Simulators for Transportation Human Factors

Research and Practice

THE HUMAN FACTORS OF SIMULATION AND ASSESSMENT

Series Editors

Michael G. Lenné

Monash University Accident Research Centre, Australia

Mark S. Young

Loughborough Design School, Loughborough University, UK

Ongoing advances in lower-cost technologies are supporting a substantive growth worldwide in the use of simulation and naturalistic performance assessment methods for research, training and operational purposes in domains such as road, rail, aviation, mining and healthcare. However, this has not been accompanied by a similar growth in the expertise required to develop and use such systems for evaluating human performance and state. Whether for research or practitioner purposes, many of the challenges in assessing operator performance and state, both using simulation and in natural environments, are common. What performance measures should be used, what technology can support the collection of these measures across the different designs, how can other methods and performance measures be integrated to complement objective data and how should behaviours be coded and the performance standards measured and defined? How can these approaches be used to support product development and training, and how can performance within these complex systems be validated? This series addresses a shortfall in knowledge and expertise by providing a unique and dedicated forum for researchers and experienced users of simulation and field-based assessment methods to share practical experiences and knowledge in sufficient depth to facilitate delivery of practical guidance.

THE HUMAN FACTORS OF SIMULATION AND ASSESSMENT

Series Editors

Michael G. Lenné

Monash University Accident Research Centre, Australia

Mark S. Young

Loughborough Design School, Loughborough University, UK

Increasing Motorcycle Conspicuity: Design and Assessment of Interventions to Enhance Rider Safety

Lars Rößger, Michael G. Lenné, and Geoff Underwood

ISBN 9781472411129 hardcover • (2015)

ISBN 9781138747647 paperback • (2017)

FORTHCOMING

Integrating Human Factors Methods and Systems Thinking for Transport Analysis and Design: Integrating Human Factors Methods and Systems Thinking for Transport Analysis and Design

Michael G. Lenné, Paul M. Salmon, Neville A. Stanton, Vanessa Beanland, and Gemma Read

ISBN 9781409463191 • (2017)

Simulators for Transportation Human Factors

Research and Practice

Edited by
Mark S. Young
Michael G. Lenné

CRC Press
Taylor & Francis Group
Boca Raton London New York

CRC Press is an imprint of the
Taylor & Francis Group, an **informa** business

CRC Press
Taylor & Francis Group
6000 Broken Sound Parkway NW, Suite 300
Boca Raton, FL 33487-2742

First issued in paperback 2019

© 2017 by Taylor & Francis Group, LLC
CRC Press is an imprint of Taylor & Francis Group, an Informa business

No claim to original U.S. Government works

ISBN-13: 978-1-4724-1143-3 (hbk)
ISBN-13: 978-0-367-87920-4 (pbk)

This book contains information obtained from authentic and highly regarded sources. Reasonable efforts have been made to publish reliable data and information, but the author and publisher cannot assume responsibility for the validity of all materials or the consequences of their use. The authors and publishers have attempted to trace the copyright holders of all material reproduced in this publication and apologize to copyright holders if permission to publish in this form has not been obtained. If any copyright material has not been acknowledged please write and let us know so we may rectify in any future reprint.

Library of Congress Cataloging-in-Publication Data

Names: Young, Mark S., editor. | Lenné, Michael G., editor.
Title: Simulators for transportation human factors : research and practice /
Mark S. Young, Michael G. Lenné, editors.
Description: Boca Raton : Taylor & Francis, CRC Press, 2017. | Series: The Human
factors of simulation and assessment | Includes bibliographical references.
Identifiers: LCCN 2017002250 | ISBN 9781472411433 (hardback : alk. paper) |
ISBN 9781315609126 (ebook)
Subjects: LCSH: Automobile driving simulators. | Flight simulators. |
Navigation--Computer simulation. | Human engineering.
Classification: LCC TL152.7.D7 S565 2017 | DDC 629.04--dc23
LC record available at https://lccn.loc.gov/2017002250

Visit the Taylor & Francis Web site at
http://www.taylorandfrancis.com

and the CRC Press Web site at
http://www.crcpress.com

Contents

Section I   Introduction

Section II   Road

Section III   Rail

Section IV   Air

Section V   Maritime

Section VI   Conclusions

Preface

Simulation continues to be a growth area in transportation human factors. From the long pedigree in aviation for pilot training, the use of simulators has since widened to cars, lorries, motorcycles, trains and ships, with applications ranging from empirical studies in the laboratory to the latest training techniques in the field.

Much of this growth seems to have occurred over the last 20 years. Two decades ago, the use of simulators in automotive human factors research was just gaining traction, with several laboratories having driving simulators of varying levels of fidelity. The first editor recalls his early experiences in this field using half a car connected to a computer running bespoke driving simulator software. (It was the front half of the car, fortunately.) Meanwhile, the rail and maritime transport modes had yet to really take advantage of simulators for research or training.

That situation was changing by 2004, when the UK Ergonomics Society (now the Chartered Institute of Ergonomics and Human Factors) convened a one-day conference that emphasised the diverse nature of simulators across the transport modes and demonstrated the popularity of this method for both researchers and practitioners. Today, simulators are being increasingly used within transport organisations to support operator training and performance assessment (particularly in rail and maritime), whilst growth in research is reflected by a noticeably increasing publication count where simulation is the primary research methodology (especially in automotive human factors).

Such progress has, of course, run in parallel with developments in both software and hardware technology. The availability of lower-cost personal computer (PC)–based technology has made (and continues to make) simulation a more widely accessible and affordable tool for both training and research worldwide. Whilst some laboratories showcase the higher-end motion-based simulators, there are many more research programmes using commercial and off-the-shelf (often PC–based) simulator systems.

Whilst the use of simulators in transportation human factors continues to grow, we feel that it is not accompanied by a similar expansion in the knowledge and expertise required to develop and use such systems. This book aims to fill that gap by drawing together current trends in simulator use for both research and training in the road, rail, air and maritime sectors.

The book is designed to appeal to a broad audience of transport researchers and practitioners across academia and industry who focus on human factors and performance assessment. The chapters, from a selection of international experts in their field, discuss traditional uses of simulators in transport (such as pilot training and automotive research) as well as less common applications (including driver training and rail research). The material focuses on simulator use rather than simulator development, but there is information in here that can help those looking to develop a functional requirements specification for a simulator. It is not intended as a handbook, but as a source of literature and case studies as well as a means for sharing experiences and lessons across the transport domains. By bridging research and practice in all of the transport modes, readers will have an accessible overview of the latest simulator techniques and studies across the gamut of transportation human factors.

Acknowledgements

We owe our gratitude to several people for making this book happen: first and foremost, to Guy Loft, the original publisher of this series and the one who kicked off this book project; Cindy Carelli, executive editor; and Renee Nakash, editorial assistant at Taylor & Francis, who have made sure that we finished it; and, secondly, of course, to all the chapter authors for their patience and tenacity throughout the whole process. Finally, we would like to thank both our families for indulging us whilst this book occupied so many evenings of our lives. It is done now.

Editors

Mark S. Young is a visiting professor at Loughborough Design School, Loughborough University, UK. He has a BSc degree in psychology and a PhD degree in cognitive ergonomics, and is a chartered ergonomist and human factors specialist with the Chartered Institute of Ergonomics and Human Factors (CIEHF). Mark has previously held visiting fellow positions in the Department of Aviation, University of New South Wales, and at Curtin University in Perth, Australia. He also served as an editor of the journal Ergonomics from 2010 to 2015. His interests focus on the human factors of transport systems, and much of his work has been based in simulators, investigating issues such as driver workload, distraction, and the effects of automation and novel technologies.

Michael G. Lenné is an adjunct professor (research) at the Monash University Accident Research Centre (MUARC), Clayton, Australia. He earned a PhD in experimental psychology from Monash University in 1998 and has spent the last 18 years working on the roles of human factors in both government and university settings. Until late 2014, Dr Lenné was a professor at MUARC, where he led the human factors research team for nearly eight years. His research used driving simulators and instrumented vehicles to study the impacts of in-vehicle technologies on driver behaviour and to measure driver responses to distraction, drowsiness and other forms of impairment. He is currently the chief scientific officer of Human Factors at Seeing Machines, where his research is centred on characterising and validating metrics for driver-state assessments in automotive, heavy vehicle and aviation sectors.

Contributors

Nora Balfe
Centre for Innovative Human
  Systems
Trinity College Dublin
Dublin, Republic of Ireland

Paul Brown
PBMaritime
Launceston, Australia

Gary Burnett
Human Factors Research Group
University of Nottingham
Nottingham, UK

Rebecca Charles
Cranfield Safety and Accident
  Investigation Centre
Cranfield University
Cranfield, UK

David Crundall
Department of Psychology
Nottingham Trent University
Nottingham, UK

Richard Donkor
Jaguar Land Rover
Coventry, UK

Richard Dunham
AMC Search Ltd
Launceston, Australia

David Golightly
Human Factors Research Group
University of Nottingham
Nottingham, UK

Rebecca L. Grant
Human Systems Integration Group
Centre for Mobility and Transport
Coventry University
Coventry, UK

Don Harris
Human Systems Integration Group
Centre for Mobility and Transport
Coventry University
Coventry, UK

Catherine Harvey
Human Factors Research Group
University of Nottingham
Nottingham, UK

Magnus Hontvedt
Department of Pedagogy
University College of Southeast
  Norway
Borre, Norway

John Huddlestone
Human Systems Integration Group
Centre for Mobility and Transport
Coventry University
Coventry, UK

Mary K. Kaiser
Ames Research Center
National Aeronautics and Space
  Administration
Mountain View, California

Barbara G. Kanki
Ames Research Center
National Aeronautics and Space
    Administration
Mountain View, California

Margareta Lützhöft
Australian Maritime College
University of Tasmania
Launceston, Australia

Dave Moore
Faculty of Design and Creative
    Technologies
Auckland University of Technology
Auckland, New Zealand

Arzoo Naghiyev
Human Factors Research Group
University of Nottingham
Nottingham, UK

Anjum Naweed
Appleton Institute for Behavioural
    Science
Central Queensland University
Adelaide, Australia

Salman Nazir
Department of Maritime Operations
University College of Southeast
    Norway
Borre, Norway

Kjell Ivar Øvergård
Department of Maritime Operations
University College of Southeast
    Norway
Borre, Norway

Andrew Parkes
Centre for Mobility and Transport
Coventry University
Coventry, UK

Nick Reed
Transport Research Laboratory
Wokingham, UK

Dale Richards
Human Systems Integration Group
Centre for Mobility and Transport
Coventry University
Coventry, UK

Sarah Sharples
Human Factors Research Group
University of Nottingham
Nottingham, UK

Paul Nikolai Smit
Department of Maritime Operations
University College of Southeast
    Norway
Borre, Norway

Linda Johnstone Sorensen
BW Gas AS
Lysaker, Norway

Alex W. Stedmon
Human Systems Integration Group
Centre for Mobility and Transport
Coventry University
Coventry, UK

Wessel M. A. van Leeuwen
Stress Research Institute
Stockholm University
Stockholm, Sweden

Peter M. T. Zaal
San José State University Research
    Foundation
San Jose, California

Section I

Introduction

1

Prologue

Mark S. Young, Michael G. Lenné and Alex W. Stedmon

CONTENTS

1.1 Advance of Transport Simulators

Simulators are vital tools in transportation human factors. From empirical studies in the laboratory to the latest training techniques in the field, simulators have been offering benefits to researchers and practitioners across the transport modes for many years.

The story of simulators in transport begins in aviation, where they originally evolved as training tools that allowed aircrew to experience aspects of flight operations and systems without actually flying real aircraft (Stedmon et al. 2012). The earliest flight simulators were developed in the interwar years as a cost-effective means of training military personnel when real aircraft were not available or were in short supply. This approach then migrated to civil aviation for commercial pilot training with the development of multimillion-pound simulators that we are perhaps more familiar with and that are now used regularly for routine and refresher training.

Nowadays, simulators are common across many transportation modes. Along with technological developments and the rise of powerful computer processing capabilities, there have been significant investment and expansion of simulators specifically for automotive, rail and maritime applications. Nevertheless, the evolution of these simulators has progressed along largely independent paths that have led to many different developments in the way that simulators are designed and used. In rail, for instance, there was a rapid growth in driver training and signalling simulators following the recommendations of Lord Cullen after the Ladbroke Grove accident in the United Kingdom (Stedmon et al. 2009). Meanwhile, simulators in the automotive sector are more typically used for researching various aspects of driver behaviour and performance. Within the maritime domain, simulators are regularly used to train bridge operations and navigation in and out of busy ports and also in fire and flood responses aboard a ship.

We will return to these applications later in the chapter, as a prelude to the rest of this book. Before we get there, though, let us consider some of the more generic aspects of, and issues with, transport simulators.

1.2 What Is a Simulator?

For anyone looking for a simple definition of a simulator, it is easy to assume that it means a high-fidelity replica of the real system. However, as we shall see, simulators cover an extensive range of different tools. At a generic level, simulators offer a level of abstraction from the real world by providing an artificial environment in which users can experience the characteristics of a real system. With this in mind, any of the following could be classed as a simulator (cf. Stedmon et al. 2009):

- Static physical mock-up (e.g. full-size or scale model or schematic; for familiarisation with a system)
- Desktop computer-based interactive media (e.g. for training rules and procedures)
- Part-task simulator (for specific aspects of the task, such as fault diagnosis)
- Full simulator (for whole-system experience and deeper levels of training that require a whole-system approach)
- Original equipment (i.e. using a real system but not in a live operational setting, such as a walk-through)

Historically, simulators have developed along technology-driven parameters that, more recently, have seen advanced visual systems, computer graphics, virtual reality and even moving platforms incorporated into their design (De Winter et al. 2009, Stedmon and Stone 2001). In the context of transport, the mention of the word *simulator* often conjures up what might be described as a replica simulator, that is, a complete, exact duplication of the real system environment, most likely with high-quality graphics and a motion platform to provide an immersive user experience. However, simulators – and simulation* – need not involve such a faithful reproduction of the real world. In some cases, it is preferable not to try and replicate every aspect of a real system, as this may cause distractions to novice users who are learning basic principles before progressing to more interactive and demanding experiences. Whilst a simulation is, by definition, a representation of reality, that representation needs to address the requirements for replicating every detail or only a subset of key features. There is no prerequisite that a simulator should attempt to mimic every aspect of the real world, although some level of interactivity is required; otherwise, what is produced is just a demonstration rather than an immersive experience.

Whilst we often think of simulators as hi-tech applications, a simulation can take place in the classroom with minimal equipment, using face-to-face role-play techniques. In these situations, a specific scenario is simulated rather than a suite of technologies or equipment. These exercises can be particularly useful for studying or training social aspects of tasks, such as communications or interpersonal relationships. They may also overlap with more conventional ideas of simulators – as described earlier – where an experimenter or an assessor acts out a particular role (e.g. of a signaller communicating by radio with a train driver) in a simulator in order to establish a more realistic operational environment.

The choice of which level of simulation to adopt should be driven by the requirements and purpose of the simulator; this then forms the basis of a design specification, which in turn will depend on the resources available. Generally speaking, the types of simulators listed earlier increase in terms of the costs involved (for both procurement and ongoing maintenance), their levels of interactivity and their fidelity. It is the last aspect, fidelity, that is probably the most crucial consideration for developing simulators that are valid and fit for purpose.

1.3 Simulator Fidelity and Validity

The concept of fidelity is closely associated with simulation. Fidelity relates to the degree to which the simulator reproduces the characteristics of the real environment and is, in itself, an important factor of simulation design.

---

* The chapters in this book tend, on the whole, to refer to simulators, but the term may be used interchangeably with *simulation*. In this chapter, we use *simulator* as meaning the tool itself (be it a computer system, mock-up or full replica), whilst *simulation* refers to the environment being reproduced (which may, for instance, be achieved during a role-play exercise).

Amongst the various aspects of simulator fidelity, three key areas emerge (Stanton 1996):

- Physical fidelity (the degree to which the simulated environment looks like the real environment)
- Functional fidelity (the degree to which the simulated environment behaves like the real environment)
- Psychological fidelity (the degree to which behaviour and performance in the simulator reflects what we would expect in the real world)

Deciding what level of fidelity is appropriate for a given simulator is often difficult but is usually driven by the application needs and requirements of the simulator. For example, perceptual–motor tasks require greater physical fidelity than cognitive tasks do (Baum et al. 1982). Whilst physical fidelity may be beneficial in training simulators (De Winter et al. 2009, Greenberg and Blommer 2011), there is evidence that functional fidelity is more important for research simulators, where physical fidelity can be compromised without affecting the transferability of results (e.g. Stammers 1986). Nevertheless, physical fidelity can help to enhance a user's sense of immersion, to convince them that they are in a realistic environment and thus to encourage realistic behaviour (Stanney and Salvendy 1996).

Traditionally, research using simulators in domains such as aviation has demanded very high levels of physical and functional fidelity (i.e. the simulator has to look like and operate like the real system does), but this may have been more to reassure the customers and users of those simulators rather than because it improves the quality of the research (cf. Stammers 1985, Stanton 1996).

Fidelity therefore also plays an important part in establishing the validity of simulators (Lenné et al. 2011, Liu et al. 2009, Wade Allen et al. 2010). High physical fidelity provides face validity – or how much the system matches user expectations in terms of look and feel (Stanton 1996).

Fundamental questions on simulator fidelity remain, such as what cues – in addition to visual displays (for which highly technical solutions are often developed) – the user might receive from audio stimuli, motion or tactile perception (Greenberg and Blommer 2011). The incorporation of motion (i.e. using a moving base) is often at the centre of debates on physical fidelity in the context of transport simulators. A moving-base simulator is often seen to represent the state of the art in technology, can greatly enhance the participants' experience and can provide crucial performance advantages when the perception of acceleration is important. However, it has been said that the use of motion 'falls victim to the "more is better" philosophy' (Lenné et al. 2011, p. 432).

Motion platforms are still rather expensive and difficult to implement effectively. If there are even slight inaccuracies in synchronising the movement of the simulator with the images displayed to the user, then this can also induce discomfort and symptoms of simulator sickness. The benefits of motion systems are rarely proportionate to the costs associated with them and are therefore usually the first feature to be sacrificed if the budget is limited. Along the different routes that transport simulation has evolved, many innovative solutions have emerged to support the user experience. One such example, to counter the issues associated with moving platforms, is the use of a low-frequency sound resonator (or a simple subwoofer) channelling engine noise into the environment that can also provide enough vibration to give a sense of movement at a fraction of the cost.

Moving-base simulators are a good example of the trade-off between the costs associated with the simulator and the degree of fidelity offered. In fact, there would appear to be a law of diminishing returns associated with physical fidelity, such that vast increases in the cost of a simulator may lead to only small increases in the quality of the data obtained (Stanton 1996). One study suggested that 80% of the benefit can often be achieved at 20% of the cost of full fidelity (Welham 1986, cited in Stanton 1996).

There is a danger, then, of overspecifying physical fidelity when it might be better to use scaled-down or part-task simulators (Rushby 2006). However, the balance between physical and functional fidelity is often linked to the desired levels of ecological validity that the simulator might be expected to achieve. Ecological validity relates to the transfer of behaviour and performance from the simulator to the real world (Wade Allen et al. 2010). It is not necessarily true that low-fidelity simulators lack ecological validity; the question is how well the simulator accurately represents the critical aspects of the experience. Modern computing power and commercial off-the-shelf PC technology offer relatively inexpensive access to often high levels of functional fidelity. With such advances, transport simulation evolves into a more credible research tool and the potential to generalise findings is enhanced.

It is usually assumed that naturalistic research in the field is the best basis for ecological validity, but that is not necessarily the case (Coolican 2005). Whilst a field study should be high in realism, if it is too specific and constrained by a particular context, then the findings will not generalise to other settings, and so it potentially has low ecological validity (Coolican 2005). The recent trend in research methods towards naturalistic, 'living-lab' approaches is further countered by the perennial argument that real-world research sacrifices experimental control and empirical rigour.

The validity of simulators in terms of accurately representing the phenomena under test is therefore an essential factor to consider in developing appropriate solutions and is discussed in many of the chapters in this book (particularly Chapters 2 through 5 and 9 through 11). Confidence in the validity of any simulator can – and should – be established by conducting validation studies. By comparing the performance in the simulator to that in the field, it is possible to quantify the transferability of other data collected in the simulator and be sure that the results will generalise to the real world. Simulators can therefore be ecologically valid if they demonstrate findings that transfer to other contexts of use and can offer potential advantages over research and training in the real world. It is to these relative advantages of simulation that we now turn our attention.

1.4 Benefits of Simulators

The previous discussion has already alluded to the healthy debate in many areas of applied investigation (and transportation human factors is no exception) over the relative merits of laboratory and field research. From a methodological perspective, there are both advantages and disadvantages to laboratory and field research. Whilst data derived from real-world contexts can be extremely valuable for investigating aspects of human–machine interaction and user behaviour (Stedmon et al. 2009), simulators offer several advantages for both research and training applications.

First and foremost is the fact that simulators can be used to put people in safety-critical situations which would be unethical or dangerous to reproduce in the real world. Examples may be testing the effects of distraction on driver performance (e.g. Young and Mahfoud 2007), investigating potential performance impairments (such as fatigue or workload) or training for abnormal or emergency events (e.g. aircraft evacuation, engine fires, train malfunctions).

Indeed, even collecting data in a live environment can incur some risk to personal or public safety, which can be avoided by using a simulator. Research into driver mental workload has often used secondary-task paradigms with participants dividing their attention between driving and an additional task (e.g. Young and Stanton 2004, 2007). Methods such as this and the use of head-mounted eye-tracking equipment are inadvisable in real road studies for safety reasons. Real-world methods often have to take a non-invasive approach so that the driver is not unduly distracted from the task of safe driving. For this reason, pre- and post-trial comparisons or non-intrusive measures are often used (e.g. psychophysiological or remote monitoring using video analysis), which can raise issues associated with retrospective data collection and lengthy post-trial analyses. Simulators therefore provide a means for investigating fundamental human factors questions and collecting crucial data that would be difficult or impossible to conduct in the field.

Simulators are also a safer option when there is a need to expose untrained or semi-experienced users to a system (with or without supervision) within a safe environment. In a simulator, users are able (and sometimes even encouraged) to make mistakes that would not be practical in a real situation. This allows users to gain experience and familiarity with a system, to repeat actions in order to learn from their mistakes and to practice their skills before working in a live environment.

Another key advantage to using a simulator is the high degree of control over the environment. The real world is highly dynamic, making it extremely difficult to control more than a few variables with which the user interacts. The myriad of extraneous factors that can influence performance may confound training, data collection and analysis. In the simulator, it is possible to hold all of these variables constant to minimise such confounds and isolate the factors of interest. Differences in performance can then be attributed to the manipulated variables with a high degree of confidence.

Similarly, the simulator offers repeatability, in that a given scenario will be experienced in exactly the same way by all users, time and time again. In the real world, no two journeys are the same, subject as they are to the vagaries of factors such as weather, traffic, lighting and other conditions. The ability to repeat the same scenario facilitates extensive data collection for research as well as consistency and reliability in training and assessment.

Timescales can also be compressed in a simulator, so that specific (rare) events can be presented on demand and in a short period, without waiting for the right conditions to arise. It is also possible to model hazards and conditions in simulators that are not easy to recreate or capture in the real world due to their infrequency and unpredictability (Stedmon et al. 2009). This capability may be useful to address specific research questions or in training (and refresher training) for infrequent or abnormal events.

A major advantage is that simulators offer a data-rich environment; the software used to develop scenarios often provides the basis for collecting a vast array of data, which is usually not possible in the real world. The performance recording and assessment capabilities of the simulator can be used for later statistical analysis in research or for objective comparison of a trainee's performance against set criteria. Meanwhile, ancillary equipment (such as eye tracking) can be used to supplement data collection.

Simulators can also offer practical benefits, as it is possible to run trials regardless of the external weather, time of day or other conditions, and trials can be aborted relatively easily. A simulator can also be used as the environment within which to test and train participants on new equipment, prototypes or procedures, in order to understand the impacts of such changes and facilitate their smooth introduction into service.

From an economic and ecological perspective, simulator studies typically consume fewer resources than real-world research does (although the capital, operating and maintenance costs of a simulator can be considerable, especially when taking into account space and facilities requirements). In most cases, the costs and logistical challenges of running a study in a simulator would be far lower than conducting the same study out in the field, where as well as the impact of using operational equipment, it might also involve disruptions to the transport network. Simulators are therefore an attractive option where the system itself is a valuable resource that is needed for operational purposes and cannot be spared for research or training activities (e.g. using a full aircraft, occupying the rail network, highways and even maritime access).

The availability and consistency of the simulated environment allow for a much higher throughput of participants as well. A simulator can be operated around the clock, if needs be, whereas real-world testing depends on the availability of operational equipment. For training and assessment in particular, this facilitates extended practice of procedural or psychomotor skills – skills which are best learned by doing rather than studying the theory and rules in the classroom. Finally, there is the potential for simulator training to increase the efficiency of training courses and potentially reduce the amount of training time required on live assets.

Nevertheless, simulators are not a panacea for research or training. There is still debate about the transfer of training from a simulator to the real world, with particular concerns over negative transfer whereby actions learned in the simulator prove inappropriate or unsafe in the live environment (e.g. where part-task simulation does not provide a holistic understanding for the user). It therefore seems unlikely that simulators will completely replace on-the-job training any time soon (see Chapter 9 for further discussion on this point).

With any simulation experience, participants usually recognise that they are not operating a real system with its associated consequences if they make a mistake (i.e. they know that they are safe from harm regardless of their actions). Some simulators, particularly those including movement, can have adverse physiological consequences in the form of simulator sickness, as discussed earlier in this chapter and elsewhere in this book (see Chapters 2, 3, 5 and 11). In addition, there will be limits as to the degree of realism that can be achieved in a simulator, even with modern computing power. Some elements of the visual scene (such as glare and contrast) are notoriously difficult to simulate, although advances in software and display technology are leading to improvements in these areas.

It is therefore important to maintain a balanced perspective on the relative merits of real-world trials against laboratory-based simulator testing. The final decision is usually based on the requirements and the complexity of the investigation. In some cases, that may mean choosing real-world studies or quasi-experimental trials over simulator studies. If the solution is to use a simulator, its specification should be developed to meet the purpose for which it is intended (e.g. training, research, product development), based on a fundamental understanding of user requirements, user expectations and the intended user experience (Stedmon et al. 2009). Ultimately, it is a case of using the right tool for the job.

1.5 Current Trends in Transport Simulators

Within transportation research, simulators have been developed as sophisticated tools to support our understanding of the complex tasks that operators perform and, as such, simulators are now regularly used in the course of research across the transport modes (e.g. Harris et al. 1999, Naweed et al. 2013, van Leeuwen et al. 2013, Young and Stanton 2002). Many of these studies have been made possible only through the empirical control, flexibility and the safe environment afforded by a driving simulator. The rest of this book captures such research in detail; here, we pick out a few examples from road and rail.

Within road transport research, simulator use has evolved along two main pathways where they have been used to investigate either aspects of driver error and underlying causes of road accidents and/or the integration of new technologies within vehicles along with the need to understand the impact that they might have on typical driver performance. There is now a relatively long pedigree of simulator research to investigate the impacts of advanced technologies such as adaptive cruise control (Nilsson 1995, Young and Stanton 2004) and active steering (Young and Stanton 2002), vision enhancement systems (Stanton and Pinto 2000) and speech recognition interfaces (Stedmon et al. 2002, 2012), and satellite navigation and driver information systems (Burnett 2008, Pettitt et al. 2007). Simulator research has also demonstrated that detrimental effects occur when car drivers eat and drink whilst driving (Young et al. 2008) or are distracted by roadside advertisements (Young and Mahfoud 2007).

Other research has used simulators to examine hypervigilance, where the ability to react to unpredictable events deteriorates in otherwise monotonous and uneventful driving tasks (Larue et al. 2011). In such situations, highway design reduces the driving task to a lane-keeping activity and the resulting lack of driver stimulation can cause driver inattention. This study illustrated that a simulator could be used to assess the effects of road design and roadside variability on driving performance, and this was the first study to correlate hypervigilance and driver performance in varied monotonous conditions.

The ecological validity of motion-based research was illustrated in one study through the use of an in-vehicle touch screen interface (Salmon et al. 2011). The results indicated that higher levels of motion were associated with inferior performance and higher subjective workload. These findings provided guidelines for the design and use of these interfaces in mobile platforms.

In rail, many human factors aspects of train driving lend themselves to simulator research (Young 2003), such as the perception of signals and signs, driver workload, in-cab interface design and ergonomics, abnormal and degraded working, train protection and automation systems and route knowledge and competence. Train driving simulators have been developed to test aspects such as these (Naweed et al. 2013, Yates et al. 2007) and to avoid many of the problems that can beset such research in the field. For instance, previous field studies into drivers' eye movements to explore issues such as visual acuity or distribution of attention (Merat et al. 2002, Naghiyev et al. 2014; see also Chapter 5) have faced logistical challenges in terms of conflicts of interest with scheduled services, train crew availability and track access.

The rail domain can be used to illustrate two areas where simulators can offer a novel approach to researching established issues. The first is in fitness for duty, particularly fatigue, which is a recurring topic in rail incidents (e.g. RAIB 2016). Some of the key questions involve the factors influencing, and those that can mitigate, fatigue. Studying fatigue in a simulator would necessitate a longitudinal study and may be difficult to coordinate. However, it is by no means impossible; such a project has already been carried out in the maritime world (van Leeuwen et al. 2013).

The second area is in verbal communications, such as between a driver and a signaller. Here, we could exploit the flexibility of the simulated environment and stage a 'Wizard of Oz' scenario, in which the experimenter plays the role of the signaller. Alternatively, and more ambitiously, it is theoretically possible to link a driving simulator with a signalling simulator and have the participants interact according to normal rules and procedures, but in a virtual world.

Such a scenario could be used in training as much as in research, and the different types of simulators come into their own with different aspects of training. For instance, a computer-based training tool might be used for rules and procedures, a part-task simulator for abnormal events or a full simulator for hands-on, interactive practice of psychomotor skills. Simulators can also be used at any stage of the training process, ab initio, through refresher and conversion training, to post-incident retraining. The capabilities of the simulator to record, store and analyse data allow trainers and assessors to evaluate performance against set criteria in order to determine competence. In any case, the use of simulators should be integrated into a training programme and based on a training needs analysis, rather than being seen as a technological bolt-on for its own sake.

In general, the evolution of simulators in transport has started in training before developing into research applications (the notable exception being in the road domain, where simulator research is relatively widespread compared to training). From a human performance perspective, as technologies became more complicated to operate, it became more necessary to train people how to use them. Although it is possible to train people using real technologies (e.g. training learner drivers on the road), there comes a point when a simulator might provide another, more beneficial, training aid.

Whilst the benefits of simulators (as reviewed earlier) are largely common to research and training, there are some key distinctions in the fundamental specifications for such simulators – distinctions which are particularly apparent if trying to adapt a training simulator for research use. Simulators for research on the whole require much greater flexibility than those for training and assessment in terms of both scenario development and data analysis. Training simulators tend to be designed for particular routes or environments and with specific performance metrics (to assess competence) in mind. This constrains both the experimental design and analysis aspects that are crucial to empirical research.

As we have already noted in this chapter, some of the earliest simulators were developed in aviation for training purposes. As in the research world, then, simulators for training have the longest history in aviation (see Chapter 9), but it is notable that their use in UK rail has grown as a result of the high-profile disasters at Southall and Ladbroke Grove in the late 1990s. The inquiries into these accidents (Cullen 2000, Uff 2000) both highlighted a role for simulators in training for drivers and signallers.

There are similar examples of training simulators in the road and maritime domains, just as there are numerous research studies using simulators in the aviation and maritime worlds. In this chapter, we have deliberately only touched upon some of these applications as a prelude to the rest of this book, where we draw together current trends in research and training simulators across all the transport modes.

1.6 Structure of This Book

So far in this chapter, we have established that there are disparate uses for simulators across transportation human factors, but that there is also a lot of common ground in terms of the advantages that they offer. This book therefore aims to map out both research and training uses of simulators across the road, rail, air and maritime domains. It offers case studies, practical examples, literature reviews and scientific research to cover the area from a functional rather than a technological perspective. This is a book of academic and operational experience, not a technical encyclopedia.

The book is accordingly divided into four main sections (Sections II through V), covering all of the transport modes, with each section containing at least two chapters (one for research, one for training). Many of the sections also contain a bonus chapter to reflect other applications of simulators within that mode (motorcycles for road transport, signalling for rail and cabin evacuation for aviation). The arrangement of sections is arbitrary and is not meant to imply any rank ordering or sequencing; readers can dip into sections and chapters as fits their interests without necessarily having to read the book from start to finish. Nevertheless, we hope that readers will find value and inspiration in learning from the experiences of those in other modes and other applications than their own domain. Finally, the concluding chapter summarises the key themes from the rest of the book and offers our own thoughts on creating a functional specification for researchers and practitioners in this area.

References

Baum, D.R., Riedel, S., Hays, R.T., and Mirabella, A. 1982. *Training Effectiveness as a Function of Training Device Effectiveness.* ARI Technical Report 593. Alexandria, VA: US Army Research Institute.

Burnett, G.E. 2008. Designing and evaluating in-car user-interfaces. In J. Lumsden (Ed.), *Handbook of Research on User-Interface Design and Evaluation for Mobile Technology* (218–236). Hershey, PA: Idea Group.

Coolican, H. 2005. *Research Methods and Statistics in Psychology* (3rd Edition). London: Hodder & Stoughton.

Cullen, Rt Hon. Lord. 2000. *The Ladbroke Grove Rail Inquiry: Part 1 Report.* Sudbury, Canada: HSE Books.

De Winter, J.C.F., de Groot, S., Mulder, M., Wieringa, P.A., Dankelman, J., and Mulder, J.A. 2009. Relationships between driving simulator performance and driving test results. *Ergonomics*, 52: 137–153.

Greenberg, J.A., and Blommer, M. 2011. Physical fidelity of driving simulators. In D.L. Fisher, M. Rizzo, J.K. Caird and J.D. Lee (Eds.), *Handbook of Driving Simulation for Engineering, Medicine and Psychology* (7.1–7.24). Boca Raton, FL: CRC Press.

Harris, D., Payne, K., and Gautrey, J. 1999. A multi-dimensional scale to assess aircraft handling qualities. In D. Harris (Ed.), *Engineering Psychology and Cognitive Ergonomics Volume Three: Transportation Systems, Medical Ergonomics and Training* (277–285). Aldershot, UK: Ashgate.

Larue, G.S., Rakotonirainy, A., and Pettitt, A.N. 2011. Driving performance impairments due to hypovigilance on monotonous roads. *Accident Analysis & Prevention*, 43: 2037–2046.

Lenné, M.G., Groeger, J.A., and Triggs, T.J. 2011. Contemporary use of simulation in traffic psychology research: Bringing home the bacon? *Transport Research Part F: Traffic Psychology and Behaviour*, 14: 431–434.

Liu, D., Macchiarella, N.D., and Vincenzi, D.A. 2009. Simulation fidelity. In D.A. Vincenzi, J.A. Wise, M. Mouloua and P.A. Hancock (Eds.), *Human Factors in Simulation and Training* (61–73). Boca Raton, FL: CRC Press.

Merat, N., Mills, A., Bradshaw, M., Everatt, J., and Groeger, J. 2002. Allocation of attention among train drivers. In P.T. McCabe (Ed.), *Contemporary Ergonomics 2002* (185–190). London: Taylor & Francis.

Naghiyev, A., Sharples, S., Carey, M., Coplestone, A., and Ryan, B. 2014. ERTMS train driving-in cab vs. outside: An explorative eye-tracking field study. In S. Sharples and S. Shorrock (Eds.), *Contemporary Ergonomics and Human Factors 2014* (343–350). Boca Raton, FL: Taylor & Francis.

Naweed, A., Hockey, G.R.J., and Clarke, S.D. 2013. Designing simulator tools for rail research: The case study of a train driving microworld. *Applied Ergonomics*, 44: 445–454.

Nilsson, L. 1995. Safety effects of adaptive cruise control in critical traffic situations. *Proceedings of the Second World Congress on Intelligent Transport Systems* (vol. 3) (1254–1259). Tokyo: VERTIS.

Pettitt, M., Burnett, G.E., and Stevens, A. 2007. An extended keystroke level model (KLM) for predicting the visual demand of in-vehicle information systems. *Proceedings of the SIGCHI Conference on Human Factors in Computing Systems 2007, San Jose, CA, 28 April–3 May* (1515–1524). New York: Association of Computing Machinery.

RAIB (Rail Accident Investigation Branch). 2016. *Two Signal Passed at Danger Incidents, at Reading Westbury Line Junction, 28 March 2015, and Ruscombe Junction, 3 November 2015*. Report 18/2016. Derby, UK: Rail Accident Investigation Branch.

Rushby, N. 2006. *How Will We Know if Games and Simulations Enable Learners to Learn Anything?* Crawley, UK: Conation Technologies.

Salmon, P.M., Lenne, M.G., Triggs, T., Goode, N., Cornelissen, M., and Demczuk, V. 2011. The effects of motion on in-vehicle touch screen system operation: A battle management system case study. *Transport Research Part F: Traffic Psychology and Behaviour*, 14: 494–503.

Stammers, R.B. 1985. Instructional psychology and the design of training simulators. In D.G. Walton (Ed.), *Simulation for Nuclear Reactor Technology* (161–176). Cambridge, UK: Cambridge University Press.

Stammers, R.B. 1986. Psychological aspects of simulator design and use. In J. Lewins and M. Becker (Eds.), *Advances in Nuclear Science and Technology* (vol. 17) (117–132). New York: Plenum Press.

Stanney, K., and Salvendy, G. 1996. After-effects and sense of presence in virtual environments: Formulation of a research and development agenda. *International Journal of Human–Computer Interaction*, 10: 135–187.

Stanton, N.A. 1996. Simulators: A review of research and practice. In N.A. Stanton (Ed.), *Human Factors in Nuclear Safety* (114–137). London: Taylor & Francis.

Stanton, N.A., and Pinto, M. 2000. Behavioural compensation with vision enhancement systems. *Ergonomics*, 43: 1359–1370.

Stedmon, A.W. and Stone, R. 2001. Re-viewing reality: Human factors issues in synthetic training environments. *International Journal of Human Computer Studies*, 55: 675–698.

Stedmon, A.W., Richardson, J.R., and Bayer, S.H. 2002. In-car ASR: Speech as a secondary workload factor. In P. McCabe (Ed.), *Contemporary Ergonomics 2002* (252–257). London: Taylor & Francis.

Stedmon, A.W., Young, M.S., and Hasseldine, B. 2009. Keeping it real or faking it: The trials and tribulations of real road studies and simulators in transport research. In P.D. Bust (Ed.), *Contemporary Ergonomics 2009* (442–450). London: Taylor & Francis.

Stedmon, A.W., Richardson, J., Bayer, S.H, Graham, R., Carter, C., Young, M., and Hasseldine, B. 2012. Speech input applications for driving: Using different levels of fidelity in simulator research. *Advances in Transport Studies*, 28: 17–34.

Uff, J. 2000. *The Southall Rail Accident Inquiry Report*. Sudbury, Canada: HSE Books.

van Leeuwen, W.M., Kircher, A., Dahlgren, A., Lützhöft, M., Barnett, M., Kecklund, G., and Åkerstedt, T. 2013. Sleep, sleepiness, and neurobehavioral performance while on watch in a simulated 4 hours on/8 hours off maritime watch system. *Chronobiology International*, 30: 1108–1115.

Wade Allen, R., Park, G.D., and Cook, M.L. 2010. Simulator fidelity and validity in a transfer-of-training context. *Transportation Research Record: Journal of the Transportation Research Board*, 2185: 40–47.

Yates, T.K., Sharples, S.C., Morrisroe, G., and Clarke, T. 2007. Determining user requirements for a human factors research train driver simulator. In J.R. Wilson, B. Norris, T. Clarke and A. Mills (Eds.), *People and Rail Systems: Human Factors at the Heart of the Railway* (155–165). Aldershot, UK: Ashgate.

Young, M.S., and Stanton, N.A. 2002. Malleable attentional resources theory: A new explanation for the effects of mental underload on performance. *Human Factors*, 44: 365–375.

Young, M.S. 2003. Development of a railway safety research simulator. In P.T. McCabe (Ed.), *Contemporary Ergonomics 2003 – Proceedings of the Annual Conference of the Ergonomics Society, Edinburgh, April 15–17, 2003* (379–384). London: Taylor & Francis.

Young, M.S. and Stanton, N.A. 2004. Taking the load off: Investigations of how adaptive cruise control affects mental workload. *Ergonomics*, 47: 1014–1035.

Young, M.S. and Mahfoud, J.M. 2007. Driven to distraction: The effects of roadside advertising on driver attention. In P.D. Bust (Ed.), *Contemporary Ergonomics 2007* (145–150). London: Taylor & Francis.

Young, M.S. and Stanton, N.A. 2007. Miles away: Determining the extent of secondary task interference on simulated driving. *Theoretical Issues in Ergonomics Science*, 8: 233–253.

Young, M.S., Mahfoud, J.M., Walker, G.H., Jenkins, D.P., and Stanton, N.A. 2008. Crash dieting: The effects of eating and drinking on driving performance. *Accident Analysis & Prevention*, 40: 142–148.

Section II

Road

2

Driving Simulators for Research

Gary Burnett, Catherine Harvey and Richard Donkor

CONTENTS

2.1 Introduction

Computer-based simulation is widely used in human factors and applied psychology research as a controlled and cost-effective means of investigating people's interactions with technology. In safety-critical contexts such as driving, simulation also offers the fundamental advantage that research questions can be addressed in an environment in which the participant(s), the research team and other road users experience no objective risk. Such benefits, together with reductions in hardware and software costs, have led to a number of research groups developing interactive driving simulator facilities. For instance, within the United Kingdom, there are now a large number of teams which possess driving simulators of varying levels of fidelity, ranging from those with single computer screens and game controller configurations, to real car cabins with multiple projections and motion systems (see Figure 2.1 for an example).

Initially developed to assess the skills and capabilities of van drivers, modern driving simulators have become tools for transport and road safety scientific research; driver and race driver training, assessment and rehabilitation; and entertainment (de Winter, van Leeuwen and Happee 2012). Driving simulators are used to study all three dimensions of driving – the driver, the vehicle and the traffic – road environment system – and their respective interactions. In terms of driving research, simulators have been used to address a wide variety of different issues, including the following:

- The effects of distraction tasks, drugs, alcohol and sleep deprivation on drivers' performance and situation awareness (Brookhuis, De Vries and De Waard 1991; Fairclough and Graham 1999; Philip et al. 2003)
- The effect of new in-vehicle technologies and displays on driver behaviour and performance (Burnett and Donkor 2010; Harvey et al. 2011; Large et al. 2016)
- Drivers' responses to intelligent transport systems and automated driving systems (Banks, Stanton and Harvey 2014; McGehee, Mazzae and Baldwin 2000; Saad 2006)

FIGURE 2.1
Example of a medium-fidelity driving simulator facility.

- Driver training regimes (Dorn and Barker 2005)
- Road and traffic design and evaluation issues, for example innovative road design and geometrics (Kelly et al. 2010) and traffic conditions (Konstantopoulos, Chapman and Crundall 2010)

The rest of this chapter is organised into three main sections: Section 2.2 covers theoretical considerations for the use of simulators for driving research studies and is split into subsections on reliability, validity, fidelity and presence. Section 2.3 covers practical considerations for the development and running of a driving simulator facility and covers safety, control and cost. Section 2.4 provides some practical guidance on how to develop a 'good' driving simulator and simulation experience for human factors research, based upon the state of the art in this field and our own expertise and experiences.

2.2 Theoretical Considerations for Simulated Driving Studies

2.2.1 Reliability

One of the major requirements of a driving simulator is the capability to create repeatable and consistent driving situations: in other words, reliability. Driving is an activity which does not happen in isolation; rather, it takes place in an environment which is continuously impacted by the actions of other people and external factors such as weather and time of the day. In a real driving environment, it is impossible to reliably recreate consistent road, traffic and environmental conditions from one study to the next and no two participants' experiences would be the same. This makes it difficult to draw conclusions about the effects of any independent variables being tested, due to the lack of consistency in these confounding factors. In a simulator study, all of these factors can be tightly controlled, meaning that not only will all participants in a single study be exposed to exactly the same conditions, but that the study can be repeated by other researchers in other simulators at any time.

Such reliability is, however, dependent on the assumption that driving behaviour and performance are measured and reported consistently enough by different researchers to allow repeatability. This is an issue addressed recently by the publication of *SAE J2944, Operational Definitions of Driving Performance Measures and Statistics* (Society of Automotive Engineers 2015). This document provides guidance on standard names of driving measures and how they should be calculated and reported, aiming to ensure consistency across publications (see also Green 2013). Reliability is therefore important for the body of driving research as a whole. Repeatability also enables the compression of experience, which allows many different situations to be tested in a short period. This enables us to study the characteristics of driving performance and behaviour, which would otherwise emerge only by chance after many hours, weeks or even months of observations of drivers in real vehicles.

Blana (1996) distinguished between physical reliability (the consistency and repeatability in the performance of physical components of a simulator) and behavioural reliability (the consistency and repeatability in performance of participants in the simulator across exposures). As behavioural reliability is much more difficult to measure than physical reliability, it is possible that factors which might temporarily influence participant behaviour (e.g. low motivation, unplanned distractions and discomfort) could lower the overall reliability of a test. The design of a simulator can contribute to the repeatability and consistency of individual subsystems within it, particularly with regard to physical reliability; however, this alone does not guarantee an accurate representation of real driving. For example the vehicle dynamics may operate consistently across all simulator trials but might not be representative of vehicle dynamics on a real road: this is an issue of validity, which will be discussed next.

2.2.2 Validity

Validity represents the extent to which a test situation measures what it is supposed to measure. There are many definitions of validity, depending on the type of research question being investigated or the assessment domain. For the purposes of using simulation for human factors research in driving, two main concepts – physical validity and behavioural validity – are pertinent (Blaauw 1982; Harms 1994).

Physical validity (sometimes referred to as face validity) refers to the degree of realism that a simulator presents to a user: in other words, the correspondence between the components, configuration and dynamics of the simulator and a real car driving environment (Blaauw 1982; Wang et al. 2010). It is important primarily not only due to its influence on a participant's motivation within the simulator, but also because of its impact on wider stakeholder acceptance, for example whether a manager (decision maker) believes in the results emerging from a simulator study.

Researchers often describe the physical correspondence between the simulator and real driving but neglect to report the behavioural correspondence (Godley, Triggs and Fildes 2002). *Behavioural validity* refers to the extent to which results obtained through a set of experiments involving specific driver populations, time periods and environments are representative of behaviour in the real world (Kaptein, Theeuwes and Van der Horst 1996; Wang et al. 2010). Behavioural validity has two components. *Absolute validity* refers to the degree that measurements of performance in the simulated environment and the real world or field test equate perfectly for the same variable (Godley,

Triggs and Fildes 2002; Kaptein, Theeuwes and Van der Horst 1996). *Relative validity* is concerned only with the extent to which the direction and magnitude of behaviours are comparable between the testing settings (Wang et al. 2010). Most researchers agree that absolute validity is difficult to prove, but that relative validity is sufficient to address many research questions in the driving domain (Kaptein, Theeuwes and Van der Horst 1996; Parkes 2012; Törnros 1998; Wang et al. 2010).

Driving simulator validity is very difficult to assess. Running equivalent road trials is time consuming, expensive and subject to various practical and ethical concerns, meaning that data against which to validate are not easy to obtain. A small number of studies have attempted to validate driving simulators against real road driving for measures including speed and lateral control (Blaauw 1982; Jamson 2001), headway maintenance (Jamson 2001; Kaptein, Theeuwes and Van der Hors 1996), braking (Hoffman et al. 2002) and steering (Toffin et al. 2003). Underwood, Crundall and Chapman (2011) recommended that these low-level vehicle control measures should be regarded as necessary, rather than sufficient, for simulator validation and that more behavioural measures are also required. In addition, previous research into the validity of driving simulators has shown that it is difficult to generalise from existing validity studies, largely because results cover only a small number of variables, are based on varying hardware and software configurations and are reported inconsistently (e.g. see Törnros 1998 compared with Blaauw 1982). Despite these difficulties, there have been some more recent attempts at validating driving simulators with what Underwood, Crundall and Chapman (2011) described as 'higher-level cognitive measures', such as hazard perception and situation awareness, as opposed to low-level vehicle control measures such as speed and lateral control. Underwood, Crundall and Chapman (2011) were able to demonstrate relative validity in a study of hazard detection in real driving and in a driving simulator, although absolute validity was not demonstrated as that would have required exactly the same hazardous scenarios to have occurred in the simulated and real driving environments. Donkor, Burnett and Sharples (2014) also found that drivers in low- and medium-fidelity simulators exhibited the same facial expressions in response to hazards as had been observed in real-world driving, demonstrating the use of emotional response as a tool for establishing simulator validity. Validity is frequently reported with reference to the level of 'fidelity' of a simulated environment and the level of 'presence' experienced by people within that environment: these important concepts are discussed in the following sections.

2.2.3 Fidelity

The concept of fidelity has been used and reported in a variety of different ways (Hays 1980; Schricker, Franceschini and Johnson 2001) and, as a consequence, is difficult to define. In the broadest terms, fidelity refers to the

'degree of similarity between the simulator and the equipment which is simulated' (Hays 1980, p. 9). This can, however, be broken down further to refer
to physical fidelity (does it look, sound, feel like the real car?), functional
fidelity (how realistic is the information, stimuli and responses provided
by the simulator?) (Hays 1980) and psychological fidelity (does it replicate
the relevant cognitive factors, such as locus of attention or decision-making
requirements?) (Hughes and Rolek 2003; Liu, Macchiarella and Vincenzi
2009). There is, however, contention over these subcategories and many more
ways that fidelity has been described and categorised.

Simulator fidelity is normally described as low, medium and high (Blana
1996; Weir and Clark 1995), although there is no generally agreed method of
defining where on this scale a particular simulator sits. As a rough guide,
a low-fidelity driving simulator would usually comprise a 'desktop' set-up
with a single screen, a gaming wheel and pedals. Medium fidelity might
describe a half-car static rig with approximately 180-degree field of view,
and a high-fidelity simulator would be likely to have a motion base and a
360-degree visual scene. Although useful, this low–medium–high scale of
fidelity tends to focus on physical, hardware components of the simulator,
neglecting information about the simulation software, the tasks performed
and more aesthetic elements of the test environment (e.g. signage, safety
measures or separation between simulator and control room). Furthermore,
some of these elements are not frequently reported in research papers, so it is
very difficult to know how they influence fidelity and, consequently, results.
There have been attempts to include a greater level of detail in classifications
of fidelity; for example Parkes (2012) attempted to classify driving simulators for training purposes in a similar manner done for flight simulators,
based largely on technical characteristics. However, the success of this sort of
approach depends on its uptake by researchers in reporting driving studies.

Fidelity is particularly important for two reasons: (a) it is positively related
to the cost of a simulator (the higher the fidelity, the higher the likely cost to
build and maintain) and (b) the fidelity of a simulation will have a considerable impact on an operator's motivation, performance and behaviour (e.g.
Burnett, Irune and Mowforth 2007). This represents a cost–benefit trade-off.
Moreover, it is not necessarily true that higher investment in a simulation
facility will produce higher-quality research; indeed, simulator fidelity may
actually exceed research requirements in some cases (Hays 1980).

2.2.4 Presence

The experience of a driver within a simulator can be largely defined by the
concept of presence. It is closely linked to fidelity, although it is certainly not
true to say that higher fidelity always results in higher presence as presence
is also influenced by the experience and attitudes of the person experiencing
the environment. Presence is an important concept in the virtual reality literature and has been defined as 'the subjective experience of being in one place

or environment, even when one is physically situated in another' (Witmer and Singer 1998, p. 225) and the illusion of 'being there' in a virtual environment (Biocca 1997, p. 18). Biocca (1997) suggested that to be fully 'present' in a virtual location, a person should cease to be conscious of the medium that took them there: for a driving simulator, this would mean that display screens, lack of real motion, awareness of the experimenter or any recording equipment within the vehicle or situated on the body should effectively 'disappear' for the participant during the experience.

Achieving presence is rarely the end goal of a driving simulator study; however, it is widely accepted as necessary as it is highly correlated with attention, motivation, learning and other responses which are being investigated as study goals (Biocca 1997). Simply put, we can assume that a greater sense of presence, allowing a person to feel like they are really in a particular environment, should evoke more realistic driving performance and behaviours along measured dimensions. This in turn should increase the validity of any results obtained within this virtual environment. The concept of presence is, however, far from simple, with definitions and terminology varying quite widely and a lack of standardised measures. A discussion of the development of presence as a concept is beyond the scope of this chapter (for more information, the reader is referred to Biocca [1997], Lee [2004] and Witmer and Singer [1994]); instead, we focus on the measurement of presence for driving simulator studies.

There has been much debate over how to measure presence, with objective and subjective measures being used by researchers and a combination of both often recommended (see Deniaud et al. 2015). Objective measures often include physiological responses, such as heart rate, skin temperature and brain activity, and the application of these is based on an assumption that an increased sense of presence will produce physiological responses which mirror those of real driving (Deniaud et al. 2015). Behavioural measures can also be used, based on the assumption that the more a participant feels involved in a virtual environment, the more similar their behaviour will be to that exhibited in a real environment: Insko (2003) gives the example of a person ducking down to avoid a virtual ball being thrown towards their head. A problem with behavioural measures for use in driving studies is that behaviours are often very subtle; for example imagine a driver bracing themselves for a sudden braking event: this would be incredibly difficult to detect.

Due to the limitations of these objective measures, it has been more common for driving researchers to employ subjective techniques to assess presence. Self-rating questionnaires are usually administered post-immersion to assess a person's overall experience of presence in a particular environment. The presence questionnaire by Witmer and Singer (1998) and the Independent Television Commission Sense of Presence Inventory (ITC-SOPI; Lessiter et al. 2001) are two of the most frequently used. The Witmer and Singer (1998) questionnaire consists of 32 questions requiring ratings on a seven-point scale and generates a single total value for overall presence between 0 and 192. Respondents are asked questions including 'How natural did your interactions with the environment seem?' and 'How completely were all of your senses engaged?' The ITC-SOPI consists of 38 questions requiring ratings on a five-point scale, the results of which are split into four factor scores: spatial presence, engagement, ecological validity and negative effects (a measure of unpleasant feelings including sickness, sleepiness and eye strain). Respondents rate their agreements with statements including 'I felt as though I was participating in the displayed environment' and 'I felt that the displayed environment was part of the real world'. The 'negative effects' aspect of the ITC-SOPI also provides an indication of simulator sickness so it can be useful for monitoring.

The Witmer and Singer and the ITC-SOPI questionnaires were used in a recent study by Harvey and Burnett (2016) on the effect of incentives and instructions on presence and were found to have a significant positive correlation, indicating that they were both measuring the same thing. Whether or not what they were both measuring was in fact presence is harder to prove as it is a purely perceptual phenomenon and two people may have different responses to identical environments (Insko 2003). A further disadvantage of this approach is that ratings cannot easily be captured during a person's immersive experience as it would be too disruptive to the experience and would in itself be expected to lower presence. Administering questionnaires post-immersion can mean that respondents might quickly forget how they were feeling in the virtual environment, leading to unrepresentative results. Having said this, there are advantages of self-rating questionnaires and reasons why they are used frequently to assess presence – namely ease, speed and low cost of administration and analysis (Insko 2003).

Both questionnaires described earlier are designed to be applicable across media and content. This is advantageous as it means that presence can be measured consistently across different domains and findings can be compared. It does, however, also mean that some questions might not be completely appropriate for all scenarios all of the time. For example the ITC-SOPI questionnaire refers to the interaction with 'characters' in the virtual environment, but in many driving studies, there will not be any characters present, and indeed, in the authors' experience, this has caused confusion for some participants. To address this issue, there have been attempts to create more domain-specific presence questionnaires: for example Donkor, Burnett and Sharples (2014) designed a driving simulator experience questionnaire specifically to enable researchers and practitioners to understand how study participants perceive the *driving* simulation environment in relation to real-world equivalent situations. Examples of questions include those relating to *strategic* ('I felt as if I had been on a journey'), *tactical* ('I felt the need to obey the rules of the road') and *control* ('I had a strong sense of physically controlling the vehicle') elements of the driving task. Items are also included relating to *social* aspects of driving, for example 'I was aware that other people were driving cars around me'. As such, the questionnaire serves as a potentially useful tool for baselining driving simulation experiences.

2.3 Practical Considerations for the Development and Running of a Driving Simulator

2.3.1 Safety

The major advantage of driving simulators is their provision of a safe environment for research, making them an ethical option for studying driver behaviour and performance. A driving simulator offers the primary benefit of assisting in addressing research questions that concern drivers and their driving environment without the safety risks to the driver, researcher and other road users that would result from on-road testing (McGehee, Mazzae and Baldwin 2000; Reed and Green 1999). Dangerous and even illegal on-road scenarios, such as driving under the influence of alcohol or whilst experiencing high levels of fatigue, can be tested in a driving simulator with no (or at least very minimal) risks to safety. In-vehicle technologies, which potentially pose a distraction risk to drivers, can be tested in a simulator without concern over the consequences of drivers spending long periods with their eyes off the road (Nilsson 1993). In the automotive sector, there has been a recent surge in interest in automated systems, many of which are designed to operate in safety-critical situations (such as collision avoidance). It will be virtually impossible, both practically and ethically, to recreate these types of scenarios in full in an on-road test, so again simulators offer a clear benefit, allowing these systems to be tested well before implementation into production vehicles.

2.3.2 Simulator Sickness

The previous section discusses the safety benefits of using a simulator over on-road studies; however, there are also negative safety aspects of simulated driving, namely the risk of inducing simulator sickness in participants. Motion sickness is a condition brought about by exposure to different forms of real or apparent motion. Motion sickness which is induced by exposure to a moving visual scene is known as visually induced motion sickness or more specifically simulator sickness when caused by simulated environments.

The issue of sickness has been a concern since the early development years of driving simulators (Casali and Frank 1988). Symptoms can range from eye strain, dizziness, headache, vertigo, nausea, mild disorientation and vomiting to severe migraines and cluster headaches. Much of the theory of the underlying causes of simulator sickness has been based on the idea that it is caused by conflicting signals from two or more senses: this is known as sensory conflict theory (Reason and Brand 1975). Certain visual images can evoke a sense of illusory self-motion (vection) in the viewer via signals from their visual system. In this situation, the brain expects to receive signals from the vestibular system to confirm the information provided by vision: that the person is moving. However, in visual environments such as a static driving simulator, the viewer is stationary in front of a display screen, resulting in an absence of corresponding signals from the vestibular system.

Research into the consequences of simulator sickness has shown that the phenomenon can have severe negative implications on a driver's simulation experience, including their perceptions of the fidelity of the simulation and the validity of results (Casali and Wierwille 1986; Kolasinski 1995). Sickness can consequently affect a driver's motivation, behaviour and driving performance. Drivers experiencing sickness might avoid tasks that are disturbing or adopt unrealistic behaviours (e.g. closing eyes whilst turning corners). They may be distracted from normal attention allocation processes and may be preoccupied with trying to resolve the conflict between senses. People may also continue to experience the symptoms of simulator sickness after the trial has finished (Frank, Casali and Wierwille 1988; Mourant and Thattacherry 2000), so researchers must ensure that procedures are in place to assess the well-being of participants before they leave a study. If participants drop out of a study due to simulator sickness, there will be associated costs of recruiting additional people and delaying study completion.

Based on what we know about the causes of simulator sickness, there are a number of practical considerations for developing a driving simulator to minimise the risk of discomfort:

- The width and depth of the windscreen out of which the driver views the projected road scene: A larger screen size is linked to increased sickness, so a smaller windscreen might be better (Harvey and Howarth 2007; Ijsselsteijn et al. 2001).
- The width and curvature of the screen and consequently the degree of peripheral vision stimulated by the projected image: Research has shown that a wider field of view and consequently greater stimulation of peripheral vision are linked to increased sickness symptoms (Brandt, Dichgans and Koenig 1973); however, these will have to be traded off against the benefits of a larger field of view, that is, greater presence, immersion and validity.
- Design of the driving scenario: Important factors are length of drive (with increased exposure likely to result in more discomfort), number of bends in road and number of braking events (Roe, Brown and Watson 2007).

- Fixed-base versus motion-based rig: In theory, motion should provide the vestibular senses with signals which correspond to visual information in the simulated environment; however, motion systems which do not accurately simulate real movement can actually increase the risk of sickness (Parkes 2012).
- Lag between driver inputs and simulated outputs, as well as frame rate: Lags/frame progressions should be imperceptible but will be reliant on technology and processing power, increasing costs.

Individual differences will also be a significant factor in simulator sickness rates, with evidence suggesting that certain characteristics can make people more susceptible than others. For example gender (Flanagan, May and Dobie 2005; Klosterhalfen et al. 2005), age (Park et al. 2006), ethnicity (Klosterhalfen et al. 2005), menstrual cycle (Clemes and Howarth 2001) and computer game use (Bigoin et al. 2007) have all been studied and/or reported anecdotally in relation to motion sickness susceptibility. The extent to which these factors influence sickness does, however, appear to vary between different studies, and some researchers have actually found no effect of some characteristics: for example age (Mourant and Thattacherry 2000) and computer game use (Mowforth 2005).

Given such contradictory perspectives, we feel that it is important not to compromise the representativeness of a sample based on an automatic assumption that certain groups of people may be more at risk of sickness symptoms than others are (e.g. older females) – leading to an avoidance of such at-risk groups in the research. More sensibly, we recommend screening of participants by asking the study volunteers to answer questions relating to their previous experiences of motion/simulator sickness and related issues (Reed, Diels and Parkes 2007).

2.3.3 Control

There are two issues of importance in relation to control: one is control over experimental design and the other is control over the configuration of a simulator rig. Having flexibility in the design of an experiment enables researchers to answer an almost infinite number of research questions just by manipulating a few variables, such as scenario design, driving tasks, participant instructions and performance measures. Scenarios (i.e. the design of the road environment in which the simulated vehicle is placed) can be developed quickly and often require little programming knowledge (although this is dependent on the software being used). This gives researchers complete control over the vehicle types, road types, traffic densities, weather conditions, behaviours of other (simulated) road users and hazards that participants are exposed to. Simulators can also be set up to provide a huge amount of very specific driving performance and behaviour data, such as longitudinal acceleration, longitudinal and lateral velocities, lane position, heading angle, steering wheel angle input and yaw rate. This provides researchers with incredibly detailed knowledge of exactly how a driver performed at milli-second intervals throughout an experiment; for example gear changes along with simultaneous acceleration and deceleration events can be used to provide a measure of eco-driving. Simulator studies can also be stopped mid-way to enable questionnaires to be administered to participants to capture perceptions almost instantaneously. Examples include the use of workload measures and presence rating scales, such as those described earlier.

The utility of all of these data is, however, reliant on correct interpretation and meaningful statistical analysis, which requires time and expertise. Concern has also been raised over the lack of standardisation and transparency between driving performance measures (Green 2013), making comparisons across the body of simulator research troublesome. It is important to keep in mind that having a large volume of data on its own is not enough to create a holistic understanding of driver behaviour (de Winter et al. 2009) and is also not a guarantee that the same behaviour would be seen under real driving conditions.

As modern-day cars become more sophisticated in design, dynamics and functionality, testing individual, combined and interdependent components can become problematic. Researchers need a way of representing a very large number of different component configurations quickly and cheaply, without requiring access to individual production vehicles. Installing prototype technologies into a simulator rig will be much more cost effective than developing them for integration into a road car, and equipment can be adapted and swapped in and out much more quickly in a simulator in the event that changes need to be made. Simple mock-ups using prototype systems enable the human–machine interface (HMI) to be assessed and problems to be identified at an early stage in the design process, that is, before effort is invested in integrating technologies into real vehicles.

2.3.4 Cost

The cost of setting up a driving simulator varies widely depending upon the fidelity required. Table 2.1 presents a general perspective of where the bulk of costs is incurred for different 'levels' of simulator and for on-road trials as a comparison.

It is impossible to provide accurate estimates of costs for these components as they are highly dependent on other factors. For example, an automotive manufacturer providing funding to a university department for driving research may donate a vehicle free of charge, eliminating a significant cost. There is also considerable variation in software costs, usually linked to the level of control offered over the simulated environment and the level of programming required to generate scenarios. There are

TABLE 2.1

Components Required for Different Levels of Simulator and On-Road Study

Desktop Simulator (Low Fidelity)	Static Base Full Simulator (Medium Fidelity)	Motion Base Full Simulator (High Fidelity)	On-Road Study
Simulation software Hardware (PC with single screen; game wheel, shifter, pedals) Space (requires at minimum a desk and chair)	Simulation software Hardware (vehicle, vehicle controls, projectors, projector screens, multiple PCs with multiple screens) Space (dedicated room to fit car and screens, researcher's control area)	Simulation software Hardware (vehicle, vehicle controls, projectors, screens/projection dome, multiple PCs with multiple screens, hydraulic motion base) Space (dedicated large area to fit car, motion base and screens/dome, researcher's control area)	Hardware (vehicle, instrumentation) Insurance, road tax, fuel

Note: PC, personal computer.

additional associated costs with running driving studies, including but not limited to the following:

- Peripheral equipment, which depends largely on what you are measuring. This could include eye-tracking equipment, physiological monitoring devices, video/audio recording equipment and technologies integrated in-vehicle (i.e. display technologies, input devices).
- Researcher time (this is likely to be slightly more for on-road studies, due to the increased duration of such experiments).
- Payment of participants (similar between simulator and on-road studies, usually linked to length of trial).

2.4 Building and Using a Driving Simulator for Research

In this section, we will draw upon the research discussed already in this chapter, current practices in driving simulator research from published literature and our own experiences with driving simulators over many years to provide some practical guidance for those intending to build and subsequently use driving simulators in research activities.

2.4.1 Building a Simulator

When developing a simulator for use in research, one must be clear on why it is needed in the first instance. This raises various other questions, including who are the intended users of the research and what types of study might be conducted. Ultimately, in building a research-capable simulator, there is a critical requirement to carefully construct the cost–benefit argument – that is, to balance fidelity with validity. As alluded to earlier, it is important to note that fidelity is *not* the same as validity. Whilst higher fidelity is often associated with more realistic performance and behaviour for participants, the validity of a simulation study is also affected by many other things, such as procedures, choice of participants, study design and what or how you measure (see Section 2.4.2).

Designing a simulator with the correct level of fidelity is a difficult task (Hays 1980), although much can be learned from previous driving studies and the experience of other researchers in the field. We undertook a survey of articles published between 2014 and 2016 from two of the top journals in our field (*Ergonomics* and *Applied Ergonomics*) and from the most recent Automotive User Interface Conference (Nottingham, September 2015) to identify how contemporary driving simulators have been used to answer different research questions and any limitations noted by the authors of those articles. The simulators used in this sample of studies ranged in fidelity from desktop arrangements consisting of a gaming steering wheel and single display screen (e.g. Baldwin and Lewis 2014; Jakus, Dicke and Sodnik 2015) to motion-based rigs with full car bodies (e.g. Dziuda et al. 2014). In some cases, the driving environments were generated with dedicated driving simulation software (e.g. STISIM Drive® – Banks and Stanton [2015], Kaber et al. [2015] or OpenDS – Haeuslschmid et al. [2015]), whilst some researchers used off-the-shelf racing games (e.g. Jagannath and Balasubramanian 2014). Other studies used a surrogate task to simulate the effects of driving, for example the occlusion technique (e.g. Pournami et al. 2015). The phenomena studied included the following:

- HMIs in driving
- Head-up displays (Bolton, Burnett and Large 2015; Haeuslschmid et al. 2015)
- Multimodal displays (Jakus, Dicke and Sodnik 2015)
- Novel navigation concepts (Bolton, Burnett and Large 2015; Kaber et al. 2015; Pertere, Meschtscherjakov and Tscheligi 2015)
- Ambient displays (Meschtscherjakov et al. 2015)
- Vehicle automation
- Effect on emergency responses and risk taking (Banks and Stanton 2015)

- Feedback and warnings for automated driving functions (Telpaz et al. 2015)
- Driver characteristics/individual differences
- Driving experience and training (Yamani et al. 2016; Zhao et al. 2014)
- Displays for drivers on the autism spectrum (Shim et al. 2015)
- Effects of age on performance with in-vehicle tasks (McWilliams et al. 2015)
- Risk-taking behaviours (Ba et al. 2016)
- Physiological effects of driving
- Visual field-of-view effects on manoeuvres (Douissembekov et al. 2015)
- Vibration and driver discomfort (Mansfield, Sammonds and Nguyen 2015)
- Driver fatigue (Jagannath and Balasubramanian 2014)
- Simulator sickness (Dziuda et al. 2014; Gálvez-García 2015)
- Psychological effects of driving
- Influence of automation on stress (Reimer, Mehler and Coughlin 2016)
- Social interactions, situation awareness and workload (Jeon, Walker and Gable 2015)

From our perspective, an overall recommendation in designing a simulation experience is to reflect on the fundamental purpose of the simulation and understand what aspects of fidelity are critical to task behaviour or performance. As an example, a simulator intended for use in the development process within a car company may require characteristics quite different from those of a simulator used in a university research lab (e.g. relating to face validity).

Examination of the reported limitations of the studies in our review provided a very useful insight into the impact of simulator design on study outcomes. Common problems reported included the lack of real risk (safety or financial) afforded by simulator studies (Walch et al. 2015) and the poor realism and unrepresentative attentional demand in some simulated scenarios (Telpaz et al. 2015). There were also more specific issues. For example, in a study of feedback for automated lane departure warnings (Itoh and Inagaki 2014), the simulated field of view was limited to 120 degrees; in their discussion, the authors reported that this could have impacted their driving performance results as this field of view did not extend to the blind spot, which would be a factor in lane-keeping performance. Itoh and Inagaki (2014) also reported that participants did not know whether the stiffening of the steering wheel motion in some conditions was a feature of the experiment or a fault in the simulator (it was in fact the former, which was being tested as a potential feedback mechanism for lane departures in semi-automated driving).

In a study of the utility of language-based displays, Politis, Brewster and Pollick (2015) reported a lack of audio feedback from the simulated scenario, which would probably have impacted on their results as there is likely to have been an interaction effect of road/vehicle noise and the audio feedback, which was the subject of the study. Finally, for a study of fatigue measures in driving, Jagannath and Balasubramanian (2014) used an off-the-shelf racing game as the driving stimulus, which could be argued to be not representative of real driving and would likely influence the onset of fatigue differently to simulation software that has been designed specifically for research purposes. These authors also admitted that the physiological measures applied in their study, including muscle fatigue, might not have been valid as there was no force feedback or vibration from the simulator. Therefore, in this case, it would have been more appropriate to use a motion rig to simulate some level of haptic feedback from the vehicle.

These examples highlight the importance of achieving a good fit between the simulator and the research questions. However, it is not appropriate to say that one type of simulator is suitable for one type of study design, as there are so many factors which influence fit and there are often multiple research questions to be addressed in a single study. We have tried to make some general recommendations for simulator requirements across broad categories of driving research, which may help researchers to design valid and successful experiments. Nevertheless, these guidelines are by no means exhaustive and researchers are encouraged to consider every individual research question carefully, on a case-by-case basis:

- HMI in driving: In-vehicle HMIs are usually investigated with regard to their potential for distraction or effects on workload, so the simulator needs to create an equivalent level of primary task demand to real driving. The simulated driving environment and scenario need to be realistic and demanding enough to create a valid dual-task situation, in which the influence of driving on secondary task performance, including the prioritisation of such tasks, is indicative of real driving behaviour. If the research focuses on a specific presentation modality, or multiple modalities, then it is particularly important that simulated feedback relating to the vehicle and driving environment in that same modality is comprehensive and realistic. This will ensure that potential intra-modal interference (e.g. auditory in-vehicle HMI masked by vehicle/road noise) can be evaluated.

- Vehicle automation: Any simulator should be capable of adequately simulating the technical characteristics of the automated technologies under investigation. It should also fully simulate the environment which informs the actions of the driver or technology; for example if participants are required to perform a manoeuvre such as a lane change which requires an over-the-shoulder glance, then the field of view in the simulator should be wide enough to allow this. Furthermore, participants need to believe that any simulated vehicle behaviours are intended and 'real' rather than faults of the simulator. This is a particular problem with novel vehicle technologies such as automation as many participants have no expectations against which to judge realism.

- Driver characteristics/individual differences: The simulator should be accessible to different user groups to allow studies into the full range of driver characteristics. For example, to study older users' driving performance, it must be possible for older participants to get in and out of the simulator and to minimise the risks of simulator sickness, as this can affect older people more than younger people. Studies of passenger behaviour will require a rig with easy access to passenger seats, possibly both front and back depending on the research question. Access will also need to be considered for participants with mobility impairments. If the research is concerned with eliciting particular characteristics from participants, then it is important that there is enough variety in the tasks and driving environment to allow sufficient observable variation in behaviour. For example, studies of risk-taking characteristics will require enough hazards to be present in the simulated road environment to ensure that risky behaviour will be observed.

- Physiological effects of driving: Investigating physiological effects often requires specialist equipment in addition to a car rig to simulate vibrations and force feedback. The amount of equipment which can be mounted safely to a vibration simulator would be limited (and may consist of a seat, pedals and steering wheel only) and would consequently lower the fidelity of the simulation; however, this needs to be traded off against the need for accurate vibration and force feedback.

- Psychological effects of driving: It is important that the phenomenon under investigation is adequately triggered, and this will be dependent on the experimental set-up. Studying situation awareness, for example, will require a scenario containing a reasonably high number of other road users in order to ensure that the participant's awareness extends out to the wider driving environment and that this environment is representative of the real world.

Irrespective of the research question being addressed, it is also very important to plan ahead, considering the longevity of a simulation facility. For example, a point often not considered in the initial purchase of a simulator is the need for continuity of simulator management. As the fidelity of a simulator increases, so does the expertise required to maintain the facility. In this respect, one must consider the skills needed (programming, engineering, electronics etc.) and the extent to which one might become reliant on others (either internally or externally) to maintain optimum simulator operation. Finally, researchers should consider future developments in driving and driving research when selecting a simulator configuration. For example, will it be capable of simulating automated driving and increasing technology in the vehicle? Maintenance of a good-quality simulator will depend upon continuing management and its use in state-of-the art research.

2.4.2 Using a Simulator

In using a driving simulator for human factors research and development, there are many aspects of procedure that need to be considered (e.g. participant selection, use of consent forms, framing of instructions). This final section describes the procedures that should be adopted when running a driving simulator research study. As such, they offer good practice principles in an aim to ensure that the results of such studies are robust and reliable. In turn, we should have greater confidence that conclusions are based on issues of interest to the research team (e.g. which interface is least distracting), rather than confounding variables (e.g. participants' desire to please the experimenter, feelings of sickness).

2.4.2.1 Participant Selection

In choosing participants, there are generic issues to consider which are independent of the environment in which a study will be conducted – for instance, the goals of the study, relevant user characteristics, resources available and stage in design process. Specific to the driving simulator environment, it will also be important to consider the following issues when recruiting participants. Pre-study questionnaires can be employed to collect this information:

- Are the participants likely to be prone to sickness in the simulator? (see earlier section on sickness)
- Will the participants be comfortable and competent in driving the simulator vehicle? (considering whether the vehicle is right-hand or left-hand drive, manual or automatic, position of indicators etc.)
- What experience do participants have of different driving simulators? An individual's behaviour might be significantly influenced by previous experience of another simulator (e.g. in steering behaviour).

2.4.2.2 Incentives for Participants

For any research in human factors, the motivations of participants will have a considerable impact on their behaviour, performance or opinions during a study (Harvey and Burnett 2016). In the context of a driving simulator (essentially an artificial environment), there is a need to carefully ensure that motivations are congruent with those expected in the real world. In particular, we need to be confident that drivers within a simulator prioritise driving tasks (steering, braking etc.) in a manner consistent with their behaviour on the roads. To a large extent, this goal can be achieved through the use of a driving simulator with high physical (face) validity, clear instructions given within information sheets and consent forms and appropriate training periods (discussed later).

In addition, some researchers (e.g. Mazzae et al. 2004) have used financial incentives as a means of motivating 'correct' behaviour in a simulator. A typical strategy might be to set a maximum payment value and then to clearly deduct money if a participant exhibits unrealistic or undesirable driving behaviour (speeding, erratic steering etc.). Whilst this approach has been shown to lead to more realistic driving behaviour in some individuals for certain studies, there is a concern that drivers can adapt their driving style in potentially negative ways, such as incessantly glancing at the speedometer (Harvey and Burnett 2016).

2.4.2.3 Information and Consent

With experimental ethics in mind, it is crucial to obtain informed consent from participants. Typically, this requires participants to (a) be given a standard set of instructions as to the nature of the study and (b) sign a form to demonstrate their understanding. Driving simulator studies specifically need to mention the potential for sickness in these forms, although it will be equally important to avoid worrying participants and/or making them overly sensitive to symptoms; thus, these forms need to be carefully worded.

In simulator studies, it may also be desirable for participants to sign a post-study consent form. There are usually two key purposes to this form. Firstly, it allows participants to declare that they believe that they are fit to drive following the study (e.g. if sickness symptoms have been experienced). Secondly, it can be used to confirm that participants have received any compensation promised at the outset of the study.

2.4.2.4 Familiarisation and Training with the Simulator

It is usually important to provide a period of training in the use of a driving simulator. Additional training may be required in other tasks to be conducted in the study (e.g. use of a new in-vehicle system) depending on the goals of the research (such as learnability versus accomplished performance). For the simulator, participants should be initially introduced to the primary controls of the vehicle and displays and given the opportunity to adjust seat and mirrors, if possible. A short familiarisation scenario (5–10 minutes of driving) is usually sufficient for drivers to become accustomed to the simulator environment. In this scenario, it is generally recommended to start with a very simple environment (straight road, little traffic) and gradually introduce some complexity (slight bends, additional traffic). If the experimental drives are likely to include specific manoeuvre types (e.g. traffic lights, roundabouts), these should also be included in the familiarisation drive. It is particularly important to monitor participants carefully in the familiarisation drive for any symptoms of simulator sickness (see Section 2.4.2.5).

A consideration in the familiarisation drive as well as the experimental drives is how the scenario concludes. In this regard, it is good practice to allow drivers to bring the vehicle to a stop themselves (e.g. in a lay-by) rather than the scenario being aborted by the researcher. This aids in the participant's sense that they really are driving a vehicle, which can subsequently improve behavioural validity.

2.4.2.5 Sickness Prediction, Monitoring and Management

Good practice principles in driving simulator research aim to minimise (a) the likelihood of symptoms developing through pre-screening and monitoring and (b) the consequences of sickness should they arise. Firstly, it is desirable for participants to be primarily selected from lists of previous participants who are known not to suffer from simulator sickness. Where this is not possible, participants will be screened for known risk factors: for example, a history of motion sickness, migraines, epilepsy, dizziness or blurred vision (Roe, Brown and Watson 2007). Those with particular susceptibility to these factors should not take part. It is important that participants are able to self-select themselves (opt-in strategy) based on these criteria to avoid any embarrassment in admitting to specific conditions (i.e. adverts or email requests for participants should clearly specify the criteria so that affected people can choose not to participate).

As noted earlier, an initial information sheet and consent form should be given to participants to make them aware of the likelihood of sickness and to sign to confirm their understanding. A simulator sickness questionnaire should be issued before the study (as a baseline in case of participants being ill), between and after experimental conditions (e.g. Kennedy et al. 1993). The questionnaire is used firstly to ensure that participants are aware of the range of symptoms and secondly to ensure that they are not suffering symptoms during or after the study. If a participant rates higher than a pre-designated value for any of the symptoms, this should be considered a red flag and they should be advised not to continue with the study. For instance, a participant may turn up to the study with a heavy cold or headache and rate high values for specific symptoms. In this situation, they should be discouraged from continuing with the study, as the likelihood of sickness being experienced within the simulator will be relatively high. Participants should be monitored by the researcher and may also be monitored during the study via in-vehicle cameras for symptoms of simulator sickness (e.g. restlessness, excessive yawning, burping). It should also be explained to the participant that they are free to end the study at any point if they feel uncomfortable.

In the instance that a participant does suffer a degree of simulator sickness, a number of contingency measures should be implemented. Firstly, participants should be removed from the simulation environment to a quiet location where they can recover. Drinking water should be made available. The researcher should ensure that a first aider is aware of the potential risk and is available to assist in case of any issues. The participant should be encouraged to remain at the study venue for a minimum of 30 minutes to allow symptoms to subside.

As a final point, it is important to note that there is a need to be more observant regarding the possibility of sickness in situations where exposures are long (e.g. when an individual trial lasts longer than 15 minutes) or an exposure is more intensive with high optic flow (such as a trial which requires several speed changes and/or turns in succession). In both cases, studies have shown increased incidences of sickness (Mourant and Thattacherry 2000; Roe, Brown and Watson 2007). Strategies for being more observant would include (a) utilising only participants who have used the driving simulator in previous studies with no symptoms of sickness, (b) providing greater familiarisation and training in the use of the simulator and/or (c) ensuring that there are several opportunities for comfort breaks in the study.

2.5 Conclusions

This chapter outlines key theoretical and practical issues associated with the design and of use of driving simulators for human factors research. Simulators are research tools that have many advantages, specifically linked to safety, control and ease of measurement. They are also increasingly affordable for research teams, enabling a strong cost–benefit argument to be made. Nevertheless, as this chapter highlights, there are considerable issues to consider when both building a simulator for a particular purpose and then using it to generate reliable, valid and ultimately useful data. In this respect, theoretical, empirical and anecdotal evidence highlights this simple fact – namely whilst fidelity is important for a good driving simulator, *how the simulator is used* is critical to ensure good results.

References

Ba, Y., Zhang, W., Salvendy, G., Cheng, A.S.K., and Ventsislavova. 2016. Assessments of risky driving: A go/no-go simulator driving task to evaluate risky decision-making and associated behavioral patterns. *Applied Ergonomics*, 52: 265–274.

Baldwin, C.L., and Lewis, B.A. 2014. Perceived urgency mapping across modalities within a driving context. *Applied Ergonomics*, 45: 1270–1277.

Banks, V.A., and Stanton, N.A. 2015. Contrasting models of driver behaviour in emergencies using retrospective verbalisations and network analysis. *Ergonomics*, 58: 1337–1346.

Banks, V.A., Stanton, N.A., and Harvey, C. 2014. What the drivers do and do not tell you: Using verbal protocol analysis to investigate driver behaviour in emergency situations. *Ergonomics*, 57: 332–342.

Bigoin, N., Porte, J., Kartiko, I., and Kavakli, M. 2007. Effects of depth cues on simulator sickness. In *ICST IMMERSCOM 2007*. Verona, Italy: ACM SIGMM Technically sponsored by Create-Net and EURASIP. 10–12 October 2007, 1–4.

Biocca, F. 1997. The cyborg's dilemma: Embodiment in virtual environments. In *Second International Conference on Cognitive Technology: "Humanizing the Information Age"*. Stoughton, WI: IEEE.

Blaauw, G. 1982. Driving experience and task demands in simulator and instrumented car: A validation study. *Human Factors: The Journal of the Human Factors and Ergonomics Society*, 24: 473–486.

Blana, E. 1996. *Driving Simulator Validation Studies: A Literature Review*. Leeds, UK: Institute of Transport Studies, University of Leeds.

Bolton, A., Burnett, G., and Large, D.R. 2015. An investigation of augmented reality presentations of landmark-based navigation using a head-up display. In *Automotive UI '15: 7th International Conference on Automotive User Interfaces and Interactive Vehicular Applications*. Nottingham, UK: ACM, 1–3 September 2015, 56–63.

Brandt, T., Dichgans, J., and Koenig, E. 1973. Differential effects of central versus peripheral vision on egocentric and exocentric motion perception. *Experimental Brain Research*, 16: 476–491.

Brookhuis, K.A., De Vries, S.C., and De Waard, D. 1991. The effects of mobile telephoning on driving performance. *Accident Analysis & Prevention*, 23: 309–316.

Burnett, G., Irune, A., and Mowforth, A. 2007. Driving simulator sickness and validity: How important is it to use real car cabins? *Advances in Transportation Studies*, 33–42.

Burnett, G.E., and Donkor, R.A. 2010. Evaluating the impact of head-up display complexity on peripheral detection performance: A driving simulator study. *Advances in Transportation Studies*, 28: 5–16.

Casali, J.G., and Frank, L.H. 1988. *Manifestation of Visual/Vestibular Disruption in Simulators: Severity and Empirical Measurement of Symptomatology*. Blacksburg, VA/Washington, DC: Virginia Polytechnic Institute and State University/National Aeronautics and Space Administration.

Casali, J.G., and Wierwille, W.W. 1986. Potential design etiological factors of simulator sickness and a research simulator specification. *Transportation Research Record*, 1059: 66–74.

Clemes, S.A., and Howarth, P.A. 2001. Changes in virtual simulator sickness susceptibility over the menstrual cycle. In *The 38th UK Conference on Human Response
to Vibration*. Farnborough, UK: QinetiQ.

de Winter, J.C.F., de Groot, S., Mulder, M., Wieringa, P.A., Dankelman, J., and Mulder,
J.A. 2009. Relationships between driving simulator performance and driving
test results. *Ergonomics*, 52: 137–153.

de Winter, J.C.F., van Leeuwen, P.M., and Happee, R. 2012. Advantages and disadvantages of driving simulators: A discussion. In *Proceedings of Measuring Behaviour
Conference, 2012*. Utrecht, Netherlands: ACM, 28–31 August 2012, 47–50.

Deniaud, C., Honnet, V., Jeanne, B., and Mestre, D. 2015. The concept of "presence"
as a measure of ecological validity in driving simulators. *Journal of Interaction
Science*, 3: 1.

Donkor, R.A., Burnett, G., and Sharples, S. 2014. Measuring the emotional validity of
driving simulators. *Advances in Transportation Studies*, 1: 51–64.

Dorn, L., and Barker, D. 2005. The effects of driver training on simulated driving
performance. *Accident Analysis & Prevention*, 37: 63–69.

Douissembekov, E., Michael, G.A., Rogé, J., Bonhoure, P., Gabaude, C., and Navarro,
J. 2015. Effects of shrinkage of the visual field through ageing on parking performance: A parametric manipulation of salience and relevance of contextual
components. *Ergonomics*, 58: 698–711.

Dziuda, L., Biernacki, M.P., Baran, P.M., and Truszczyński, O.E. 2014. The effects of
simulated fog and motion on simulator sickness in a driving simulator and the
duration of after-effects. *Applied Ergonomics*, 45: 406–412.

Fairclough, S.H., and Graham, R. 1999. Impairment of driving performance caused by
sleep deprivation or alcohol: A comparative study. *Human Factors*, 30: 201–217.

Flanagan, M.B., May, J.G., and Dobie, T.G. 2005. Sex differences in tolerance to
visually-induced motion sickness. *Aviation, Space and Environmental Medicine*,
76: 642–646.

Frank, L., Casali, J., and Wierwille, W.W. 1988. Effects of visual display and motion
system delays on operator performance and uneasiness in a driving simulator.
Human Factors, 30: 201–217.

Gálvez-García, G. 2015. A comparison of techniques to mitigate simulator adaptation
syndrome. *Ergonomics*, 58: 1365–1371.

Godley, S.T., Triggs, T.J., and Fildes, B.N. 2002. Driving simulation validation for
speed research. *Accident Analysis & Prevention*, 34: 589–600.

Green, P. 2013. Standard definitions for driving measures and statistics: Overview
and status of recommended practice J2944. In *Automotive User Interfaces and
Interactive Vehicular Applications: AutoUI*, Eindhoven, Netherlands: ACM, 27–30
October 2013, 184–191.

Haeuslschmid, R., Schnurr, L., Wagner, J., and Butz, A. 2015. Contact-analog warn-
ings on windshield displays promote monitoring the road scene. In *Automotive
UI '15: 7th International Conference on Automotive User Interfaces and Interactive
Vehicular Applications*. Nottingham, UK: ACM, 1–3 September 2015, 64–71.

Harms, 1994. Driving performance on a real road and in a driving simulator: Results
of a validation study. Vision in vehicles, Vol. V. North Holland: Elsevier.

Harvey, C., and Burnett, G. 2016. The influence of incentives and instructions on
behaviour in driving simulation studies. In *European Conference on Cognitive
Ergonomics*. Nottingham, UK: ACM, article no. 17.

Harvey, C., and Howarth, P.A. 2007. The effect of display size on visually-induced motion sickness (VIMS) and skin temperature. In *First International Symposium on Visually Induced Motion Sickness, Fatigue, and Photosensitive Epileptic Seizures, VIMS 2007.* Hong Kong: HKUST Publishing Technology Centre, 10–11 December 2007.

Harvey, C., Stanton, N.A., Pickering, C.A., McDonald, M., and Zheng, P. 2011. To twist or poke? A method for identifying usability issues with the rotary controller and touch screen for control of in-vehicle information systems. *Ergonomics,* 54: 609–625.

Hays, R.T. 1980. Simulator fidelity: A concept paper. Alexandria, VA: US Army Institute for the Behavioral and Social Sciences.

Hoffman, J.D., Lee, J.D., Brown, T.L., and McGehee, D.V. 2002. Comparison of driver braking responses in a high-fidelity simulator and on a test track. *Transportation Research Record,* 1803: 59–65.

Hughes, T., and Rolek, E. 2003. Fidelity and validity: Issues of human behavioral representation requirements development. Paper read at *Proceedings of the 2003 Winter Simulation Conference,* 7–10 December 2003, at New Orleans, LA.

Ijsselsteijn, W., De Ridder, H., Freeman, J., Avons, S.E., and Bouwhuis, D. 2001. Effects of stereoscopic presentation, image motion, and screen size on subjective and objective corroborative measures of presence. *Presence,* 10: 298–311.

Insko, B.E. 2003. Measuring presence: Subjective, behavioral and physiological methods. In G. Riva, F. Davide and W.A. Ijsselsteijn (Eds.), *Being There: Concepts, Effects and Measurement of User Presence in Synthetic Environments.* Amsterdam, Netherlands: IOS Press, 109–119.

Itoh, M., and Inagaki, T. 2014. Design and evaluation of steering protection for avoiding collisions during a lane change. *Ergonomics,* 57: 361–373.

Jagannath, M., and Balasubramanian, V. 2014. Assessment of early onset of driver fatigue using multimodal fatigue measures in a static simulator. *Applied Ergonomics,* 45: 1140–1147.

Jakus, G., Dicke, C., and Sodnik, J. 2015. A user study of auditory, head-up and multimodal displays in vehicles. *Applied Ergonomics,* 46: 184–192.

Jamson, H. 2001. Image characteristics and their effect on driving simulator validity. In *First International Driving Symposium on Human Factors in Driver Assessment, Training and Vehicle Design.* Aspen, CO: University of Iowa, 14–17 August 2001.

Jeon, M., Walker, B.N., and Gable, T.M. 2015. The effects of social interactions with in-vehicle agents on a driver's anger level, driving performance, situation awareness, and perceived workload. *Applied Ergonomics,* 50: 185–199.

Kaber, D., Pankok Jr., C., Corbett, B., Ma, W., Hummer, J., and Rasdorf, W. 2015. Driver behavior in use of guide and logo signs under distraction and complex roadway conditions. *Applied Ergonomics,* 47: 99–106.

Kaptein, N., Theeuwes, J., and Van der Horst, R. 1996. Driving simulator validity: Some considerations. *Transportation Research Record,* 1550: 30–36.

Kelly, S.W., Kinnear, N., Thomson, J., and Stradling, S. 2010. A comparison of inexperienced and experienced drivers' cognitive and physiological responses to hazards. In L. Dorn (Ed.), *Driver Behaviour and Training.* Farnham, Surrey, UK: Ashgate, 23–36.

Kennedy, R.S., Lane, N.E., Berbaum, K.S., and Lilienthal, M.G. 1993. Simulator sickness questionnaire: An enhanced method for quantifying simulator sickness. *The International Journal of Aviation Psychology,* 3: 203–220.

Klosterhalfen, S., Kellermann, S., Fang, P., Stockhorst, U., Hall, G., and Enck, P. 2005. Effects of ethnicity and gender on motion sickness susceptibility. *Aviation, Space and Environmental Medicine,* 76: 1051–1057.

Kolasinski, E.M. 1995. *Simulator Sickness in Virtual Environments.* Alexandria, VA: Army Research Institute for the Behavioural and Social Sciences.

Konstantopoulos, P., Chapman, P., and Crundall, D. 2010. Driver's visual attention as a function of driving experience and visibility: Using a driving simulator to explore drivers' eye movements in day, night and rain driving. *Accident Analysis & Prevention,* 42: 827–834.

Large, D.R., Crundall, E., Burnett, G., Harvey, C., and Konstantopoulos, P. 2016. Driving without wings: The effect of different digital mirror locations on the visual behaviour, performance and opinions of drivers. *Applied Ergonomics,* 55: 138–148.

Lee, K.M. 2004. Presence, explicated. *Communication Theory,* 14: 27–50.

Lessiter, J., Freeman, J., Keogh, E., and Davidoff, J. 2001. A cross-media presence questionnaire: The ITC-sense of presence inventory. *Presence: Teleoperators and Virtual Environments,* 10: 282–297.

Liu, D., Macchiarella, N., and Vincenzi, D.A. 2009. Simulation fidelity. In D.A. Vincenzi, J.A. Wise, M. Mouloua, and P.A. Hancock (Eds.), *Human Factors in Simulation and Training* (61–73). Boca Raton, FL: CRC Press.

Mansfield, N., Sammonds, G., and Nguyen, L. 2015. Driver discomfort in vehicle seats – Effect of changing road conditions and seat foam composition. *Applied Ergonomics,* 50: 153–159.

Mazzae, E.N., Ranney, T.A., Watson, G.S., and Wightman, J.A. 2004. Hand-held or hands-free? The effects of wireless phone interface type on phone task performance and driver preference. *Proceedings of the Human Factors and Ergonomics Society Annual Meeting,* 48: 2218–2222.

McGehee, D., Mazzae, E., and Baldwin, G. 2000. Driver reaction time in crash avoidance research: Validation of a driving simulator study on a test track. *Proceedings of the Human Factors and Ergonomics Society Annual Meeting,* 44: 320–323.

McWilliams, T., Reimer, B., Mehler, B., Dobres, J., and Coughlin, J.F. 2015. Effects of age and smartphone experience on driver behavior during address entry: A comparison between a Samsung Galaxy and Apple iPhone. In *Automotive UI '15: 7th International Conference on Automotive User Interfaces and Interactive Vehicular Applications.* Nottingham, UK: ACM, 1–3 September 2015, 150–153.

Meschtscherjakov, A., Döttlinger, C., Rödel, C., and Tscheligi, M. 2015. ChaseLight: Ambient LED strips to control driving speed. In *7th International Conference on Automotive User Interfaces and Interactive Vehicular Applications: Auto UI.* Nottingham, UK: ACM, 1–3 September 2015, 212–219.

Mourant, R.R., and Thattacherry, T.R. 2000. Simulator sickness in a virtual environments driving simulator. *Proceedings of the Human Factors and Ergonomics Society Annual Meeting,* 44: 534–537.

Mowforth, A. 2005. *An Investigation into the Role of Presence as a Means of Validating Driving Simulators.* Nottingham, UK: University of Nottingham.

Nilsson, L. 1993. Behavioural research in an advanced driving simulator: Experience of the VTI system. *Proceedings of the Human Factors and Ergonomics Society Annual Meeting,* 37: 612–616.

Park, G.D., Allen, R.W., Fiorentino, D., Rosenthal, T.J., and Cook, M.L. 2006. Simulator sickness scores according to symptom susceptibility, age, and gender for an older driver assessment study. *Human Factors and Ergonomics Society 50th Annual Meeting*, 50: 2702–2706.

Parkes, A. 2012. The essential realism of driving simulators for research and training. In N. Gkikas (Ed.), *Automotive Ergonomics* (133–154). Boca Raton, FL: CRC Press.

Pertere, N., Meschtscherjakov, A., and Tscheligi, M. 2015. Co-navigator: An advanced navigation system for front-seat passengers. In *Automotive UI '15: 7th International Conference on Automotive User Interfaces and Interactive Vehicular Applications*. Nottingham, UK: ACM, 1–3 September 2015, 187–194.

Philip, P., Taillard, J., Klein, J., Sagaspe, E., Charles, A., Davies, W., Guilleminault, C., and Bioulac, B. 2003. Effect of fatigue on performance measured by a driving simulator in automobile drivers. *Journal of Psychosomatic Research*, 55: 197–200.

Politis, I., Brewster, S., and Pollick, F. 2015. Language-based multimodal displays for the handover of control in autonomous cars. In *Automotive UI "15: 7th International Conference on Automotive User Interfaces and Interactive Vehicular Applications*. Nottingham, UK: ACM, 1–3 September 2015, 3–10.

Pournami, S., Large, D.R., Burnett, G., and Harvey, C. 2015. Comparing the NHTSA and ISO occlusion test protocols: How many participants are sufficient? In *Automotive UI '15: 7th International Conference on Automotive User Interfaces and Interactive Vehicular Applications*. Nottingham, UK: ACM, 1–3 September 2015, 110–116.

Reason, J.T., and Brand, J.J. 1975. *Motion Sickness*. London: Academic Press.

Reed, M.P., and Green, P.A. 1999. Comparison of driving performance on-road and in a low-cost simulator using a concurrent telephone dialling task. *Ergonomics*, 42: 1015–1037.

Reed, N., Diels, C., and Parkes, A.M. 2007. Simulator sickness management: Enhanced familiarisation and screening processes. In *First International Symposium on Visually Induced Motion Sickness, Fatigue, and Photosensitive Epileptic Seizures, VIMS 2007*. Hong Kong: HKUST Publishing Technology Centre, 10–11 December 2007, 156–162.

Reimer, B., Mehler, B., and Coughlin, J.F. 2016. Reductions in self-reported stress and anticipatory heart rate with the use of a semi-automated parallel parking system. *Applied Ergonomics*, 52: 120–127.

Roe, C., Brown, T., and Watson, G. 2007. Factors associated with simulator sickness in a high-fidelity simulator. Education, 251, 5A.

Saad, F. 2006. Some critical issues when studying driver behavioural adaptations to new driver support systems. *Cognition, Technology & Work*, 8: 175–181.

Schricker, B.C., Franceschini, R.W., and Johnson, T.C. 2001. Fidelity evaluation framework. In *34th Annual Simulation Symposium*. Seattle, WA: IEEE, 22–26 April 2001.

Shim, L., Liu, P., Politis, I., Regener, P., Brewster, S., and Pollick, F. 2015. Evaluating multimodal driver displays of varying urgency for drivers on the autistic spectrum. In *Automotive UI '15: 7th International Conference on Automotive User Interfaces and Interactive Vehicular Applications*. Nottingham, UK: ACM, 1–3 September 2015, 133–140.

Society of Automotive Engineers. 2015. *SAE J2944, Operational Definitions of Driving Performance Measures and Statistics*. Warrendale, PA: SAE.

Telpaz, A., Rhindress, B., Zelman, I., and Tsimhoni, O. 2015. Haptic seat for automated driving: Preparing the driver to take control effectively. In *Automotive UI '15: 7th International Conference on Automotive User Interfaces and Interactive Vehicular Applications*. Nottingham, UK: ACM, 1–3 September 2015, 23–30.

Toffin, D., Reymond, G., Kemeny, A., and Droulez, J. 2003. Influence of steering wheel toque feedback in a dynamic driving simulator. In *Driving Simulation Conference – North America*. Dearborn, MI: DSC, 8–10 October 2003.

Törnros, J. 1998. Driving behaviour in a real and a simulated road tunnel – A validation study. *Accident Analysis & Prevention*, 30: 497–503.

Underwood, G., Crundall, D., and Chapman, P. 2011. Driving simulator validation with hazard perception. *Transportation Research Part F: Traffic Psychology and Behaviour*, 14: 435–446.

Walch, M., Lange, K., Baumann, M., and Weber, M. 2015. Autonomous driving: Investigating the feasibility of car-driver handover assistance. In *Automotive UI '15: 7th International Conference on Automotive User Interfaces and Interactive Vehicular Applications*. Nottingham, UK: ACM, 1–3 September 2015, 11–18.

Wang, Y., Mehler, B., Reimer, B., Lammers, V., D'Ambrosio, L.A., and Coughlin, J.F. 2010. The validity of driving simulation for assessing differences between in-vehicle informational interfaces: A comparison with field testing. *Ergonomics*, 53: 404–420.

Weir, D., and Clark, A. 1995. *A Survey of Mid-level Driving Simulators (SAE Technical Paper 950172)*. Warrendale, PA: Society of Automotive Engineers.

Witmer, B.G., and Singer, M.J. 1994. *Measuring Presence in Virtual Environments*. Alexandria, VA: US Army Research Institute for the Behavioural and Social Sciences.

Witmer, B.G., and Singer, M.J. 1998. Measuring presence in virtual environments: A presence questionnaire. *Presence*, 7: 225–240.

Yamani, Y., Samuel, S., Knodler, M.A., and Fisher, D.L. 2016. Evaluation of the effectiveness of a multi-skill program for training younger drivers on higher cognitive skills. *Applied Ergonomics*, 52: 135–141.

Zhao, N., Chen, W., Xuan, Y., Mehler, B., Reimer, B., and Fu, X. 2014. Drivers' and non-drivers' performance in a change detection task with static driving scenes: Is there a benefit of experience? *Ergonomics*, 57: 998–1007.

3

Driver Training

Andrew Parkes and Nick Reed

CONTENTS

3.1 Introduction

The case for simulator training of vehicle operators is compelling. Simulators can safely provide trainees with virtual experiences of risky or uncommon challenges that might be encountered in real situations. The use of computers to deliver the simulated experience means that the training scenario is a data-rich environment. These data can be exploited to develop targeted performance evaluation metrics and feedback. Furthermore, simulated training scenarios are precisely repeatable, such that performance improvement by a trainee can be demonstrated by repetition of the same scenario and/or the performance of that trainee can be compared to that of another in a robust manner.

In recent years, the areas of use for simulators in operator training have spread considerably from its origins in flight and there are now simulators for car, bus and truck driver training. In this chapter, we explore the research undertaken to examine the use of driving simulators for the training of drivers of road vehicles, with a particular focus on the use of simulation for the training of truck drivers, and we consider some of the challenges that must be overcome in the successful delivery of such training.

3.2 Types of Simulator Technology for Driver Training

Simulators used for driver training are a subset of synthetic driver training systems. They attempt to convey aspects of the real driving situation to the trainee, who, usually with the support of a qualified trainer, acts on the available controls and experiences the resultant outcomes in prepared scenarios. Learning is supported through these experiences and any real-time or post hoc feedback, whether through objective measures from the system or subjective comments from the trainer. This view of a simulator as a device to provide a trainee with aspects of the real driving situation permits a wide range of systems to be legitimately referred to as 'simulators'.

Driving simulators come in a wide range of configurations with a similarly wide cost base and offering very different levels of fidelity of simulation. The most basic, it could be argued, are free-to-download driving simulator smartphone applications that offer an entertainment experience in the context of driving training (e.g. 'School Driving 3D' for iPhone operating system [iOS]/ Android [Ovidiu Pop 2016], 'Dr. Driving' for iOS/Android [SUD Inc. 2016]). The most sophisticated research driving simulator may be the National Advanced Driving Simulator (NADS) in the United States, which, in addition to being a multimillion dollar system to develop, requires a full-time support team of specialists to maintain and develop the system and create new driving scenarios.

Lang et al. (2007), as part of the TRAIN-ALL project funded by the European Commission, suggested a five-tier incremental classification of driving simulators based on a range of parameters, similar in spirit to the classification used by the US Federal Aviation Administration. The classification ranges from level A (the most basic) to level E (the most complex). A simulator classified as level A would consist of a single display screen, mock vehicle controls and simple training scenarios with no kinaesthetic feedback and very limited flexibility in the vehicle models available or driving environments. Progressing through the simulator classifications adds greater sophistication in a number of areas: more realistic control interfaces, larger field of view, greater flexibility in the development of training scenarios, more realistic behaviour of other road users and the addition of motion cueing. Level E simulators (such as the NADS) have large-scale, high-resolution graphic displays, advanced motion cueing systems and complex, highly interactive training scenarios. However, no current driving simulator dedicated to driver training would be classified as falling within level E.

3.3 Commercial Driver Training

In 1996, a report from the US Federal Highway Administration (FHWA 1996) detailed a scoping study on commercial motor vehicle driving simulator technology. It cited an earlier 1991 special issue of *Heavy Duty Trucking* that claimed the following: 'Cost-effective training simulators are becoming technologically possible – there have been astounding leaps in computer graphics and realism – at the same time the driver shortage and the Commercial Driver License (CDL) are forcing the trucking industry to seek more effective methods for driver training, selection and screening'. Given the sheer size of the trucking industry in the United States and Europe, some may consider it surprising that there are relatively few commercial truck simulators in existence and little consensus on the content of any curriculum delivery.

Indeed, the intervening period since the FHWA study has seen continued technological development in simulators, particularly in visual database rendering, but commercial uptake and development of simulation facilities for commercial truck driver training have been inconsistent and not widely implemented. There appear to be three fundamental reasons for the relatively slow adoption of simulation as a key component of professional truck driver training:

- A lack of documented evidence showing a clear benefit of simulation training over traditional on-road and test track methods
- A concern over the economics of providing high-technology facilities and the attendant high costs of entry to the area
- A concern from the drivers that such training will be additional to, rather than replace parts of, the current requirements

One might conclude that, to date, genuine benefits to simulator training have been hard to identify and there has been limited direction or compulsion from regulations to push the industry towards the use of simulators for driver training and contrast simulation training in the commercial truck sector with that in the very different military ground vehicle and aviation sectors where the presence of cost–benefit models, accreditation and certification bodies and agreed curricula is evident (see e.g. Chapter 9).

The European Union Directive 2003/59/EC provided the haulage industry in Europe with a mandate that had the potential to support the adoption of simulator training. It requires all professional bus, coach and lorry drivers to hold a Drivers' Certificate of Professional Competence (Driver CPC) in addition to their vocational driving licence. The objective of the Driver CPC is to improve the knowledge and skills of large goods vehicle and driving passenger-carrying vehicle (PCV) drivers upon entering the sector, and their key target is to ensure professional driver development throughout a career. The Driver CPC directive was introduced into an industry that has typically been sceptical towards driver training, particularly in the context of diminishing profit margins and escalating fuel costs. There is awareness that within the wider community, the need is to ensure minimum compliance with the Driver CPC. However, some companies are embracing the opportunity and introducing training programmes that can be recognised as delivering wider change management. The Driver CPC is simply the first step within the EU to ensure that all individuals entering the profession understand that they have a career of continuous skill development, which in the longer term may be related to the type of vehicle and the loads and environment in which they are able to drive.

The directive applied to all member states from 10 September 2009 and made 35 hours of periodic training obligatory every five years for truck drivers in order to maintain their employability. Directive 2003/59/EC of the European Parliament and the Council of 15 July 2003 on the initial qualification and periodic training of drivers of certain road vehicles for the carriage of goods or passengers amends Council Regulation (European Economic Community [EEC]) no. 3820/85 and Council Directive 91/439/EEC and repeals Council Directive 766/914/EEC; it also allows for the use of simulation for both for heavy goods vehicle and PCV training (EU Commission 2003).

The periodic training is not prescribed. It may include computer-based training and workshops, be classroom based or 'in-vehicle', depending on which suits the needs of the individual driver and/or their employer. In the United Kingdom, the training centre and course must be approved to deliver Driver CPC training by the Joint Approvals Unit for Periodic Training.

The regulation shows where simulation, and synthetic training in general, could provide a valuable role but does not prescribe exactly which elements may be suitable, nor proscribe those that are unsuitable. Furthermore, drivers can fulfil their training requirements under the CPC regulations by completing relatively low-cost, commoditised classroom-based training rather than more expensive but potentially more valuable simulator-based training. In this context, whilst simulation training is widespread, successful and necessary in aviation or military ground vehicle applications, it has yet to become similarly well accepted and suitable for truck driver training.

3.4 Fidelity and Validity

A review by Williges et al. (1973) pondered the 50-year history of flight simulation and concluded that '… many issues concerning ground based flight simulators and trainers remain unanswered'. Many concerns remain in aviation, and most remain to be addressed at all in a systematic fashion in the driver training industry.

The prospective benefits of simulation are clear. There is potential for control of the training environment, repeatability of specific combinations of features, objective performance scoring, cost reduction and consistent online tutorial delivery. The training environment has the potential to be more effective than the real world due to the ability to remove unessential elements from any particular scenario and safer due to the lack of physical risk, no matter how catastrophic the performance failure is.

However, to make rational judgements about the use of simulators for cost-effective driver training, potential operators require answers to the following questions:

- What can training simulators really do?
- How well can they do it (and what supporting evidence is there to prove this)?
- How much will they cost (initial, ongoing, upgrade and disposal costs)?
- What new skills will trainers need (and/or what new trainers will be needed)?
- How will the simulators, trainers and training scenarios be accredited?
- How should simulators be used within a wider curriculum?

The problem at present is that, whilst there are several convincing high-fidelity driving simulation systems available, there are very few answers available to the last of the previous questions. There is little known in relation to driving, and little that is directly transferable from aviation, that can inform discussion of what should be delivered in a simulation training package, nor how the costs and benefits might compare to real road training.

Figure 3.1 attempts to demonstrate the current dilemma. We might expect that there would be a clear (even possibly linear) relationship between the cost of a particular simulation system and the value of the training transfer that could be derived (line A). In reality, the relationship is likely to be less straightforward.

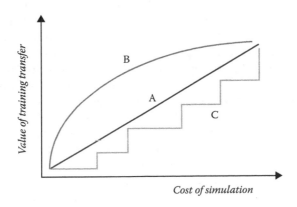

FIGURE 3.1
Models of fidelity versus cost of simulation. (From Parkes, A. M., in Gkikas, N. (ed), *Automotive Ergonomics*, CRC Press, Boca Raton, FL, 2013, pp. 133–154.)

There is certainly a strong suspicion that it may become increasingly expensive to add ever greater fidelity, and added expenditure may result in diminishing returns on investment (line B). With real-life procurement, there are many go/no go decisions to be made in simulator specification, and so, a step function (line C) may be more realistic. Decisions such as whether to include a motion system to provide acceleration cues and thereby increase perceived realism, to include multiple channel projection systems for a panoramic visual projection or to include sophisticated three-dimensional sound rendering for a convincing soundscape all require jumps in technology provision that have substantial cost implications.

So, what degree and fidelity of simulation are necessary for effective training of truck drivers? For maximum face validity of a driving simulator, it would be necessary to specify the highest degree and fidelity available within a particular budget, but if the training has to be cost effective when compared to traditional real-world training, budgets will be constrained and result in a compromise between what might be desired and what can be provided.

At present, there is little information available to enable perfect choices between expenditure on a particular motion system instead of on a particular visual system or even sound and vibration systems. It is not possible to state unequivocally at present whether motion is necessary for successful training or, more specifically, for what types of driver training it is or is not necessary. The aviation literature provides a range of views. Some have suggested that because experienced pilots often rely on motion rather than instrument readings, motion becomes more important as experience level increases (Briggs and Wiener 1959, cited in Williges et al. [1973]). Similarly, it might be argued that experienced truck or car drivers rely more on motion, sound and vibration rather than dashboard displays to judge the performance of the vehicle, whereas novice drivers might derive substantial benefit from systems that focus on instrument displays. Thus, some training lessons appropriate for novice drivers might be conducted on simple part-task trainers, but advanced skill-based lessons would require a motion component. If we decide that the motion is important, then fidelity must be addressed. Poor motion systems might not only have a negative transfer of training to real-world situations; they will also lead to increased levels of simulator sickness.

Similar arguments might surround the fidelity of visual databases. The simple view is that they need to be as realistic as possible. However, from a training perspective, that might not be correct. Certainly, in terms of resolution, field of view, brightness, contrast and refresh rates, there seems value to having higher fidelity. However, it might be argued that the content of the visual scene itself does not have to be high fidelity (if that means close to photo-realistic representation of a real scene). There may be value in taking unimportant elements out of a visual scene, allowing the driver to concentrate on elements salient to the training objective without distraction. Anyone involved in visual database (traffic scene) development knows that there is a distinct law of diminishing returns (line B in Figure 3.1) to further expenditure beyond a certain point. Williges et al. (1973) proposed the notion of essential realism, relating not to what might be regarded as essential for improved face validity but, instead, essential to the particular training requirements under consideration.

There are three important elements that should drive decisions on simulation provision within the training process:

- The efficiency and acceptability of the learning experience in the simulator
- The transfer of the learning to the real world
- The retention of what was learned

Welles and Holdsworth (2000) reviewed features necessary to successful training in a range of commercial simulators and concluded that '... data to date, although sketchy, anecdotal or very preliminary, provides strong suggestion that driving simulators ... can reduce accidents, improve driver proficiency and safety awareness, and reduce fleet operations and maintenance costs'. They refer to hazard perception training with a particular police force leading to reductions in intersection accidents of around 74% and an overall accident reduction of around 24% in a six-month period following training. In another report, Helman et al. (2010) argued that it is hazard perception training that has the most easily demonstrated benefit for novice drivers in terms of direct transfer of skills and subsequent collision involvement rates.

3.5 Development of a Truck Driving Simulator

The UK Department for Transport commissioned TRL (Transport Research Laboratory) to investigate the feasibility of a truck driving simulator tailored to the needs of the UK road haulage industry. An initial scoping study (Lester et al. 2003) engaged with industry stakeholders to determine what features would be desired from a truck driver training simulator. Based on the findings of this study and funded by the UK Department for Transport, TRL procured a full-mission high-fidelity simulator, with appropriate bespoke UK road databases and courseware (Parkes 2003. Parkes and Rau 2004).

The full system became operational in October 2003 at the TRL headquarters in Berkshire, England. During the period November 2003 to March 2004, over 600 drivers took part in training and validation trials. The main focus was to provide quantitative analysis of the efficiency and acceptability of training exercises provided within the synthetic environment.

The presentation of training exercises was balanced so that sufficient numbers of students were exposed to each one. The emphasis in this phase of the project was not on the general improvement of each individual student, but rather on the perceived pedagogical value of the exercise itself.

This research phase had two main objectives: first, to expose a large number of professional drivers and freight companies to the potential of synthetic training and, second, to inform the relevant UK government authorities how synthetic training could best be integrated into programmes of training and testing to support compliance with the EU directives and help support such an important industry.

Trainees were put into pairs, with one trainer to each pair of trainees. A curriculum was provided that covered driving theory, hazard perception training as an interactive PC programme, simulation familiarisation on a part-task trainer and then training exercises in the full-mission simulator, followed by detailed feedback at a separate review station.

The need to expose as many drivers as possible to the truck simulator in the available time meant that there was a practical limit to the time available for each driver in the full-mission simulator. Each pair of drivers had around one hour in the simulator followed by around 30 minutes at the debrief station. This was insufficient to provide tailored training for the individual but provided time in which the student could gain a significant appreciation of the benefits and drawbacks of the simulator and the courseware provided. The whole session for the trainee lasted around four hours.

In terms of the various training exercises, it was those that focused on the tactical level of driving (fuel efficiency, poor weather driving techniques, adapting to different dynamic load conditions such as liquid or hanging loads or responding to emergency situations such as a tyre blowout) that were deemed the most effective. Exercises that focused more on the operational control level, such as low-speed parking manoeuvres, were seen to have less of an advantage over traditional training in a real vehicle and would require more development before being seen as a viable alternative. It is instructive that tactical rather than operational level driving tasks were rated as being more effective. These require greater relative dependence on drivers having a well-developed understanding of the driving situation and manipulating the vehicle controls to ensure that safe and effective progress is maintained. This understanding is relatively more important than the simulation delivering the exact sensory feedback that would be experienced in a real vehicle. Conversely, for operational driving, accurate sensory feedback may be vital in giving the driver a sense of the size, mass and movement of the vehicle, whilst accurate tactile feedback through the control interfaces may be important for the driver to develop a sense of how to manoeuvre a large vehicle in a confined environment.

In an extension of the truck simulator research programme, a longitudinal cohort study (Reed et al. 2009) was conducted to investigate transfer of training effects. In this study, 60 professional truck drivers were recruited from

11 different companies to participate in a simulator training programme designed to improve fuel efficiency. Participants attended the simulator on three occasions for training over a period of six months. In each training visit, the trainees completed a familiarisation drive in the simulator followed by two drives on a test route featuring a mix of rural and urban driving environments. Between the two drives, drivers were given instructions by a professional trainer in how to complete the route fuel efficiently. After their second drive on the test route, the trainees were given feedback by the trainer on their performance. This included a report generated by the simulator data which measured their performance on a number of criteria (e.g. speed, torque, engine revolutions per minute [RPM], accelerator use, brake use, time taken, number of gear changes) and gave a red/amber/green assessment with a computer-generated comment on how to improve. The trainer used this sheet to deliver guidance to the trainee on the elements of their driving upon which they should concentrate in order to achieve the biggest improvements in fuel efficiency.

Fuel efficiency records from real vehicles for each of the 60 drivers recruited to participate in training were requested from their employers for the week before and the week after each of their three simulator training sessions. This permitted observation of both the immediate impact of the training on fuel efficiency and the resilience of any training effects over time. To act as a control group, the same real-world fuel efficiency records for a matched driver (who did not receive any training) were sourced from the same company for each of the 60 drivers participating in the training. The matched driver was specified as having driving duties similar to those of the trained driver for whom they were the control. This control group reduced the impact of natural variation in fuel efficiency over the year due to changes in traffic and climatic conditions.

Complete records for all simulator training data and from all pre- and post-training real vehicle fuel efficiency periods were available for 37 of the 60 drivers that participated in the study. For this group, the results demonstrated that the drivers experiencing simulator training showed a significant real-world improvement in fuel efficiency, suggesting that techniques trained in the simulator had transferred to their real vehicles. The mean improvement in fuel efficiency across the 37 drivers was 15.7%. This represents a big potential saving for haulage companies as margins are tight and fuel is a significant proportion of their costs.

Despite all efforts made to mitigate its effects, six drivers failed to complete the first training session due to symptoms of simulator sickness. Interestingly, no drivers who successfully completed session 1 dropped out in either session 2 or 3. This suggests either that this first training session filtered out all the drivers who were susceptible to simulator sickness and/or that one cause of simulator sickness might be that the stress/anxiety of participating in training heightened drivers' sensitivity to simulator sickness – having successfully completed one session, anxiety was reduced with an associated reduction in symptoms of simulator sickness.

The first company in the United Kingdom to use the truck simulator of TRL to provide training for their truck drivers on a commercial basis was Allied Bakeries (AB). This commenced with a pilot study, supported by the Welsh Assembly government, in which AB committed six drivers to a simulator training programme, similar to that described in the cohort study but in which drivers visited TRL for training on two occasions (rather than three). AB were able to keep detailed fuel consumption records for the six drivers involved in the study and to minimise other factors that may affect fuel efficiency. For example, AB ensured that each driver always drove the same vehicle and always drove on the same routes throughout the period of the study. The drivers were chosen to represent a range of driving styles and to cover different normal driving environments.

Training was provided by an approved driving instructor for all forms of commercial vehicles. Participants visited TRL for two training sessions. In addition to this, all participants visited TRL for a familiarisation visit in order to see the simulator facility and to have a brief drive of the simulator to help reduce any feelings of anxiety that they might have about training on the simulator at TRL. This familiarisation visit was conducted in April 2008. The first training visit was conducted in July 2008; the second was conducted in September 2008.

Drivers were asked to operate a simulated Mercedes Actros 2541 rigid (6 × 4 axle configuration) lorry unit with 50% load (estimated gross vehicle weight 17 tonnes). This vehicle type was selected to be most similar to the type of truck and load typically driven by the AB drivers. The training route was a mixture of rural and urban driving with a number of events designed to challenge the driver and provide opportunities to display fuel-efficient driving practices. Drivers received feedback on their performance by using a revised version of automated assessment system, tailored to suit the type of simulated vehicle driven in this training programme.

A potential problem that may affect all training providers was difficulty in scheduling drivers for training sessions. Training on the simulator required coordination of trainees to all attend the facility on specified dates. Issues such as illness, injury, departure and annual leave all affected driver bookings for the training programme. However, on completion of the training programme, results demonstrated that trainees produced a 25.6% improvement in simulated fuel efficiency. Exploration of the changes in behaviour indicated that this was accompanied by a reduction in RPM (reduced by 33.4%), resulting in greater engine torque (increased by 50.5%) when accelerating and a reduction in fuel wastage when slowing the vehicle. Drivers also made significantly fewer gear changes in the simulator after training (down by 26.4%). This also contributed to greater torque values and better vehicle sympathy but may have been due partially to increased familiarity with the simulator gearbox.

Results in the real world showed that simulator training was associated with a mean fuel efficiency improvement of 7.3%. Biggest improvements were seen for drivers who completed mixed driving routes, whereas the smallest improvements were seen for drivers who usually drove in the urban environment. The observed real-world fuel efficiency improvements were applied to annual fuel cost and CO_2 emission calculations, resulting in an average per driver saving of over £1691.27 in fuel costs and over 4 tonnes of CO_2, if a driver were able to sustain this fuel efficiency benefit over one year. Based on that assumption, the improved fuel efficiency of the trained drivers would provide a return on investment in a little over four months. This is based on the training provided to AB by TRL at £300 per driver per training session and fuel costs at 110 p per litre.

This study supported the results of the previous simulator fuel efficiency training programme conducted at TRL, demonstrating that simulator training can achieve significant and pragmatic fuel efficiency savings on a commercial basis.

3.6 Transfer of Training

More recently, Hirsch and Bellavance (2014) reported on the results of a survey study of a cohort of novice drivers for whom a proportion of the time that would ordinarily be spent training on-road in a real vehicle was substituted by training in a driving simulator as part of a long-term study into the effectiveness of transfer of simulator training to real-world driving. In their study, one driving simulator hour could replace one on-road hour for up to 50% of the 15 hours of mandatory on-road lessons. The survey investigated the extent to which their sample of 229 learner drivers perceived the simulator training to be effective relative to on-road lessons. Results indicated that simulator training was perceived to be either more efficient than or as equally efficient as on-road lessons for 13 of 15 specific driving skills. The two skills for which the learner drivers gave the lowest ratings for efficiency of training in the simulator were parking and speed control. This aligns neatly with the findings from the truck simulator of TRL whereby the training of tactical rather than operational skills was found to be more effective.

Hirsch and Bellavance (2014) noted a possible issue with self-selection bias in their sample. It is likely that only novice drivers well disposed to simulator training would have volunteered to participate in such a study, and so, their ratings of the simulator may be more positive than a random sample of novice drivers.

These varied studies, when taken together, present an intriguing picture, with precise handling skills being less easily trained and having less evidence of direct transfer to real-world performance than higher-order tactical skills. Unfortunately, the evidence base lacks any form of controlled studies that systematically compared the effectiveness of delivery of the same curriculum content across different media (simulator, classroom, real road, e-learning and so on) nor has there been research that has looked in detail at where in the learning course simulator (synthetic) training is best positioned. We do not yet know if driving simulation can or should completely replace some on road elements of a curriculum. If we assume that the notion of blended learning is appropriate (see Oliver and Trigwell 2005), then we would need to know whether the tactical components of a safe and efficient driving curriculum are best interspersed with on-road experience or in a block after a certain level of operational skill (vehicle control) has been demonstrated.

Although the majority of evidence supports the transfer of training from the simulated to the real environment, some studies have shown that training in a synthetic environment does not always improve performance in the real task and can sometimes cause task decrement. In reference to simulator training, Boldovici (1987) suggested that there are three types of training transfer: positive – learning is improved due to training via the simulator; negative – training in the simulator somehow interferes with performance in the task; and neutral – training in the simulator has no discernible effect. Research has demonstrated that simulators can cause all three. There are numerous examples of positive transfer (e.g. Atkins et al. 2002, Carretta and Dunlap 1998, Dolan et al. 2003, Reed et al. 2009, Williams and Flexman 1949). However, despite these positive results, some research has found no evidence of transfer of training. Kozak et al. (1993) found no significant transfer-of-training benefits from training on a manual pick-and-place task in a virtual environment. Meanwhile, negative transfer of training in virtual environment training for submariners was shown by Golas et al. (1996), which the authors attributed to distracting artefacts in the visual presentation.

Many of the risks of negative training transfer that are found with real-world driver training are also true for simulators such as changed attitudes to risk, a mismatch between training scenarios and the real situation or differences between the performances of the training equipment relative to the real vehicle. However, simulator training carries some additional risks. It can be envisaged that the incidence of symptoms of simulator sickness may cause drivers to behave in such a way that either causes them to drive faster such that they finish the training scenarios sooner or that they adapt their driving style in ways that are intended to mitigate the sickness symptoms. Either of these driving style adaptations may restrict the training benefit achieved or even cause negative training effects.

3.7 Simulator Sickness

A potential problem with the implementation of a programme of training in a simulated environment is the incidence of simulator sickness amongst trainees. The symptoms of simulator sickness are similar to those of motion sickness and include nausea, dizziness and headache or eye strain. The most widely accepted theory to explain motion sickness is the sensory conflict theory (e.g. Reason and Brand 1975, Regan 1993) and has been applied to the symptoms that can arise with simulator use. It proposes that the symptoms occur as the result of conflict between three sensory systems – the visual system, the vestibular system and non-vestibular proprioception. Kennedy et al. (1992) found that whilst sufferers of motion sickness tend primarily to experience nausea and vomiting, the symptoms of simulator sickness are typically slightly different and include sweating, headache, disorientation and postural instability. Simulator sickness is a concern because it can potentially degrade the training effectiveness and the well-being of trainees.

The incidence and severity of simulator sickness vary widely depending on the simulator used, the type and duration of the exercise performed and the population from which the participants are drawn. Studies suggest that 80–95% of individuals exposed to virtual environments experience some symptomatic response due to exposure (e.g. Cobb et al. 1996, Kennedy and Stanney 1997). Bles and Wertheim (2000) suggested that 5% of the population may be particularly prone to simulation sickness and unable to adapt. However, sickness effects may cause the user only some mild discomfort and might not interfere with the training or research in progress.

There are many characteristics associated with simulation that have been found to be potentially implicated in causing simulator sickness. Clearly, if conflict between visual, vestibular and proprioceptive senses is at the root of the problem, then anything that serves to heighten this conflict could worsen symptoms of sickness. Although simulators with a wide field of view may provide the user with a more immersive experience, the tendency to cause simulator sickness has been found to increase with the size of the field of view (Kennedy et al. 1989). This may be because visual stimuli detected by the periphery of the retina are subject to the increased motion sensitivity of the cells in that region, and as a result, any sensory conflict may be more conspicuous.

Raisler and Lampton (1996) documented a study of simulator sickness in tank driver trainers. The simulator used in the study was the M1 Tank Driver Trainer (TDT). The TDT programme in the United States is a good example of how computer-based simulators can provide training that is less expensive, safer and more flexible than training conducted with operational equipment. The study was commissioned to establish if the TDT was being affected by simulator sickness and if there were ways to alleviate it.

The report describes how simulator sickness may degrade training effectiveness despite the absence of severe symptoms such as vomiting. Discomfort in the simulator may distract the trainee and may lead to a negative transfer of training in that the trainees may develop driving techniques that mitigate sickness in the simulator rather than those that would be most effective in real-world driving conditions. Such effects may be detrimental if they are transferred to the real world.

The results from this study showed that the majority of TDT trainees reported few or no symptoms. However, some trainees experienced significant levels of discomfort that meant that their training was compromised. The results also indicated that simulator sickness decreased over the numerous sessions. This suggests that trainees may adapt to the visual and movement display of the simulators and may even learn to avoid certain head movements within the virtual environment that lead to simulator sickness. This supports the work of Uliano et al. (1986), who also found that pilots showed decreasing symptoms of simulator sickness with repeat exposure to a flight simulator. However, Kennedy and Frank (1983) and Regan (1993) suggested that by adapting to the simulated environment, a user may experience greater post-simulation after-effects. Raisler and Lampton (1996) also suggested that certain scenarios, such as routes involving rapid left and right turns in rapid succession, may produce more simulator sickness than others. This has implications for the development of both the synthetic environments for training and the exercises conducted within that environment. The environment and exercise design should avoid including regions that would lead to the trainee performing manoeuvres that cause sickness. This must be balanced against the needs of the training. An ideal virtual environment for the avoidance of simulator sickness might include no sharp bends and no steep gradients, but this could be interpreted as unrealistic by the trainee and restrict transfer of training.

Based on a review of literature and adapting guidance applied in other forms of vehicle simulator, Ward and Parkes (1996) developed a set of recommendations for alleviating simulator sickness in driving simulators. They were as follows:

1. Subjects should be briefed that symptoms are not abnormal to relieve anxiety, particularly for those not accustomed to simulation.
2. Persons who are experts at a task in the real environment, particularly those unaccustomed to simulation of that task, are most at risk. Such individuals have stronger expectations about sensory associations that may be foiled by simulation and may well have no adaptation to sensory associations within the simulated environment.
3. Simulator sickness may be 'contagious'. In the case of simulations with both driver and passengers, distress evident in one person may increase symptoms or willingness to express symptoms in others. Persons in distress should be excluded at first signs.
4. Only use fit, healthy, well-rested participants. Illness, fatigue, emotional upset and hangovers may make persons more susceptible and exacerbate the effects of simulator sickness.

5. Adaptation to the simulation can lead to resistance to symptoms. A minimum of one day and a maximum of seven days between simulator sessions are advised.

6. Simulator sessions should not be followed by substantial driving sessions. This is to avoid any danger of interference from after-effects or delayed symptoms with driving in the real environment.

7. Sessions should be less than two hours to limit the accumulation of symptoms, particularly fatigue related (e.g. eye strain, headache, dizziness) and after-effects.

8. Include rest breaks at logical points in complex simulation tasks.

9. Turn simulation off and other lights on before the subject enters, during breaks and after the session.

10. Complex and rapid changing of direction should be avoided. Such actions are most conducive to compelling visual fields and visual–vestibular conflicts.

11. Avoid freezing of the simulated scene during trials, unless preceded by opportunity for the driver to achieve stable and straight control of the vehicle.

12. Do not reposition the subject within the virtual world whilst simulation is active. This may lead to scene distortion.

13. Use as narrow a field of view as required by the task. Reduce the field of view in simulation that is known to be nauseogenic.

The items described indicate how significant care and attention must be applied in the suitable design and implementation of simulator training schemes. In a research setting, many of these points can be addressed through careful design of the simulator scenarios and selection of appropriate recruitment criteria. However, when delivering driver training using a simulator, scenarios are necessarily more prescriptive in order to train drivers in specific skills. Item 2 on the list is an issue for the training of qualified drivers. Those that are highly experienced may be more likely to suffer with symptoms of simulator sickness. Presumably, their greater experience means that they are more attuned to visual and kinaesthetic differences between simulated and real driving resulting in simulator sickness.

Item 4 relates to the physical and mental state of participants when they arrive at the simulator facilities. Guidance can be given to prospective trainees on these matters prior to arrival. However, the reality is that the training organisation must manage the experience of whoever has paid to receive the training and in whatever state that they choose to arrive – whether well rested/nourished/hydrated or not.

Another key difference between the use of driving simulators for research and their use for training is that trainees (or their employer) will have paid for their experience, whereas research participants are likely to receive compensation for their involvement in a study. Paying for simulator training raises expectation that a positive, beneficial experience will be achieved. If simulator sickness curtails the training session, then either alternative training must be offered or the fees wholly or partially refunded.

If a trainee does experience simulator sickness, a further complication is that the symptoms can persist for minutes or even hours after termination of the simulated driving (Stanney et al. 1999). With symptoms including nausea, dizziness and blurred vision, there is an important role for the simulator provider in confirming that trainees are in a fit state to leave the training facilities and begin their onward journey – either driving a vehicle or using public transport.

3.8 Conclusion

Experiences with driving simulators indicate that they can deliver a commercially viable training package. However, a consistent, integrated simulator-based training package that successfully complements and builds upon traditional training techniques has yet to emerge. There are many possible reasons why this may be the case. Firstly, training costs in this relatively low-margin activity are critical. Simulator training programmes to date have typically been delivered by specialist organisations to low numbers of students. A commoditised training product that could be delivered at a low cost has not been established. In higher-value, higher-risk industries such as aviation, military and Formula One, the cost and practicality of simulation make the costs easier to justify. For more mundane driver training, where on-road techniques can be delivered at relatively low cost, the relative value proposition of simulator training is more challenging.

Secondly, there is inertia within the training industry and the training market towards changing from traditional training techniques. Simulation at the lower end of the fidelity spectrum is seen as 'just a game', whilst higher-fidelity systems are seen as too costly or too likely to cause issues of simulator sickness.

Thirdly, little has been done to demonstrate the genuine benefit of simulator training as compared to or in addition to standard training techniques. Several studies have indicated that significant fuel efficiency benefits can be achieved (e.g. Dolan et al. 2003, Reed et al. 2009), but even in those studies, it was not possible to measure the long-term benefit of the training. To our knowledge, no studies have identified an optimum frequency of training. Identifying a safety benefit is even more difficult to achieve. Rosenbloom and Eldror (2014) attempted to evaluate the relative safety of simulator-trained drivers by using proxy measures captured using telematic systems, finding no advantage (or disadvantage) for those drivers trained in a simulator. Furthermore, it is very difficult to compare studies of the effectiveness of simulator training when there are so many variables in the way that training is delivered; these include software, hardware, scenarios, trainer involvement, trainee briefing and feedback. All could contribute significantly to the effectiveness (or otherwise) of the simulator training.

Driving simulation for training is far from being new and has been discussed for many years; given the potential benefits in terms of road safety and fuel efficiency that equate to substantial savings at a societal level, it is surprising that there has not been more rigorous evaluation or direct comparison to other forms of learning.

It is highly likely that driving simulators can enable learning or experienced drivers to develop and enhance driving skills in a manner that results in improved driving performance on real roads. The non-trivial challenge to the simulator training industry is to prove that this is the case and to deliver the training in a cost-effective manner that is readily accepted by the market and complements existing training provisions. There are pockets of the industry where these challenges are being addressed, but more work is required to make simulator training a truly mass-market proposition.

References

Atkins, R.J., Lansdowne, A.T., Pfister, H.P., and Provost, S. 2002. Conversion between control mechanisms in simulated flight: An ab initio quasi-transfer study. *Australian Journal of Psychology*, 54: 144–149.

Bles, W., and Wertheim, A.H. 2000. Appropriate use of virtual environments to minimise motion sickness. In *RTO HFM Workshop: What Is Essential for Virtual Reality Systems to Meet Military Human Performance Goals?* (P7-1–P7-9). The Hague, Netherlands: RTO MP-058, 13–15 April 2000.

Boldovici, J.A. 1987. Measuring transfer in military settings. In S.M. Cormier and J.D. Hagman (Eds.), *Transfer of Learning: Contemporary Research and Applications* (239–260). San Diego, CA: Academic Press.

Briggs, G.E., and Wiener, E.L. 1959. *Fidelity of Simulation: I. Time Sharing Requirements and Control Loading as Factors in Transfer of Training*. Tech. Rep. NAVTRADEVCEN-TR-508-4. Port Washington, NY: Office of Naval Research, Naval Training Device Center.

Caretta, T.R., and Dunlap, R.D. 1998. *Transfer of Training Effectiveness in Flight Simulation: 1986–1997*. Report no. AFRL-HE-AZ-TR01998-0078 (DTIC no. ADA362818). Mesa, AZ: Air Force Research Laboratory.

Cobb, S., Nichols, S., Ramsey, A., and Wilson, J.R. 1996. Health and safety implications of virtual reality: Results and conclusions from an experimental programme. *Proceedings of FIVE 1996 Conference, Framework for Immersive Virtual Environments* (158–166). Pisa, Italy, 21 December.

Dolan, D.M., Rupp, D.A., Allen, J.R., Strayer, D.L., and Drews, F.A. 2003. Simulator training improves driver efficiency: Transfer from simulator to real world. *Proceedings of Second International Driving Symposium on Human Factors in Driver Assessment, Training and Vehicle Design.* Park City, UT.

EU Commission. 2003. Directive 2003/59/EC of the European Parliament and of the Council of 15 July 2003 on the initial qualification and periodic training of drivers of certain road vehicles for the carriage of goods or passengers, amending Council Regulation (EEC) no. 3820/85 and Council Directive 91/439/EEC and repealing Council Directive 76/914/EEC European Parliament (2003) PE 330.876.

FHWA. 1996. *Commercial Motor Vehicle Simulation to Improve Driver Training, Testing and Licensing Methods.* Report no. FHWA-MC-96-003. Washington, DC: US Department of Transportation, April 1996.

Golas, K., Royse, S., and Anderson, J. 1996. *Training Media Analysis (TMA) for SMQ(D) Course.* San Antonio, TX: Southwest Research Institute.

Helman, S., Grayson, G., and Parkes, A.M. 2010. *How Can We Produce Safer New Drivers? A Review of the Effects of Experience, Training, and Limiting Exposure on the Collision Risk of New Drivers.* TRL Insight Report (INS005). Crowthorne, UK: Transport Research Laboratory.

Hirsch, P., and Bellavance, F. 2014. *Novice Learner Driver Perceptions of the Efficiency of Driving Simulator-Based Training in a Natural Setting in Quebec.* Washington, DC: Transportation Research Board 93rd Annual Meeting.

Kennedy, R.S., and Frank, L.H. 1983. A review of motion sickness with special reference to simulator sickness. Paper presented to the National Academy of Sciences/National Research Council Committee on Human Factors, Monterey, CA.

Kennedy, R.S., Lilienthal, M.G., Berbaum, K.S., Baltzley, D.R., and McCauley, M.E. 1989. Simulator sickness in U.S. Navy flight simulators. *Aviation, Space, and Environmental Medicine,* 60: 10–16.

Kennedy, R.S., Lane, N.E., Lilienthal, M.G., Berbaum, K.S., and Hettinger, L.J. 1992. Profile analysis of simulator sickness symptoms: Application to virtual environment systems. *Presence,* 1: 295–301.

Kennedy, R.S., and Stanney, K.M. 1997. Aftereffects of virtual environment exposure: Psychometric issues. *Advances in Human Factors/Ergonomics,* 897–900.

Kozak, J.J., Hancock, P.A., Arthur, E., and Chrysler, S. 1993. Transfer of training from virtual reality. *Ergonomics,* 36: 777–784.

Lang, B., Parkes, A.M., Cotter, S. et al. 2007. *Benchmarking and Classification of CBT Tools for Driver Training.* TRAIN-ALL (European Union FP6 Project) Deliverable 1.1.

Lester, T., Rehm, L., and Vallint, J. 2003. Research for the development and implementation of a purpose built Truck Simulator for the UK – Phase I. PR/T/122/02. Crowthorne, UK: Transport Research Laboratory.

Oliver, M., and Trigwell, K. 2005. Can "blended learning" be redeemed? *E-Learning and Digital Media,* 2: 17–26.

Ovidiu Pop. 2016. School Driving 3D. Retrieved from: https://play.google.com/store/apps/details?id=com.ovilex.schooldriving3d&hl=en_GB.

Parkes, A.M. 2003. Truck driver training using simulation in England. *Proceedings of Second International Driving Symposium on Human Factors in Driver Assessment, Training and Vehicle Design* (59–63). Park City, UT.

Parkes, A.M., and Rau, J. 2004. *An Evaluation of Simulation as a Viable Tool for Truck Driver Training.* London: ITEC.

Parkes, A.M. 2013. Essential realism of driving simulation for research and training. In N. Gkikas (Ed.), *Automotive Ergonomics* (133–154). Boca Raton, FL: CRC Press.

Raisler, R.B., and Lampton, D.R. 1996. Simulator sickness in tank trainer drivers. *Proceedings I/ITSEC*. Orlando, FL.

Reason, J.T., and Brand, J.J. 1975. *Motion Sickness*. London: Academic Press.

Reed, N., Cynk, S., and Parkes, A.M. 2009. From research to commercial fuel efficiency training for truck drivers using TruckSim. In L. Dorn (Ed.), *Driver Behaviour and Training, Volume IV* (257–268). Aldershot, UK: Ashgate.

Regan, E.C. 1993. Side-effects of immersion virtual reality. Paper presented at the International Applied Military Psychology Symposium. Cambridge, UK, 26–29 July 1993.

Rosenbloom, T., and Eldror, E. 2014. Effectiveness evaluation of simulative workshops for newly licensed drivers. *Accident Analysis and Prevention*, 63: 30–36.

Stanney, K.M., Kennedy, R.S., Drexler, J.M., and Harm, D.L. 1999. Motion sickness and proprioceptive aftereffects following virtual environment exposure. *Applied Ergonomics*, 30: 27–38.

SUD Inc. 2016. Dr. Driving. Retrieved from: https://play.google.com/store/apps/details?id=com.ansangha.drdriving&hl=en.

Uliano, K.C., Lambert, E.Y., Kennedy, R.S., and Sheppard, D.J. 1986. *The Effects of Asynchronous Visual Delays on Simulator Flight Performance and the Development of Simulator Sickness Symptomatology*. NAVTRASYSCEN 85-D-0026-1. Orlando, FL: Naval Training Systems Center.

Ward, N.J., and Parkes, A.P. 1996. The impediment of driving simulator sickness: Prevalence, process, and proposals. *International Conference on Traffic and Transport Psychology*, Valencia, Spain, 22–25 May.

Welles, R.T., and Holdsworth, M. 2000. Tactical driver training using simulation. *I/ITSEC 200 Conference*. Orlando, FL, 30 November 2000.

Williams, A.C., Jr., and Flexman, R.E. 1949. Evaluation of the School Link as an aid in primary flight instruction. *University of Illinois Bulletin*, 46 (71) (Aeronautics Bulletin 5).

Williges, B.H., Roscoe, S.N., and Williges, R.C. 1973. Synthetic flight training revisited. *Human Factors*, 15: 543–560.

4

Motorcycle Simulator Solutions for Rider Research

Alex W. Stedmon, David Crundall, Dave Moore and Mark S. Young

CONTENTS

4.1 Motorcyclists as Vulnerable Road Users

Around the world, motorcyclists represent one of the most vulnerable forms of road user. In the United Kingdom, motorcyclists are generally considered to be around 51 times more likely to be killed or seriously injured (KSI) than car drivers (Department for Transport 2007). Recent data also indicate that motorcyclists are involved in 9.5% of all recorded collisions regardless of severity (Department for Transport 2012a) and account for 21.5% of KSI statistics (Department for Transport 2012a). These figures are set against a backdrop over the last decade, where motorcycles have contributed to only around 1% of the 300 billion vehicle miles travelled in any single year (Department for Transport 2012b), further compounding the underlying vulnerability of these road users when compared with other forms of transport. Whilst many of these incidents may be due to the behaviour of other road users (Clarke et al. 2007; Crundall et al. 2012a), a large proportion of motorcycle accidents can be attributed to rider error, such as losing control of the motorcycle on a bend in the road (Clarke et al. 2007) or an inappropriate choice of speed or overtaking manoeuvre for the prevailing road or traffic conditions (Carroll and Waller 1980; Lynham et al. 2001; Mannering and Grodsky 1995).

Previous research (Clarke et al. 2004) has shown that the three most common types of motorcycle accidents include the following:

- Right-of-way violations – These are characterised by the problem of other road users not seeing motorcyclists and assuming their own right of way. In many observation failure cases of this kind, the motorcycle that the driver had failed to see was so close to a junction that there appeared to be no explanation as to why it had not been seen (Clarke et al. 2004). This phenomenon is also referred to as a 'looked but did not see' accident (Brown 2002) or the more colloquial 'sorry, mate, I didn't see you' accident. If these accidents were to be eliminated, there would be an approximate 25% fall in the total UK motorcycle accident rate (Clarke et al. 2004).
- Loss of control on a bend, corner or curve – These accidents are usually regarded as primarily the fault of the motorcyclist rather than that of the other road users, with accidents associated more with riding for pleasure as they tend not to involve other traffic (Clarke et al. 2004).
- Motorcycle manoeuvrability accidents – A subgroup of accident cases appear to be related to the way that motorcyclists manoeuvre their motorcycles. Taking all accident cases where riders were judged as blameworthy, 16.5% involved a motorcyclist overtaking other vehicles and causing an accident (Clarke et al. 2004). These at-fault riders tend to be younger and riding higher-engine capacity motorcycles than other accident-involved riders. However, motorcycle accidents also occurred when riders took the opportunity to pass alongside slow-moving or stationary traffic, which is often referred to as *filtering* (Clarke et al. 2004).

Whilst some of these accidents are caused by vehicle, road or environmental factors, a larger proportion of them can also be attributed to skill and attitudinal gaps in rider behaviour (Clarke et al. 2007). Recent studies have attempted to interpret the heightened crash risk of motorcyclists by looking for potential advantages associated with riding experience or advanced levels of training (Crundall et al. 2012b; Crundall et al. 2013a; Hosking et al. 2010; Liu et al. 2009; Shahar et al. 2010; Stedmon et al. 2011; Vidotto et al. 2011).

Although there has been research into the causes of these types of road traffic accidents, virtually no human factors research has been conducted specifically looking at rider behaviour (Stedmon et al. 2011). Motorcycle ergonomics and rider human factors are therefore an emerging research area, and with interest and support from manufacturers, road safety organisations, motorcycle media and the riders themselves, there is a growing need to investigate key factors of rider behaviour and performance in order to gain a deeper understanding of the vulnerabilities associated with this mode of transport (Robertson et al. 2009; Stedmon et al. 2012). The factors that need to be considered include furthering our understanding of the 'rider experience' that encompasses rider behaviour, road design and more detailed investigations of the complex interactions between the rider, motorcycle and environment.

4.2 Human Factors and Research Requirements for Motorcycle Simulation

From a human factors perspective, the rider and motorcycle can be understood in terms of an interactive system operating within a very demanding and safety-critical environment (Stedmon et al. 2012). Across the different forms of road transport that exist, all users share the same physical road space and are bound by the same road laws. Taking a systems approach, all transport modes contribute to the overall safety and performance of the road transport system, but each will have different needs and requirements (Helander 1976). As motorcycles interact with others in the wider transport system in unique ways compared to other modes of transport (Lee et al. 2007; Martin et al. 2001; McInally 2003; Robertson 2003) and utilise the physical road space differently from the way other types of vehicle do (Robertson 2003), research should therefore address aspects such as how riders process the vast array of information around them (directly or indirectly from the motorcycle and environment) and what impact rider experience and training have on their ability to control the motorcycle and interact with the ever-changing road environment.

Real-world data are valuable in understanding complex issues within transport research. However, it is not always possible to conduct detailed experiments or studies in situations where personal and public safety may be compromised. As a consequence, simulators are often developed and employed where real-world data collection is particularly challenging. Around the world, much of the attention on development and research of transport simulation has focused on cars and aircraft. Whilst it is apparent that safety-critical aspects of flying lend themselves to simulation and in many countries that conduct research the majority of road users drive cars, motorcycle simulation has not developed in the same way. It is easy to overlook the differences between cars and motorcycles, not just in the fundamental principles of human–machine interaction, but also in the unique challenges that each mode of transport presents for developing specific simulators. However, as increased traffic levels present an increased danger for motorcyclists and major limitations in conducting research on the road, simulators offer a means of investigating many of the issues surrounding motorcycle riding without subjecting people to the real dangers of riding. It is with these issues in mind that this chapter considers the current status of motorcycle simulation for investigating rider behaviour. We review and compare the known motorcycle simulators along with the issues associated with developing simulators for rider research. Particular attention is given to the underlying methodological issues associated with motorcycle research in relation to issues of simulation fidelity and ecological validity.

4.3 Review of Motorcycle Simulation Research

Simulators have a well-established pedigree in driving research (see Chapter 2). A relatively new development in transport simulation is the focus on motorcycles. Within the field of motorcycle ergonomics and rider human factors, it is now possible for motorcyclists to ride identical computer-generated scenarios in order to control many of the variables that would affect consistent data collection in the real world in order to investigate isolated variables independently. In comparison to driver behaviour, motorcycle simulation research is a less developed area of investigation, and the research that has been conducted has focused primarily on experiential differences between novice, inexperienced and experienced motorcyclists. There is considerable research examining these issues in driving, which provides useful context before discussing the studies of motorcyclists.

Driving simulators have been used to examine many aspects of driver performance such as the effectiveness of training on reaction times to hazards. For example in a study that investigated the effects of experience and training on young drivers' performance, inexperienced drivers who had undergone hazard perception training drove differently from the way untrained inexperienced drivers did (Fisher et al. 2002). Research using driving simulators has also had some success in identifying performance differences in reaction to hazards across apparently homogenous groups.

Another driving simulator research illustrated differences in cognitive workload levels between inexperienced and experienced drivers with trained and experienced drivers being able to automate the driving task more effectively than inexperienced drivers did (Patten et al. 2006). It appeared that experienced drivers had more spare mental capacity and were therefore able to focus spare resources on assessing peripheral information. This argument was used to support the finding that experienced drivers spotted more hazards than the inexperienced drivers (Patten et al. 2006). This research provided a basis for understanding the differences between experienced and professionally trained drivers and drivers who have little or only modest driving experience, but who were not unqualified or novice drivers (Patten et al. 2006).

Driving simulators have also been used to investigate how drivers assess the potential risk of turning into the path of a motorcyclist. Using the trajectory

of approach of a motorcycle, participants driving a high-fidelity driving simulator were instructed to indicate when gaps were safe enough for them to
turn at an intersection. Through careful manipulation of the size of the gaps
and the type of oncoming vehicle, the results indicated that drivers were
more likely to turn in front of an oncoming motorcycle when it travels in the
left-of-lane position than when it travels in the right-of-lane position (Sager
et al. 2014). Simulator studies have also helped to investigate and explain the
theoretical background of safety interventions that have been shown to work
effectively in the real world (Jamson et al. 2010; Lewis-Evans and Charlton
2006). The introduction of hazard marker posts that accentuated vanishing
point information throughout a bend on a rural road in Buckinghamshire
(United Kingdom) appeared to reduce a previously high motorcycle KSI rate
observed for the previous eight years to zero in the three years following the
introduction of the hazard marker posts (James 2005).

Similar differences have been demonstrated where experienced riders crashed less often, achieved better rider performance scores after
each hazardous event and were more likely to approach hazards at an
appropriate speed than inexperienced or novice riders did (Liu et al. 2009;
although Shahar et al. 2010 failed to replicate these results by using the
same type of simulator and software). Furthermore, young motorcycle riders tended to be overconfident in their hazard perception abilities, but this
did not translate into better performance in a hazard perception task (Liu
et al. 2009). This finding builds on earlier motorcycle simulator research
in which experienced riders had superior hazard perception skills than
those of novice riders (Bastianelli et al. 2008). From this study, it was suggested that novice riders both are unable to allocate sufficient cognitive
resources to visual search activities and have an inadequate mental model
for detecting traffic hazards (i.e. a cognitive concept for real-world hazards). It was also suggested that increased skill might be associated with
an increase in the ability to acquire and filter information from traffic
events and develop more specific mental models of a variety of hazards
(Bastianelli et al. 2008).

More recently, research has investigated hazard perception and visual
scanning patterns by using a motorcycle simulator (Hosking et al. 2010).
Results indicated that experienced motorcyclists were faster to respond to
hazards than inexperienced riders and that the faster response times might
be due to experienced riders having a visual search pattern that is more flexible than that of inexperienced riders (Hosking et al. 2010). Within this study,
prior car driving experience led to some improvements in hazard perception
skills of inexperienced riders, illustrating that some of these skills learned
in car driving seem to transfer across to motorcycle riding (Hosking et al.
2010). More specifically, inexperienced motorcyclists who were experienced
car drivers were faster to respond to hazards and had more flexible search
patterns than riders who were inexperienced users of both forms of transport. Nevertheless, further improvements would still be required in order

for the hazard perception skills of these inexperienced riders to reach the levels attained by experienced riders (Hosking et al. 2010).

Another research investigated how far a motorcycle simulator could improve hazard avoidance skills in teenagers. In a simulator study involving 410 participants, the experimental group after training avoided more hazards than a control group who received passive training based on a road safety lesson. These results indicate that training within a simulator increased the number of avoided accidents in the virtual environment based on higher levels of hazard perception skills (Vidotto et al. 2011). Similar research involving teenage riders of mopeds (typically low-capacity powered two-wheelers that are a first step in motorcycle riding or mobility for younger adults) investigated the role of attention in enhancing the skills required to ride a moped simulator. Two experiments were conducted where participants were required to ride safely and avoid hazards. In the first experiment, hazards were designed so that a shift in attention was required to escape the danger. In the second experiment, participants not only were required to conduct the same tasks but were also required to attend to an additional attention-based task. The results from the first experiment indicated that when no attentional shift was required, more hazards were avoided. In the second experiment, the results indicated that the additional attention-based task did not impair hazard performance and that it also produced an improvement in the ability to shift attentional focus, preserving performance efficiency. From this, the authors concluded that poor attentional shift plays a role in accounting for accidents and that attentional training can provide benefits for improved processing efficiency that may account for fewer accidents (Tagliabue et al. 2013).

A motorcycle simulator was also used to evaluate an intelligent curve warning system that has been designed to give riders support when negotiating a curve. A study was conducted where participants performed three rides: one without the system (baseline) and two experimental rides using a curve warning system provided by either a force feedback throttle and/or a haptic glove. The results indicated that warnings provided by both systems produced earlier and stronger responses in relation to motorcycle dynamics through a curve than when the riders did not use the system. Furthermore, the haptic glove system led to a reduction of critical curve events. Whilst subjective workload levels were not affected by the use of either system, the force feedback throttle required increased attention. Overall, there was a preference for the haptic feedback glove, which could then be developed and tested in a real road environment, having been evaluated in a simulator (Huth et al. 2012).

In other research works, a motorcycle simulator was used to conduct one of the first in-depth rider behaviour investigations of its kind (Stedmon et al. 2009; Stedmon et al. 2011). Using the STISIM Drive software as part of an integrated experimental paradigm, the motorcycle simulator offered a unique tool to explore rider behaviour across different experience and training levels. The results indicated clear differences in the riding styles of different rider groups (e.g. novice, experienced and advanced trained riders) as well as how they interpreted potential hazards on the road and the road positions that they took prior to and once they had encountered hazards. This was possible only by using simulated riding scenarios that allowed for detailed data collection and comparisons in order to understand these differences. A key finding was that the advanced trained riders generally travelled faster through bends but did so in a smoother manner and with a more defensive riding style (i.e. measured by lateral position and speed as they entered the bend and as they travelled through the bend) (Crundall et al. 2012a; Crundall et al. 2012b; Crundall et al. 2013b; Stedmon et al. 2009; Stedmon et al. 2011).

From this work, a further area of investigation, for which the simulator provided detailed data, focused on rider behaviour through bends. As a considerable number of fatalities and injuries are due to the actions of the rider themselves (i.e. losing control on a bend when no other vehicle is involved), this aspect of the research analysed differences in the experience and training levels of novice, experienced and advanced trained riders for a range of bends in a simulated environment. In a comparison of 60 and 40 mph zones, advanced riders rode more slowly in the 40 mph zones and had greater variation in lane position than the other two groups. In the 60 mph zones, both advanced and experienced riders had greater lane variation than novices. Across the whole ride, novices tended to position themselves closer to the kerbside. In a second analysis across four classifications of curvature (straight, slight, medium, tight), advanced and experienced riders varied their lateral position more so than novices, although advanced riders had greater variation in lane position than even experienced riders in some conditions. These results indicated that experience and advanced training led to changes in behaviour compared to novice riders that can be interpreted as having a potentially positive impact on road safety (Crundall et al. 2014).

In a different approach, a motorcycle simulator was used to investigate the effects of blood alcohol concentration (BAC). Motorcyclists tend to be overrepresented at low BAC levels, and so, a simulator study was designed to assess differences in hazard response across three BAC levels (sober, 0.02% BAC and 0.05% BAC). Equal numbers of novice and experienced participants completed simulated rides in both urban and rural settings whilst also responding to a safety-critical peripheral detection task. The results indicated that under the 0.05% BAC condition, there was a significant increase in the standard deviation of lateral position in the urban scenario and peripheral detection task reaction time in the rural scenario when compared with the sober condition (i.e. no alcohol). Participants were most likely to collide with an unexpected pedestrian in the urban scenario at 0.02% BAC, with novice participants at a greater relative risk than experienced riders. Novices rode faster than experienced participants did in the rural setting regardless of their BAC level (Filtness et al. 2013). This study illustrated one of the key benefits of simulator trials in general: being able to conduct research that is highly challenging (and potentially illegal) on real roads.

4.4 Evolution of Motorcycle Simulators

Whilst most modes of transport simulation have developed considerably over the last 20 years, motorcycle simulation has not received similar attention. As a result, motorcycle simulation technologies are still evolving. There are perhaps fewer than 10 different systems in use and in the public domain around the world. In many ways, this could be due to motorcyclists in many areas of the world (e.g. Europe and the United States) representing niche road user groups. Whilst other areas of the world (e.g. South America, Africa and the Far East) rely on motorcycles more as a common means of transport, research in these regions is still limited. However, the vulnerability of motorcyclists on roads around the world is still a major concern for government departments and rider safety is still a strategic area of road safety campaigns. In 2012, the Parliament of Victoria Road Safety Committee conducted an international review of motorcycle safety and visited Europe to meet with a range of road safety industry and policy advisors. The output of this work was Parliamentary Paper no. 197, *Inquiry into Motorcycle Safety*,* which made a specific recommendation to investigate the potential for simulation in rider training and safety.

More recently, and in response to the increased importance of safeguarding vulnerable road users, the European Union has supported a specific Cooperation in Science and Technology Action (COST Action) focused on safety for powered two-wheelers (Safe2Wheelers: TU 1407). This COST Action provides a unique means for European researchers, engineers and scholars to jointly develop their own ideas and new initiatives through trans-European networking of nationally funded research activities. Safe2Wheelers operates across a range of working groups specifically focused on accidentology; rider behaviour; traffic environment; technical solutions; primary and secondary safety; and policy, integration and dissemination. In 2016, the COST Action ran a dedicated Accidentology and Motorcycle Simulator Workshop that attracted over 80 international participants. It was one of the first times that representatives from a range of established and pioneering motorcycle simulators around Europe were all present at the same meeting.

Interestingly, motorcycle manufacturers do not seem to have built simulators to develop their products in the same way that the automotive industry has embraced driving simulators. Notable exceptions to this include Honda Motorcycles, who developed a full-size motorcycle simulator† for rider research back in the late 1980s (Ferrazzin et al. 2001). This simulator was developed through a number of iterations with a more recent training simulator (Honda Riding Trainer‡), which was a useful sales and demonstration

---

* http://www.parliament.vic.gov.au/57th-parliament/rsc/inquiries/article/1409.
† http://world.honda.com/motorcycle-technology/sim/p3.html.
‡ http://world.honda.com/news/2013/c130327Updated-Software-Riding-Trainer/index.html.

tool in showrooms for new riders to familiarise themselves with typical riding hazards. This is the only major motorcycle manufacturer that has developed a simulator for public use and which has then been provided to research institutions as a tool to support rider research. The simulator operates as a computer-based system with the rider facing a screen whilst sitting on a small bench seat. Interaction is through realistic controls and a simulated riding environment. A major benefit is that the simulator is a mobile simulation platform that is easily transported between locations. However, the riding scenarios cannot be easily altered (they were not originally designed to be so), data cannot be easily manipulated and the physical size and steering action of the motorcycle compromise the face validity of the simulator. That said, it is currently one of the most widely used simulators for rider research in training and assessing novice riders (Vidotto et al. 2008), comparing driving experience with hazard perception and hazard reaction between novice and experienced motorcyclists (Liu et al. 2009), assessing rider mental workload as they were exposed to hazardous situations (Di Stasi et al. 2009) and investigating differences between riders and drivers on the basis of their responses to hazards (Shahar et al. 2010) as well as more recent research comparing riding performance between returned, continuing and new riders (Symmons and Mulvihill 2011).

Elsewhere, simulators have been developed by individual institutions. Early motorcycle training and traffic safety simulators were developed, but little information remains available for them (e.g. Doz Nadal 2015). Even as recently as 2000, another simulator was built at Tokyo University (Ferrazzin et al. 2001), but again, little information is available for this. The California Superbike School* has built a simulator that is used to train MotoGP riders. It is a full-size moving platform with realistic controls used to train riders on the correct postures to adopt when riding and racing but is not linked to any dedicated simulation scenario software. This simulator acts as a real-size platform to mimic the lean angles of a racing motorcycle so that riders can learn to position themselves more appropriately to maintain better control of the motorcycle. Around Europe, the Institut für Zweiradsicherheit e.V. (IFZ[†]) simulator in Germany and the Motorcycle Rider Simulator (MORIS) simulator (Ferrazzin et al. 2001) in Italy both allow users to operate realistic controls and are linked to simulation software. The IFZ simulator has been built using a Bayerische Motoren Werke (BMW Motorrad) motorcycle, whereas the MORIS simulator uses components of a Piaggio moped mounted on a motion platform. These simulators appear to have been developed more for engineering, two-wheeled dynamics and demonstration purposes rather than for specific rider research or training. A more recent development is the Motorcycle Dynamics Research Group (MDRG[‡]) simulator from Padova

---

* http://www.superbikeschool.co.uk.
† http://www.ifz.de/.
‡ http://www.dinamoto.it.

University in Italy. In many ways, this is similar to the MORIS simulator but uses motorcycle components mounted on a bespoke rig that provides pitch and roll actions. Unlike the previously mentioned simulators, the MDRG simulator uses a three-screen projection system and a 5.1 surround sound audio system to provide an enhanced immersive experience for the rider that would therefore support rider behaviour research. This simulator has been designed to develop and test electronic devices for improving rider safety and vehicle performance, investigate different design factors on vehicle dynamics and train riders and study their behaviours in different riding scenarios (Cossalter et al. 2011).

As part of the European Union-funded SAFERIDER* project, the French Institute of Science and Technology for Transport, Development and Networks (IFSTTAR) motorcycle simulator has been used as a demonstrator and research tool. The IFSTTAR simulator consists of a motion platform, computer software displaying rider environments on a projection screen and a sound system. The simulator uses components of a small-capacity (125 cc) motorcycle, including steering/handlebars, fuel tank, seat, footrests, throttle, front and rear brakes and gear shifting mechanism, to support a realistic rider experience in the simulator. The front fork utilises two lateral electric actuators (for pitch and roll movements), whilst the rear of the motorcycle is connected to a powered horizontal track to facilitate yaw movements (Arioui et al. 2010; Benedetto et al. 2014).

Further afield, the Monash University in Australia has also developed a motorcycle simulator. Within the Monash University Accident Research Centre (MUARC), a dedicated simulator uses a full-size static motorcycle positioned in front of a wide field-of-view projection screen to provide the rider with an immersive riding experience of the simulated scenario before them. There is little information available on how the motorcycle interfaces with the simulation software, but it appears to be similar to other simulators that have been developed in the United Kingdom and in Europe.

In the United Kingdom, the development of MotorcycleSim has occurred alongside the emergence of these other simulators but has taken a user-centred approach from the outset in order to capture the underlying user requirements of a simulator before it was built. MotorcycleSim was designed and built with the main aim of conducting rider human factors research. The simulator consists of the hardware (real motorcycle and user input controls) providing data to the STISIM Drive software, which is then used to provide visual feedback via a projected riding scenario directly in front of the user. The simulator was built using a full-size and fully equipped Triumph Daytona 675 motorcycle (kindly supplied by Triumph Motorcycles Ltd). The existing motorcycle controls (throttle, brake lever, brake pedal, gear selector, clutch lever) were modified or adapted to work with the simulation software in a realistic manner. Surround sound speakers provide engine noise

---

* http://www.saferider-eu.org/.

feedback. Visual feedback can be manipulated to enhance acceleration and braking effects (by altering the degree of dynamic pitch in the visual scene as the motorcycle accelerates and slows down), tilting the scenery as the rider steers in a particular direction (to enhance the perception of leaning into a corner) and increasing the rider's field of view to take account of peripheral visual cues. The simulator has been designed as a static platform that provides a high degree of stability. Unlike some other simulators that offer little flexibility for research purposes, the STISIM Drive simulation software can be used to build interactive riding scenarios with different weather and traffic conditions and typical hazards such as vehicles pulling out at junctions (Stedmon et al. 2011). As it is possible to model all the scenario attributes, this allows for strictly controlled experimental repeatability in a laboratory setting (Figure 4.1). In addition, the STISIM Drive software is able to record a variety of different measures for rider behaviour research (e.g. motorcycle speed, road position, acceleration, braking effort, closure speeds/times, vigilance, number of accidents, red light violations, speeding offences).

Another recent development in motorcycle simulation involves a partnership between the Würzburg Institute for Traffic Sciences (WIVW) and BMW Motorrad. This simulator has been built as a tool to investigate riding technologies and aspects of rider workload (Will and Schmidt 2015). A particular focus is on the fundamental understanding of the complex riding task, motorcycle ergonomics and rider assistance systems. With the

FIGURE 4.1
MotorcycleSim using the STISIM Drive software.

combination of a restricted and limited workspace and demanding riding environment, the use of a simulator for evaluating control and display concepts is an extremely valuable and cost-effective tool. The simulator has been designed to emulate a real-size motorcycle through the use of a rig that houses the front fairing, handlebars, seat and tank of a BMW motorcycle. The rig is housed on a motion platform and positioned in front of a wide field-of-view projection screen to provide an immersive environment for the rider (Figure 4.2).

Different scenarios are built using their in-house SILAB software, which supports simulators at various stages of development, from a single PC with a gaming steering wheel and pedals to more sophisticated simulators with multichannel field of view, real vehicle rigs and a motion platform system. The virtual world is presented to the rider via a modern graphics engine that can mimic specific times of day, lighting models and weather and road conditions. Sound is generated for the simulated vehicle and of other road users by a three-dimensional sound model so that riders can also assess their environment and other road users (e.g. their type, distance and speed) through peripheral sound cues.

A further development has been the incorporation of a prototype 'G-Vest' developed by researchers at BMW Motorrad (Doz Nadal 2015) in order to address the lack of any mechanisms to present acceleration forces and

FIGURE 4.2
WIVW motorcycle simulator.

feedback to rider in motorcycle simulators. The rider wears the G-Vest over their usual riding kit, and this is then attached via dedicated cables to a controller at the rear of the motorcycle simulator. Actuators then create a backward force to stimulate the somatosensory system by producing pressure variations to simulate acceleration effects. Initial evaluations have shown that inertial and airflow-induced forces can be represented by a surface pressure on the torso, and the perception of acceleration is realistic without conflicting with the vestibular system. This particular example illustrates the need for innovative solutions to enhance the simulation experience.

4.4.1 Comparisons of Motorcycle Simulators

It is apparent from the previous review of simulators that many have been developed for slightly different purposes, and this makes it difficult to compare them side by side. However, by using some basic metrics, it is possible to make some distinctions between them. Table 4.1 illustrates the key

TABLE 4.1

Simulator Characteristics and Fidelity Factors

	Simulator Characteristics				
	Platform, Controls, Environment (Yes/No)			Fidelity (Low/Medium/High)	
	Real Motorcycle	Real Controls	Simulated Environment	Physical Fidelity	Functional Fidelity
Training simulator of Dahl	Yes	Yes	No	High	Low
Research simulator of Born	Yes	Yes	Yes	Medium	High
Tokyo University simulator	n/a	n/a	n/a	n/a	n/a
Early Honda simulator	Yes	Yes	Yes	Medium	Medium
Honda Riding Trainer	No	Yes	Yes	Low	Low
California Superbike School	No	Yes	No	Medium	High
MORIS simulator	No[a]	Yes	Yes	Medium	Medium
IFZ simulator	Yes	Yes	Yes	High	Medium
MDRG simulator	No	Yes	Yes	Medium	High
IFSTTAR simulator	Yes	Yes	Yes	Medium	High
MUARC simulator	Yes	Yes	Yes	High	Medium
MotorcycleSim	Yes	Yes	Yes	High	Medium
WIVW simulator	No	Yes	Yes	Medium	High

Note: n/a = not available.

[a] Not a real motorcycle but is based on a real moped (i.e. small-capacity powered two-wheeler).

characteristics of these simulators based on whether they incorporate a real motorcycle and real controls or provide the opportunity for the rider to experience riding in a simulated environment. From this, it is then possible to consider aspects of physical and functional fidelity of the overall simulator solution.

What Table 4.1 illustrates is that across the range of simulators, there are many different configurations and levels of physical and functional fidelity. These two dimensions of fidelity are considered as having the greatest impact on the validity of simulator research (Stanton 1996). Physical fidelity relates to the extent that a simulator looks like a real system, and functional fidelity relates to the extent that a simulator acts like the real system. One thing that emerges from this classification of the simulators is that whilst it is necessary to have a real motorcycle to develop high levels of physical fidelity, it is not required for high levels of functional fidelity.

Whilst physical fidelity is important in training simulators (Greenberg and Blommer 2011), it has been argued that it can be compromised in research simulators (assuming that functional fidelity is maintained or enhanced) without affecting the transferability of results (Stammers 1986). From previous research that captured detailed user requirements for the development of a motorcycle simulator, riders identified characteristics associated with functional fidelity, such as the simulator leaning like a real motorcycle, having realistic acceleration and braking effects and the controls working in a realistic manner, as more important than characteristics associated with physical fidelity, such as using a real motorcycle, or whether the simulator was aesthetically pleasing (Stedmon et al. 2012).

From additional comments made during focus group exercises, potential users also emphasised the importance of realism within the simulation software and how other traffic and road conditions need to be incorporated into riding scenarios (Stedmon et al. 2012). Most of the simulators with medium to high levels of functional fidelity present the rider with a simulated environment. This can help enhance the functional fidelity of the simulator by providing a responsive experience for the rider based on their inputs to the motorcycle. This factor perhaps marks a distinction between those simulators that have been built (particularly early in the relative history of such simulators) that perform specific functions and those that have been developed for investigating factors in the wider context of motorcycle riding. One exception to this is the California Superbike School simulator, which does not present a simulated environment but produces a high level of functional fidelity, albeit for a very specific subset of riding activity (i.e. race training and rider posture).

It is interesting that all of the simulators use real controls, but not all of them use a real motorcycle. This can affect the face validity of the simulator. If the simulator rig does not look like a motorcycle, then there is a danger that riders will not assume that the simulator will necessarily behave like a real motorcycle (or a particular motorcycle that the simulator is trying to mimic).

However, as with driving simulators, having a real vehicle for the simulator can then be a limiting factor for researchers looking to investigate different configurations on/in the vehicle where some form of modular/configurable rig would have advantages. The use of real controls also supports the concept of functional fidelity so that riders are interacting with actual motorcycle controls with which they are more familiar.

4.5 Design Considerations and Future Applications for Motorcycle Simulators

Whilst most of the simulators that are known to exist have been built by institutions and university departments, few appear to have been developed primarily from a rider behaviour perspective (Stedmon et al. 2012). As the development of motorcycle simulators has been rather ad hoc, there have been no precedents for how to build a simulator and they are not available to buy 'off the shelf' in the same way as driving simulation packages. With no real design specifications for building a simulator, this probably explains why the simulators that exist today have evolved in their own innovative ways to address key issues of replicating a faithful rider experience. Rather than setting down a list of requirements, this section highlights some of the considerations for developing a motorcycle simulator by raising awareness of the unique nature of this mode of transport.

Compared to most modes of road transport that operate using traditional steering wheel and pedal interfaces, motorcycles are unique and have fundamentally different user controls in the way that they are distributed and operated: the clutch is a lever, usually operated by the left hand; the front brake is a lever, usually operated by the right hand; the throttle is a twist grip operated by the right hand (this is logical as the throttle is usually closed before the front brake is applied); the gearbox is operated by a foot pedal by using the left foot, and the rear brake is also a foot pedal, operated by the right foot. Lights, indicators, horn and ancillary equipment are usually operated by the fingers and thumbs without removing the hands from the handgrips (Stedmon et al. 2012).

Riders may also adopt a variety of postures to control the motorcycle, counter environmental effects (e.g. crosswinds, inclement weather), adjust to the speed of the motorcycle or gain a better view of the road (McInally 2003; Robertson and Minter 1996; Stedmon et al. 2012). The rider therefore controls the motorcycle through their body position, the physical activity of putting force onto the handlebars and the unique action of countersteering (Robertson et al. 2009). By comparison, car drivers are protected from the environment by sitting inside the vehicle and have very little influence over the behaviour of the vehicle beyond their inputs to the direct driving controls.

There are considerable engineering challenges faced in building simulators to emulate realistic motorcycle riding (e.g. the forces exerted on riders as they lean the motorcycle, braking and acceleration effects; environmental factors such as side and head winds, turbulence and buffeting; wind noise that riders experience at different speeds). In addition, research has been conducted to evaluate whether motion platforms or static, fixed-base simulators provide specific benefits (Stedmon et al. 2012); roll motion parameters, speed cues and environment design (Shahar et al. 2014); and the trajectory control modality and the leaning rendering (Benedetto et al. 2014). When developing any motorcycle simulator for rider research, it is necessary to capture user requirements from the outset in order to specify and build a solution that is fit for research purposes and provides the desired levels of fidelity associated with real-world riding to support user acceptance.

One of the underlying issues with motorcycle simulators is how they might handle the delicate balance of control interfaces that the rider uses throughout a ride. In real riding, the motorcyclist is constantly adapting their inputs to the prevailing conditions (e.g. surrounding traffic, weather, wind turbulence). It is important that the simulator interacts with the simulation software through the existing motorcycle controls (that may require adaptations). The mechanical movements of the throttle, braking, gear selection and steering angle of the motorcycle need to be interfaced with some form of control unit.

A key factor in motorcycle simulation is designing the motorcycle brakes to operate in a realistic manner. On motorcycles, using both the front and rear brakes together or independently is a key riding skill, and so, solutions need to achieve these inputs, especially if adapting existing driving software where a single braking input is expected. Signals from the front and rear brakes of the motorcycle need to communicate with the simulation software with distributed ratios so that the front brake produces a greater braking force than that produced by the rear brake. This is important in preserving the functional fidelity of the brakes where a rider must operate both brakes for maximum braking effort (as would be expected on a real motorcycle).

From a practical perspective, it is also important that riders wear their motorcycle helmet and riding kit to mimic their experience of real riding. This can be particularly advantageous as the helmet naturally restricts extremes of peripheral vision and therefore can help focus the rider's attention on the simulated environment that they are negotiating. There is also anecdotal evidence from driving simulators that when other cues are present (e.g. seatbelts, doors, handbrake mechanisms), there is a greater likelihood that users will exhibit other real behaviours (such as checking if it is safe before leaving their vehicle). In motorcycle simulation, behaviours that have been observed include rider safety checks (i.e. looking over their shoulder for potential hazards), looking round vehicles prior to an overtaking procedure and seeking to use mirrors to check the rear view.

Motorcycle simulation is set to become more widespread. As different simulators evolve, the applications that they will be used for will also expand. In the field of motorcycle ergonomics and rider human factors, research will continue to explore areas such as motorcycle design and rider equipment (e.g. testing prototype designs and the impact of motorcycle technologies such as Satnav systems, intercoms or advanced head-up/helmet-mounted display systems); road safety (e.g. scenarios built to test the interplay of roadside furniture, signage and natural features on riding behaviour); driver education (e.g. using carefully constructed riding scenarios or real video footage, an innovative use of the simulator could be in educating non-motorcyclists on what it is like to ride a motorcycle); accident investigation and forensics (e.g. where there is potential to use the simulator for reconstructing crash scenes by using known data about accidents in order for investigators to analyse what might have happened in a particular incident); and rider skills, attitudes and behaviour (e.g. issues such as workload, situation awareness, vigilance and spare mental capacity, decision-making, risk taking and perceptions of danger, as well as allowing riders to train for unlikely events in the same way that aircraft pilots train for emergency situations so they are better prepared if they happen for real).

4.6 Conclusion

Whilst the evolution of motorcycle simulation has lagged behind other forms of transport simulation, it is now developing as a research application in its own right for a number of engineering, research and design purposes. Of the simulators that are known to exist, each has unique characteristics and each provides the user with different levels of physical and functional fidelity. This chapter presents an overview of the evolution of motorcycle simulation technologies and identifies their unique place within the wider arena of transport simulation. A key factor in developing a motorcycle simulator is the balance between physical and functional fidelity alongside the integration of rider controls with appropriate simulation software that enhances the rider experience.

The potential of motorcycle ergonomics in all its guises is vast, and both real road and simulators will continue to provide a valuable research and training tool in understanding road user behaviour within wider interactive transport systems. Motorcycle simulation is an evolving field and will feature on the emerging research agendas in road transport just as they do in automotive, aviation and rail research. With the recent developments in motorcycle simulator technology and application, this is an area that can contribute to the safety, performance and experience of these vulnerable road users in future.

References

Arioui, H., Nehaoua, L., Hima, S., Seguy, N., and Espie, S. 2010. Mechatronics, design and modeling of a motorcycle riding simulator. *IEEE/ASME Transactions on Mechatronics, Institute of Electrical and Electronics Engineers* 15: 805–818.

Bastianelli, A., Spoto, A., and Vidotto, G. 2008. Visual attention while riding a motorcycle: Improving traffic hazard perception using a riding simulator. *Perception* 37: 135.

Benedetto, S., Lobjois, R., Faure, V., Dang, N.-T., Pedrotti, M., and Caro, S. 2014. A comparison of immersive and interactive motorcycle simulator configurations. *Transportation Research Part F: Traffic Psychology and Behaviour* 23: 88–100.

Brown, I. D. 2002. A review of the "look but failed to see" accident causation factor. *Behavioural Research in Road Safety, XI Seminar*: 116–124. London: Department of Transport, Local Government and the Regions.

Carroll, C. L., and Waller, P. 1980. Analysis of fatal and non-fatal motorcycle crashes and comparisons with passenger cars. *Proceedings of the International Motorcycle Safety Conference*, Maryland: 18–23 May 1980, 3: 1153–1178. Washington, DC: Motorcycle Safety Foundation.

Clarke, D. D., Ward, P., Bartle, C., and Truman, W. 2004. *In-depth study of motorcycle accidents*. Road Safety Research Report No. 54. London: Department for Transport.

Clarke, D. D., Ward, P., Bartle, C., and Truman, W. 2007. The role of motorcyclist and other driver behaviour in two types of serious accident in the UK. *Accident Analysis and Prevention* 39: 974–981.

Cossalter, V., Lot, R., Massaro, M., and Sartori, R. 2011. Development and validation of an advanced motorcycle riding simulator. *Proceedings of the Institution of Mechanical Engineers, Part D: Journal of Automobile Engineering*, 225(6): 705–720.

Crundall, D., Crundall, E., Clarke, D., and Shahar, A. 2012a. Why do car drivers fail to give way to motorcycles at t-junctions? *Accident Analysis and Prevention* 44: 88–96.

Crundall, E., Crundall, D., and Stedmon, A. 2012b. Negotiating left-hand and right-hand bends: A motorcycle simulator study to investigate experiential and behaviour differences across rider groups. *PLoS ONE* 7: e29978.

Crundall, D., van Loon, E., Stedmon, A., and Crundall, E. 2013a. Motorcycling experience and hazard perception. *Accident Analysis and Prevention* 50: 456–464.

Crundall, E., Stedmon, A., Saikayasit, R., and Crundall, D. 2013b. A simulator study investigating how motorcyclists approach side-road hazards. *Accident Analysis and Prevention* 51: 42–50.

Crundall, D., Stedmon, A., Crundall, E., and Saikayasit, R. 2014. The role of experience and advanced training on performance in a motorcycle simulator. *Accident Analysis and Prevention* 73: 81–90.

Department for Transport 2007. *Transport Statistics Bulletin, Road Statistics 2006: Traffic, Speed, Congestion*. London: Department for Transport.

Department for Transport 2012a. Road traffic (vehicle miles) by vehicle type in Great Britain: Table TRA0101. https://www.gov.uk/government/statistical-data-sets/tra01-traffic-by-road-class-and-region-miles (accessed 14 September 2016).

Department for Transport 2012b. Reported road casualties Great Britain 2011. https://www.gov.uk/government/uploads/system/uploads/attachment_data /file/9280/rrcgb2011-complete.pdf (accessed 14 September 2016).

Di Stasi, L., Alvarez-Valbuena, V., Canas, J., Maldonado, A., Catena, A., Antoli, A., and Candido, A. 2009. Risk behaviour and mental workload: Multimodal assessment techniques applied to motorbike riding simulation. *Transportation Research Part F: Traffic Psychology and Behaviour* 12: 361–370.

Doz Nadal, A. 2015. Evaluation of acceleration sensation induced by proprioception on a motorcycle simulator. Master Thesis, Technische Universität München, Germany. http://upcommons.upc.edu/handle/2117/81067 (accessed 14 September 2016).

Ferrazzin, D., Salsedo, F., Barbagli, F., Avizzano, C. A., Di Pietro, G., Brogni, A., Vignoni, M., and Bergamasco, M. 2001. The MORIS motorcycle simulator: An overview. *Proceedings of the Small Engine Conference Technology and Exhibition*, Pisa, Italy: November 2001. http://www0.cs.ucl.ac.uk/staff/a.brogni/public /download/2001_Ferrazzin_TMM.pdf (accessed 14 September 2016).

Filtness, A. J., Rudin-Brown, C. M., Mulvihill, C. M., and Lenné, M. G. 2013. Impairment of simulated motorcycle riding performance under low dose alcohol. *Accident Analysis and Prevention* 50: 608–615.

Fisher, D. L., Laurie, N. E., Glaser, R., Connerney, K., Pollatsek, A., and Duffy, S. A. 2002. Use of a fixed-base driving simulator to evaluate the effects of experience and PC-based risk awareness training on drivers' decisions. *Human Factors* 44: 287–302.

Greenberg, J., and Blommer, M. 2011. Physical fidelity of driving simulators. In *Handbook of Driving Simulation for Engineering, Medicine, and Psychology*, eds. D. L. Fisher, M. Rizzo, J. K. Caird, and J. D. Lee, 7.2–7.24. Boca Raton, FL: CRC Press/Taylor & Francis.

Helander, M. 1976. *Vehicle Control and Highway Experience: A Psychophysiological Approach*. Gothenburg, Sweden: Chalmers University of Technology Division of Highway Engineering.

Hosking, S. G., Liu, C. C., and Bayly, M. 2010. The visual search patterns and hazard responses of experienced and inexperienced motorcycle riders. *Accident Analysis and Prevention* 42: 196–202.

Huth, V., Biral, F., Martín, O., and Lot, R. 2012. Comparison of two warning concepts of an intelligent curve warning system for motorcyclists in a simulator study. *Accident Analysis and Prevention* 44: 118–125.

James, M. H. 2005. Where you look is where you go: A fresh approach to the treatment of bends. Collision Investigation and Analysis Team, Buckinghamshire County Council. http://www.motorcycleguidelines.org.uk/wp-content/uploads /2013/08/WYLIWYGbooklet.pdf (accessed 14 September 2016).

Jamson, S., Lai, F., and Jamson, H. 2010. Driving simulators for robust comparisons: A case study evaluating road safety engineering treatments. *Accident Analysis and Prevention* 42: 961–971.

Lee, T. C., Polak, J. W., and Bell, M. G. H. 2007. Do motorcyclists behave aggressively? A comparison of the kinematical features between motorcycles and passenger cars in urban networks. *Proceedings of the 11th World Conference on Transport Research*. http://www.tsc.berkeley.edu/newsletter/fall2007/C3/1400/wctr_submit20070501 .doc (accessed 14 September 2016).

Lewis-Evans, B., and Charlton, S. G. 2006. Explicit and implicit processes in behavioural adaptation to road width. *Accident Analysis and Prevention* 38: 610–617.

Liu, C. C., Hosking, S. G., and Lenné, M. G. 2009. Hazard perception abilities of experienced and novice motorcyclists: An interactive simulator experiment. *Transportation Research Part F: Traffic Psychology and Behaviour* 12: 325–334.

Lynham, D., Broughton, J., Minton, R., and Tunbridge, R. J. 2001. *An analysis of police reports of fatal accidents involving motorcycles*. TRL Report No. 492. Crowthorne, UK: Transport Research Laboratory.

Mannering, F. L., and Grodsky, L. L. 1995. Statistical analysis of motorcyclists' perceived accident risk. *Accident Analysis and Prevention* 27: 21–31.

Martin, B., Phull, S., and Robertson, S. 2001. Motorcycles and congestion. *Proceedings of the European Transport Conference*. Association for European Transport, published as a CD-RoM (ETC2001). Available from PTRC Training for Transport (http://www.ptrc-training.co.uk).

McInally, S. 2003. R3: A riding strategy formulation model for the risk averse motorcyclist. In *Contemporary Ergonomics 2003*, ed. P. T. McCabe, 406–410. London: Taylor & Francis.

Patten, C. J. D., Kircher, A., Östlund, J., Nilsson, L., and Svenson, O. 2006. Driver experience and cognitive workload in different traffic environments. *Accident Analysis and Prevention* 38: 887–894.

Robertson, S. A., and Minter, A. 1996. A study of some anthropometric characteristics of motorcycle riders. *Applied Ergonomics* 27: 223–229.

Robertson, S. 2003. Motorcycling and congestion: Behavioural impacts of a high proportion of motorcycles. In *Contemporary Ergonomics 2003*, ed. P. T. McCabe, 417–422. London: Taylor & Francis.

Robertson, S., Stedmon, A. W., Bust, P., and Stedmon, D. 2009. Motorcycle ergonomics: Some key themes in research. In *Contemporary Ergonomics 2009*, ed. P. Bust, 432–441. London: Taylor & Francis.

Sager, B., Yanko, M. R., Spalek, T. M., Froc, D. J., Bernstein, D. M., and Dastur, F. N. 2014. Motorcyclist's lane position as a factor in right-of-way violation collisions: A driving simulator study. *Accident Analysis and Prevention* 72: 325–329.

Shahar, A., Poulter, D., Clarke, D., and Crundall, D. 2010. Motorcyclists' and car drivers' responses to hazards. *Transportation Research Part F: Traffic Psychology and Behaviour* 13: 243–254.

Shahar, A., Dagonneau, V., Caro, S., Israël, I., and Lobjois, R. 2014. Towards identifying the roll motion parameters of a motorcycle simulator. *Applied Ergonomics* 45: 734–740.

Stammers, R. B. 1986. Psychological aspects of simulator design and use. *Advances in Nuclear Science and Technology* 17: 117–132.

Stanton, N. 1996. Simulators: A review of research and practice. In *Human Factors in Nuclear Safety*, ed. N. Stanton, 117–140. London: Taylor & Francis.

Stedmon, A. W., Hasseldine, B., Rice, D., Young, M., Markham, S., Hancox, M., Brickell, E., and Noble, J. 2009. MotorcycleSim: An evaluation of rider interaction with an innovative motorcycle simulator. *The Computer Journal* 54: 1010–1025.

Stedmon, A. W., Crundall, D., Crundall, E., Irune, A., Saikayasit, R., van Loon, E., Ward, P., and Greig, N. 2011. Developing a collective test battery to investigate rider behaviour: An integrated experiment approach. *Advances in Transportation Studies*: 63–78.

Stedmon, A. W., Brickell, E., Hancox, M., Noble, J., and Rice, D. 2012. MotorcycleSim: A user-centred approach in developing a simulator for motorcycle ergonomics and rider human factors research. *Advances in Transportation Studies* A27: 31–48.

Symmons, M., and Mulvihill, C. 2011. A simulator comparison of riding performance between new, returned and continuing motorcycle riders. *Proceedings of the Sixth International Driving Symposium on Human Factors in Driver Assessment, Training and Vehicle Design.* 27–30 July 2011, Lake Tahoe, CA.

Tagliabue, M., Da Pos, O., Spoto, A., and Vidotto, G. 2013. The contribution of attention in virtual moped riding training of teenagers. *Accident Analysis and Prevention* 57: 10–16.

Vidotto, G., Bastianelli, A., Spoto, A., and Sergeys, F. 2008. Enhancing hazard avoidance in teen-novice riders. *Accident Analysis and Prevention* 43: 247–252.

Vidotto, G., Bastianelli, A., Spoto, A., and Sergeys, F. 2011. Enhancing hazard avoidance in teen-novice riders. *Accident Analysis and Prevention* 43: 247–252.

Will, S., and Schmidt, E. A. 2015. Powered two wheelers' workload assessment with various methods using a motorcycle simulator. *IET Intelligent Transport Systems* 9: 702–709.

Section III

Rail

5

Train Simulators for Research

Arzoo Naghiyev and Sarah Sharples

CONTENTS

5.1 Introduction

Train simulators have become an integral part of rail system design and planning and driver training since the Ladbroke Grove accident in 1999. The subsequent inquiry into the accident criticised the limited use of simulators in the UK rail industry and recommended that further human factors research be carried out to develop the understanding of train driving (Cullen 2001). Research train simulators have been used to understand the nature of train driving and associated cognitive constructs such as workload, situation awareness and vigilance. They also allow for the design and evaluation of new rail automation and control technologies, as well as changes to the rail infrastructure, *before* these changes are implemented, thus aiding prediction of the potential impact that they may have on train drivers.

Alternatives to the use of rail simulators to support implementation of changes include computational simulation of rail systems and the capture of behaviour in a real-world context. Both of these approaches offer significant benefits, such as the opportunity to manipulate a range of variables (computational simulation) and capturing the complexity of a specific part of track or group of drivers (real-world data capture). However, train simulators are less time consuming than computational simulations, which do not always show the effect on the driver's actual behaviour (Hamilton and Clarke 2005); research in the real-world environment can be very time consuming and is above all limited by the number of uncontrollable variables. Furthermore, the infrastructure that needs to be evaluated may not exist or may be difficult to compare in a naturalistic environment, due to low internal validity.

Despite the abundance of train driving simulators for training purposes, they cannot always be used for the purposes of research. They are excellent tools for train drivers, helping them improve their route knowledge, experience abnormal working and offering opportunities to record and monitor driving performance in detail. However, due to their inflexibility, they cannot always be used for research where experimenters may want to change parts of the simulated visual scene or parts of the physical simulator itself. Nevertheless, many training simulators have been used on occasions for the purposes of research, due to the unavailability of a research-specific simulator that would address the needs of the research question and the cost and time needed to develop a train simulator for the purpose of research.

In comparison to simulators in other transportation domains, train driving research simulators are very uncommon and sparsely used. Despite the Ladbroke Grove accident inquiry criticising the limited use of rail simulators and the need to develop understanding of train driving, there has not been a large shift by the rail industry to support the development of train driving research simulators. Typically, they have been built to a low-fidelity specification in an academic environment, instead of being commissioned by the rail industry. However, there are some exceptions where train driving research simulators have been designed for the purposes of research and have been commissioned and financially supported by the rail industry.

This chapter discusses the small volume of research conducted using train driving research simulators. It then considers the requirements that should be considered when designing a train driving simulator for the purposes of research and the collection and analysis of the different types of data that could be captured. In addition, issues of validity and fidelity in train simulators for research are discussed. There are some national and global differences in train driving, but this chapter predominantly focuses on the UK rail system.

5.2 Research Using Train Driving Simulators

Over the years, a variety of research questions have been examined using train simulators of varying levels of fidelity. These questions have included studies of driver workload and performance as well as the impact of introducing new technologies, including automation, on the driving task. Much simulator research has taken place within academic institutions, in conjunction with the rail industry, and has also been undertaken by the rail industry directly. Dunn and Williamson (2012) used a low-fidelity simulator (Microsoft Train Simulator software on a PC connected to an LCD monitor and a desktop train controller) to investigate the effect of task demand on driver performance on monotonous routes. Marumo et al. (2010) used a higher-fidelity train simulator to analyse the braking behaviour of train drivers to detect unusual driving. Thomas et al. (1997) also used a high-fidelity simulator to explore the impact of work schedule on train driver performance. These and other studies are discussed further throughout the chapter.

Simulators have also been used to explore the use of new technology in train driving, for example examining eye movements of drivers by using the European Rail Traffic Management System (ERTMS) interface and a driver advisory system (DAS) in train driving (Arenius et al. 2012) and the effect of different types of DAS and task demand on performance and workload (Large et al. 2014). Additional equipment and technology has also been introduced in train simulators for the purpose of research, with the aim of informing the design of future in-cab systems. An example of this is a roof-mounted head-up display (HUD) on a train simulator that has been used to examine the feasibility of HUDs in train cabs (Davies et al. 2007). Kojima et al. (2005) used physiological testing to capture brain activity by using cerebral blood flow as an indicator of monotony for drivers in both automatic train operation and manual control.

5.3 Developing Requirements for a Train Driving Simulator

Before assessing the requirements of a train driving simulator for research, it is crucial to have a good understanding of the nature of train driving, including the types of tasks and activities involved and the associated demands. There are several train driving models that have been developed by researchers (Hamilton and Clarke 2005; McLeod et al. 2005; Roth and Mullter 2007; Naweed 2014; Zoer et al. 2014) that can be used to further one's understanding. In addition, Buksh et al. (2013) conducted a comparative cognitive task analysis of some of the different forms of train driving in the UK rail system.

Train simulators are an ideal tool for conducting research that would not necessarily be viable in a real-world context, due to either cost or the safety-critical nature of the rail environment. Examples of such include examining costly changes in the infrastructure or investigating unsafe scenarios such as trespassers or other hazards in the rail environment. In particular, research simulators are an ideal method to use where a researcher wishes to apply more intrusive methods that may either take time to apply (e.g. physiological monitoring devices) or distract the driver in some way (e.g. questioning whilst driving). They also enable us to explore the impact of implementing new infrastructure or technologies and address potential needs such as scenario flexibility (e.g. examining a range of different scenarios, contexts or tasks) and data analysis (logging in detail the interactions of the driver with the cab and external environment) (Young 2003). However, when commissioning or designing a rail simulator, a major consideration that should be addressed is whether the simulator will meet the required research purposes. Therefore, a full set of requirements is needed – these should be elicited from relevant experts and specified in a usable form that will support system purchase and design.

When developing a train simulator for research, it is essential to have relevant stakeholder and subject matter expert (SME) engagement in eliciting the user requirements, development and evaluation stages. A stakeholder might be the simulator owner (e.g. a network infrastructure controller or train-operating company), the simulator users (e.g. researchers), simulator participants (drivers) or transport policymakers (e.g. safety or standard groups). The needs of these groups may differ, and the relative priority given to requirements from different groups should be considered. For example, a rolling stock manufacturer may be particularly interested in the way in which the cab interior within the fleet should be designed; therefore, their focus may be on a realistic physical interior to the cab. Conversely, a policymaker may be interested in the impact of a change in regulations or process on train driver behaviour more generally; in this case, the focus may be on the need to simulate a range of different driving contexts and routes.

Furthermore, cross-cultural differences in train driving both nationally and globally may need to be considered when establishing simulator requirements, if relevant. For example, the requirements for a new freight-based system in a low-population density country (e.g. Sweden) would be different from those for a simulator used to support the design of a multi-country system that might be implemented in mainland Europe. In addition, the regulatory context and varied rail infrastructure standards and guidelines that need to be simulated should be identified.

In order to identify priorities for design, it is therefore appropriate to apply a structured process for requirement elicitation. The following section describes a previously applied process of requirement elicitation for a research rail simulator and presents the thematic groupings of different elements of driving simulation that emerged.

5.4 Eliciting Requirements for a Train Driving Simulator

Different types of complex work modelling, description and analysis methods can be used to understand the nature of the train driving task, which can in turn be used to understand the requirements of a simulator. Commonly used human factors approaches include work domain analysis (the first phase of a cognitive work analysis [Vicente 1999]), critical decision method (CDM) (Klein et al. 1989) and applied cognitive task analysis (Militello and Hutton 1998). The data collected for these frameworks can come from a variety of sources, such as documentation, interviews, observations, use of training simulators and cab rides, with the input of a variety of people from the rail industry. In developing a train simulator for human factors–focused academic research, Yates et al. (2007) elicited user requirements from human factors experts in the rail industry. The main themes that arose from this research are shown in Figure 5.1.

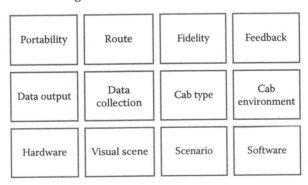

FIGURE 5.1
Themes elicited from user requirement gathering. (From Yates, T. K. et al., in Wilson, J. R. et al. (eds), *People and Rail Systems: Human Factors at the Heart of the Railway*, Ashgate, Farnham, UK, 2007, pp. 155–165.)

TABLE 5.1

Summary of Train Simulator Classification Categories and Subcategories

Train Simulator Classification Category	Subcategories
Internal cab environment	Cab type
	Controls
	Olfactory control
	Temperature control
Input control interface	Modular design
	Operational instrumentation
	Cab controls
	Gauge/dials
	Haptic feedback
Audio	Internal audio
	External audio
	Weather-related audio
Motion base	Fixed base
	Motion base
	Vibration effects
Train dynamics	Braking characteristics
	Acceleration characteristics
	Terrain-related characteristics
	Train type and loading characteristics
	Degraded situation characteristics
Visual representation of environment	Geotypical/geospecific
	Rail infrastructure
	Non-rail infrastructure items
External environmental effects	Weather
	Low visibility
	Emergency scenarios
Communications in cab	Communications
	Scenarios involving communications
System lag	Time delay
Visual display system	Type of display
	Ambient lighting
	Sun glare
Data collection	Data collection capabilities
	Sampling rate
Other	Other functionality not mentioned above
	Other information

Source: Yates, T. K., *Design and development of a virtual reality train simulator for human factors research*, PhD thesis, University of Nottingham, Nottingham, UK, 2008.

When designing a research train simulator, these themes can be used as a framework to elicit specific user requirements. Normally, when developing a research-focused simulator, it would be advisable to engage a broad range of expertise across several domains, including rail, human factors and computer science, as well as train drivers.

Tichon (2007) also elicited user requirements for a train simulator, using train drivers to understand the nature of the driving task under degraded working and to design appropriate scenarios. Tichon (2007) used a CDM (Klein et al. 1989), which is a form of cognitive task analysis that uses structured questioning to identify interaction requirements in specific points in decision-making, to focus on the decision-making elements of train driving under degraded modes as opposed to the entire task. This method has been used to elicit user requirements for simulators across a variety of other domain, and a detailed evaluation of the reliability and validity of this method can be found in the paper of Hoffman et al. (1998).

On the basis of the overall topic areas that emerged in SME interviews and informed by technical components of typical simulators, Yates (2008) generated a detailed simulator classification for train simulators. Table 5.1 shows a summary of these classifications.

Table 5.1 highlights the many factors that need to be considered and evaluated when designing a train simulator for research. Elements of these categories will be discussed in turn in detail.

5.5 Refinement of Simulator Requirements: Fidelity versus Cost

It is reasonable to assume that the designer of a research simulator will need to make compromises in design, due to time or financial constraints. Even a multimillion-pound/dollar simulator will include some compromise, in the number of routes or cab interior configurations that can be simulated for example, and more typically, a research simulator will be developed on a much lower budget. Therefore, the designer needs to think very carefully about which elements are most important to simulate and to what level of complexity.

Throughout the following sections, we discuss the concept of fidelity. Fidelity refers to the extent to which something is a faithful representation of its real-world equivalent – a high-fidelity simulator will be a closer replication of the real world than a low-fidelity simulator. Often, a single simulator will combine elements that are high and low fidelity; for example, a researcher may have access to a real train cab and thus a high-fidelity interior, but only a low-resolution or low-complexity (and thus lower-fidelity) external display. It is vital to evaluate the level of fidelity of the simulator in the design stages, prior to the implementation stage, to assess if performance in the simulator will reflect that in the real world. Once the requirements of the simulator have been gathered, it is advisable to conduct a cost–benefit analysis of the requirements for the research simulator.

Closely related to the concept of fidelity is that of validity. Other texts (e.g. Wilson and Sharples 2015) cover validity in more depth, but the essence of validity is the extent to which a measure or tool measures what it purports to measure. In the context of simulators, two types of validity are particularly important: *face validity* (the extent to which the simulator 'looks' like the real environment) and *behavioural validity* (the extent to which behaviour in a simulator is the same as that in the real world). When considering behavioural validity, it is often the case that we can make inferences with confidence about *relative validity* (i.e. the extent to which the difference in behaviour between two conditions or contexts is the same as that would be expected in the real world) and are sometimes less concerned about *absolute validity* (the ability to directly transfer behaviour metrics, such as speed, from the simulator to the real world).

In addition to validity, constructs such as immersion and presence, which are closely linked, need to be considered when designing a train simulator for research. One way of defining immersion can be the 'degree to which a person feels absorbed into an experience' (Pausch et al. 1997). It has been suggested that presence is situated immersion (Alexander et al. 2005). Nichols et al. (2000) went further to define presence as 'a sense of being there, reflected by engrossment with, and intuitive behaviour, in the virtual environment'. There are many factors which influence presence. These can include the fidelity of the sensory input, the exclusion of information from the immediate environment and the level of realism in the virtual environment (Witmer and Singer 1998). There have been many attempts to classify these factors; for example, Witmer and Singer (1998) described them as control factors, sensory factors, distraction factors and realism factors. It is important when examining train simulators to consider these key constructs. Fidelity is important for immersion and for engaging the participant and for acceptance of the train simulator by the participant, particularly with experienced train drivers.

Research into the benefits of increased fidelity in simulated environments has produced mixed results. Yates (2008) found that an increase in scene complexity (i.e. functional fidelity) had no overall benefits. However, Waller et al. (1998) suggested that an increase in fidelity would improve the performance and reactions of a participant in a simulator. When talking to train drivers, we have noted that they may express that 'everything' in the visual scene is essential when simulating the driving task. However, what is important is to identify which elements of the simulator are necessary if our primary goal is to elicit behavioural validity and presence. High levels of realism in a research simulator can increase psychological fidelity as these help to replicate stress and arousal found in degraded modes of train driving. However, there needs to be a balance between the level of fidelity and the cost, in order to have a cost-effective simulator. Unfortunately, in comparison to other transportation domains, these issues are severely underresearched in the rail domain and require further study for a comprehensive understanding.

5.6 Elements to Consider When Generating Train Simulator Requirements for Research

5.6.1 Internal Cab Environment and Input Controls

When specifying the requirements for the internal cab environment, the cab type and the train controls need to be considered. A real commuter-class train cab can be used with the standard or typical in-cab controls (e.g. throttle, brake, alarm warnings and communication interfaces) to help achieve a high level of physical fidelity. This level of physical realism is not always crucial for a research simulator, and high functional fidelity may be more desirable; however, more research is required in this area when considering train driving simulators for research. It needs to be decided at a very early stage whether a cab-based simulator or a desktop-based simulator is to be developed. This decision is usually influenced by the budget set aside for the simulator but may also be affected by requirements such as the need for a simulator to be portable between different physical locations. High physical fidelity (and thus face validity) can influence the expert user's buy-in and acceptance of the simulator and influence how motivated the user is to perform as they normally would (Alexander et al. 2005). It has been suggested that incorporating other factors such as the right temperature and appropriate smell in a cab should also be considered in order to achieve a high level of sensory fidelity (Yates et al. 2007). (However, from personal experience of train cab rides, there does not appear to be any differing cab 'smells'.) Yates (2008) found that using a full cab simulator offered little or no benefit over using a laptop- or desktop-based simulator on the level of the presence reported by experienced train drivers who used the simulators.

If a cab simulator is to be used, then the layout of controls should replicate that of a real cab. An example of this is provided by Davies et al. (2007), who closely based their cab simulator on a United Kingdom-based class 390 Pendolino. This simulator included full driving controls, an automatic warning system (AWS), a driver's vigilance system, a train management system, an air-suspended driver's seat and train protection and warning system. It should also be considered which cab type is to be used and if elements of that cab type (e.g. design or layout of controls) are specific to a particular class of train. This may influence the way that a driver engages with the research. For example, the use of an urban commuter train cab would be inappropriate to use with a high-speed long-distance route, but it may be completely acceptable to use a typical or familiar type of urban commuter cab for comparing different urban routes.

A variety of input controllers can be used in the train simulator. These must include the basic throttle and brake (Dunn and Williamson 2012). To increase the flexibility of use of the simulator, the controls could be adaptable to simulate different classes of train. For example, the number of notches for the throttle or brake could be altered depending on the type of cab that was being simulated. When considering user requirements, it should be assessed if the train simulator would be used to research freight train driving or if the simulation would require additional activities such as coupling and decoupling of carriages (and thus whether these less frequently used elements of the cab controls need to be included). Furthermore, any safety devices used in trains should also be included as they are a crucial part of the driving task. An example of this type of device is the driver safety device (DSD) in the UK rail system.

In addition, the inclusion of new technology in the cab such as an ERTMS interface or DAS should be considered if appropriate. It is important to consider the implementation of future train technologies, both new in-cab technologies and new signalling systems, when designing a train driving simulator for research that will have long-term use. When testing concepts prior to the design and build of a new train technology, a Wizard-of-Oz approach can be used. For example, Large et al. (2014) used this approach to evaluate two prototype DASs by using Microsoft PowerPoint on an interface in the cab. The experimenter updated the prototype DAS at predetermined points in the journey to imitate the DAS. The use of this approach was a cost- and time-effective method for evaluating different concepts in the early design stages.

5.6.2 Field of View

Drivers typically look predominantly at the field of view in front of them and rarely look at the side window, unless they are performing a specific task such as checking the platform area before setting off and using train driver-operated level crossings. In other simulation contexts, such as automobile driving simulation, the influence of the field of view is strong, and many simulators have wide fields of view, but this can be limiting due to motion sickness (Greenberg and Blommer 2011). The use of visual information when driving a train is arguably different for a number of reasons. Firstly, train drivers do not have mirrors or other aids to help extend their field of view, as their primary task involves focusing on a narrow field of view in front of them. Secondly, the speed at which they are often driving, coupled with the perpendicular position of the windows next to them, means that in many contexts the information from the side windows is much more limited and not so actively used in driver decision-making. Some trains are fitted with forward-facing cameras, and some train-operating companies allow drivers to use these to capture events out on the tracks. These types of cameras may be considered when designing simulators to research issues related to incidents or accidents. Yates (2008) recommended that a small flat projection screen should be used to fit the view of the cab. Not only is this cost effective, but it has been shown that participants using a higher-fidelity 160-degree curved widescreen made more mistakes in a train simulator than those using a low-fidelity small cab-based flat screen (Tichon and Wallis 2010). It was suggested by Tichon and Wallis (2010) that the narrower field of vision may provide a visible frame of reference similar to that of a real train. Therefore, in the case of train simulators designed for the purpose of research, a higher level of perceptual immersion could be detrimental to the validity.

5.6.3 Audio

Operational sounds should be included as they are a part of the driving task. This could be the horn that drivers are required to use when they see a whistle board, AWS feedback, DSD alert or alarms from the ERTMS. Furthermore, speakers can be installed into the cab to replicate realistic engine noise. The noise should be realistic as poor sensory fidelity could distract the user's attention away from the simulator task (Tichon et al. 2006). Soundproofing of the simulator cab from the surrounding environment is recommended, as it will also improve the level of presence in the simulator (however, this soundproofing can be achieved by including reasonably loud 'background' noise, as well as physically enclosing the simulator space – it is often more a problem with rail simulators that the noise from the simulator travels out of the simulation space, as opposed to external noises disturbing the simulator participants!). Foerstberg (2000) used train seats and internal train noise to improve the fidelity of the simulator. Hellier et al. (2011) found that when presented with simulated low levels of engine noise, participants drove at faster speeds and were more likely to commit driving violations – obviously with implications for the absolute validity of behavioural data from such a system. If simulating weather conditions such as heavy rain, then it should be considered if this should be included in the simulation of the audio.

5.6.4 Speed Perception

Speed perception is a very important issue to consider when designing train simulators, because the perception of speed is influenced by a variety of sensory cues such as optic flow, proprioception, distance perception and auditory cues (Wallis et al. 2007). Train drivers use speed perception and knowledge of gradients and landmarks to calculate braking distances. Unlike car driving, it can take a long time to stop a high-speed train and if miscalculated can cause problems such as overrunning a train station. One contributor to speed perception that has been the subject of much discussion within the automotive simulation sector (Kemeny and Panerai 2003; see also Chapter 3) is the simulation of motion within the cab. The use of a motion platform tends to be very costly, can contribute to increased levels of motion sickness amongst participants and has not been consistently demonstrated to greatly influence the face validity or behavioural validity of simulators (see Sharples et al. 2014). A more efficient way of influencing speed perception is, of course, to directly communicate the speed to the driver, as what would happen in the real-world context. This can be provided by a working speedometer in the cab simulator, or in a lower-fidelity simulator, a speedometer could be placed at the bottom corner of the virtual environment to aid speed perception. Auditory cues such as speed-modulated rumbles of the train could help to add extra information to the environment (Wallis et al. 2007).

5.6.5 Motion and Train Dynamics

Motion is a large part of train driving in which drivers feel the momentum of the train and experience changing gradients. Speed and motion fidelity are factors that need to be considered when designing a research simulator, as they may increase the level of sensory fidelity and presence. A simple solution is to introduce a vibration unit at the base of a cab-based simulator to simulate the sensation of an engine. Yates (2008) set this vibration unit at a constant speed; however, this can be modulated in relation to acceleration. An alternative approach to simulating vibrations is to produce visual images that shake on the simulator screen. Marumo et al. (2010) achieved this by designing a simulator that calculated the vibration of the virtual cab from the track irregularities and a vehicle dynamics model. This allowed the sensation of the virtual cab to be vibrating without the use of a motion base. Kojima et al. (2005) simulated longitudinal motion by using driving force, braking force and train resistance. They calculated driving force speed–tension properties of the specific type and class of train. They calculated the braking force depending on the type of brakes (electric or air brakes). Finally, they simulated train resistance by using starting resistance, running resistance, curve resistance and gradient resistance. In order to make the simulation more realistic, they also created slip and skid properties (track adhesion) during scenarios with rainy weather. In normal slippery conditions, a train driver would release sand onto the tracks and the integration of a sand release button could be taken into consideration in the cab design.

It is possible to develop a tilting-train simulator, using a motion system to help replicate motion. This can be achieved using a motion base, a motion controller and software containing motion generation algorithms (Kim 2007). Currently there are some Pendolino tilting trains in the United Kingdom; however, they are restricted to certain services. Therefore, if the focus of the research was not on tilting trains, then it may not be cost effective to include this function in the simulator and it would not necessarily demonstrate relevant functional fidelity. Foerstberg (2000) used a moving-vehicle simulator by using a combination of horizontal and roll accelerations to examine ride comfort and motion sickness in tilting trains. Thomas et al.'s (1997) simulator was mounted on a six-axis motion base to replicate the motion of a train. They computationally modelled the track geometry and train suspension characteristics to achieve this. The simulator was able to produce heave and sway associated with rail infrastructure such as passing over points and bridges.

5.6.6 Simulation Lag

Simulation lag is crucial to consider in terms of presence and also to reduce simulation sickness. It is recommended that the lag does not exceed 100 milliseconds, as this is when users notice latencies (Huang and Gau 2003). It was suggested by Tichon et al. (2006) that introducing cab vibration could reduce the risk of possible simulation sickness as it reduces any visual vestibular discrepancies – although the increased vestibular stimulation that may result from this approach may exacerbate symptoms in those who are already experiencing sickness due to sensory conflict, the difference in information being presented to the visual and vestibular senses (see Sharples et al. 2008 for further discussion). Using vibrations in the cab could also improve the sensory and physical fidelity of the simulator. Therefore, it could be beneficial to include a vibration unit into the base of the simulator. Another benefit of using a vibration unit, as noted earlier, is that it can provide physical feedback to the user which acts as a cue for speed.

5.6.7 Portability

Portability of the simulator is another factor that needs to be assessed as a requirement. It is difficult to access train drivers as participants for a study (as discussed later in this chapter), and making the simulator portable could help to address this issue. A portable system could enable the researcher to take the simulator along with them to conduct studies at train depots. A portable system may be more appropriate to use if a full cab simulator may not be viable due to logistics of transportation and space for storage and use, or if the topic of the research is focused on comparison of different routes or contexts, rather than measuring behaviour in response to specific cab-based elements. Another constraint is that if the simulator was to be in a fixed location, then it may be problematic if drivers need to be sampled from a large geographic area.

5.6.8 Feedback

Feedback for the participants must be considered due to the nature of train driving. Experienced train drivers will expect engine noise when the train is stationary or in motion. Kojima et al. (2005) generated sounds by recording an actual train and then using the recordings in their simulation. To enhance the realism, they adjusted the frequency and sound volume based on information from the actions of the train operator and simulated environment. Drivers calculate gradients, traction and abnormal working based on the speed of the train in relation to how they are using the controls. Therefore, if the train driver's actions are not accurately portrayed by the simulator, then the driver might not accept the simulator as an accurate representation of train driving. This in turn could have an impact on the behavioural validity of the study. This should also be considered when assessing the accuracy of the speedometer being used in the train simulator.

5.6.9 Software

When building a new train driving simulator, a decision must be made in the early stages on whether off-the-shelf train simulator software is to be used or whether designing and commissioning an entirely new simulation is an option. The choice of software used for the virtual environment can vary from virtual reality software, such as Vizard virtual reality toolkit, or package software such as Microsoft Train Simulator or RailSimulator.com's Train Simulator 2013. Train simulator packages can allow bespoke scenarios to be created, providing flexibility for research purposes, as well as high levels of fidelity. However, this is an area in train driving research simulators that has been underresearched and needs further development.

5.6.10 Visual Scene and Routes

The visual scene is a very important aspect of a train simulator. It has been suggested (Yates 2008) that the visual scene should be rendered using high-quality images to provide a high level of presence. This should include a high refresh rate of the simulation (at least 30 frames per second), good resolution of the image, an appropriate field of view and the render quality (Wallis et al. 2007). Presence has been suggested to be positively related to performance in an immersive simulator environment (Witmer and Singer 1998). Therefore, a high level of presence is crucial in the virtual environment of a train simulator as this will provoke realistic psychological responses, which will increase the psychological fidelity of the train simulator. In return, this will increase its behavioural validity, as the performance scores will more accurately represent the performance construct in the real world (Li et al. 2003).

In a medium- or high-fidelity simulator, the visual scene should include rail infrastructure in line with the relevant national rail group standards, including styling and dimensions. High fidelity can be achieved by replicating critical parts of the environment (Jentsch and Bowers 1998). It has been suggested that for signal sighting (identification of how visible an external signal is from the cab at various points along the track), a high level of realism is required in the simulation, particularly for rail infrastructure such as signal gantries and signal structures (Yates 2008). There are also local and national standards regarding design, maintenance and positioning of infrastructure that must be adhered to by the rail industry. These standards can be used as guidelines when developing virtual environments for a train simulator, in particular the positioning of signals and signal spacing. Data can be acquired by video footage of routes or route layout plans and maps from the rail industry.

The important core elements of the rail environment that should be considered when designing a virtual route based on a UK rail system include the rails, AWS magnets, signs, signals, mile markers, overhead lines, gantries and stations (Yates 2008). Non-rail-related objects should also be included as they can help a train driver determine their speed (Yates and Sharples 2005). Young (2003) and Li et al. (2006) suggested elements that should be present in the visual scene of a train simulator. These included the following:

- AWS magnets on the track
- Signals
- Speed boards
- Maintenance boxes
- Whistle boards
- Level crossings
- Viaducts
- Trackside buildings
- Overbridges
- Stations
- Tunnels

A significant part of the driving task is to monitor the environment, and drivers will examine parts of the infrastructure such as overhead power cables and points. Therefore, it is recommended that there should be some elements of the rail infrastructure replicated to support this part of the driving task. In addition, including track workers into the visual scene could improve the fidelity of the simulator or help investigate particular research questions (the UK and some other rail systems are not 'closed' systems, and drivers will encounter track workers along the routes).

The length and duration of the route should be considered when designing the route. Typically, train drivers work long shifts and can drive very long routes; therefore, a short trial may not replicate the true nature of train driving. The driving task that is to be investigated will also influence the virtual route that is to be used. Marumo et al. (2010) included four stations in their experimental route, as they were interested in braking behaviour. Sometimes the length of a route can be determined early on in the design process by the research question. For example, Dunn and Williamson (2012) used a 240 km route lasting three hours in order to understand task demand in monotonous driving. They further simulated monotony in the route by using constant speeds on straight tracks or by using night-driving. In addition, the length of the track used and the duration of the study will also depend on cross-cultural differences as well as the research question. Thomas et al. (1997) based their study in the Midwest of the United States, which consists of long-distance routes, and they used a route of 190 mi, lasting 10 hours in total. However, there were 10 stops within the route with breaks. When investigating the effects of different infrastructure on train driver behaviour, it is important to assess whether the flexibility of the visual scene simulation could be a limiting factor for research purposes (Young 2003).

Route complexity can be increased using signal density, gradient changes or speed restrictions, or by more unlikely events such as signal reversal, obstructions or equipment failure. The use of a microworld is an alternative method that can be used which does not try to replicate the complexity of the real world but focuses on the more abstract functional relationships (Canas and Waern 2005). This does not focus on the use of high visual fidelity and a complex visual scene, but instead depends on high abstract functional fidelity (Naweed et al. 2013). However, this method would lack ecological validity and should be considered only if there is a low budget and little input from the rail industry. The level of visual fidelity is a crucial part of the design of a train simulator and is further discussed earlier in this chapter.

5.6.11 Geospecific versus Geotypical Routes

There is much debate and a lack of agreement about whether the routes used in a simulator should be geospecific (i.e. direct representation of a particular route) or geotypical (not a representation of an actual route but has generic characteristics such as long distance or urban driving) (Yates 2008). One factor to consider when deciding between geospecific and geotypical routes is that train drivers that work on particular routes will have very in-depth route knowledge. For example a train driver will report that if they are driving a familiar route, then they not only are familiar with the track and track infrastructure, such as signs and signals, but also use a number of external landmarks to support good driving (e.g. position of braking points in different weather conditions) and can even 'drive with their eyes shut' (in other words, recognise where they are on a journey from the noises and movements along a track). Therefore, if a geospecific route is to be used, then all driver participants must either be experienced drivers on that route (i.e. have verified route knowledge of the environment presented) or have never driven that route (i.e. have no route knowledge to influence their performance). If it is a route that driver participants have been trained on, then there is also the risk that they will be more critical of any minor discrepancies in the representation of the simulated route (such as exact positions or colours of lineside objects or landmarks). This needs to be considered in the context of the 'willing suspension of disbelief' – in many cases, if there is enough information to make an environment familiar and we understand the context in which we are working (i.e. taking part in a research study), then we can often forgive or overlook minor omissions or differences in the simulated environment, as there are sufficient cues for our automatic or practiced behaviour to be employed.

5.6.12 Scenarios

The types of task scenarios used would be dependent on the research question and also if there is a type of technology that is being examined. Different types of scenario requirements need to be considered in the early design stages. The option to change weather conditions or specific operating scenarios should also be included to enable more versatile usage of the simulator (Tichon et al. 2006). The inclusion of such stress-inducing events has been suggested to contribute to psychological validity (Williams 1980). Previous studies have included unrealistic tasks such as including a mathematical calculation in the place of a speed board, in order to increase cognitive demand (Dunn and Williamson 2012). However, in a high-fidelity simulator, cognitive demand could be increased by using more context-relevant scenarios such as poor weather conditions (e.g. snow and fog), by simulating poor adhesion on the tracks (e.g. in leaf fall season) or by simulating many gradient and speed changes. This would make the simulation more comparable to the real world. Other potential unusual scenarios can include degraded working or simulating a large animal or road vehicle incursion on the tracks. Trespassers on the tracks or unusual passenger behaviour on stations could be used to further investigate trespassing and suicides in the rail domain from the driving perspective.

5.6.13 Communication Devices

The train cab and the driver are part of a dynamic open-loop system consisting of other operators (e.g. signallers, controllers, station staff), and communicating with them, particularly the signaller, is a part of the train driver's role. It may be necessary for future research to include a form of communication device in the simulator to enable research into communication or distraction. This may be achieved by a low-fidelity Wizard-of-Oz approach where the experimenter may play the role of the signaller via a communication device or a simple radio set-up. A high-fidelity approach could involve linking the train driving research simulator to a signalling research simulator. This would enable the researcher to investigate distributed cognition (as seen in the maritime domain with full-mission ship-handling simulators; see, e.g. Chapter 12). The inclusion of communication devices such as cab-secure radio or the more recent Global System for Mobile Communications – Railway would be important if the research question involved more realistic communication tasks and demands.

5.6.14 Hardware

Various additional hardware components, such as video cameras and eye-tracking equipment, could be implemented into a train driving simulator; however, it is important to question their purpose from a research context.

Video cameras could be mounted in the cab, to record participants' behaviours, including their use of the controls if it is not already being recorded by the simulator (Yates 2008). Eye-tracking equipment can be built into the simulator, or drivers could wear a device such as eye-tracking glasses to measure eye movements. Eye-tracking glasses have been used on train drivers in the field and were not reported to be intrusive to the driving task (Naghiyev et al. 2014). In addition, equipment for recording other physiological data, such as skin conductance or heart rate, could be considered.

5.6.15 Additional Considerations

There are some additional considerations to be made when developing a train simulator for research. It is beneficial to discuss and question any secondary task that may be used alongside the primary driving task. A secondary task could be a communication task, or it could be utilised to increase the participant's mental workload. Mental calculation problems are one method of introducing a secondary task to investigate mental workload (Marumo et al. 2010).

Another consideration may be whether or not to include the times of day in the simulation. Thomas et al. (1997) examined the effects of work schedule on driver performance, and it was therefore beneficial for them to simulate day and night video projection capabilities that corresponded to the time of day outdoors. They achieved this by manipulation of video special effects, which included a field of view at night dominated by a simulated headlamp illumination.

5.7 Collection and Analysis of Data with Research Train Simulators

The type of data required from the research train simulator should be considered in the design stages, in particular how the data are to be collected and analysed. The type of data that are to be collected should be consistent with the research questions that need to be investigated. This section discusses the choice of participants and expertise, training, performance measures, subjective measures, physiological measures and data extraction.

5.7.1 Choice of Participants and Expertise

Due to the skills and knowledge required in train driving, it is normally preferable to use experienced train drivers as participants when conducting simulator-based rail research (unless, for instance, the study is specifically examining novice drivers and training methods). The majority of studies conducted in this domain have used train drivers; however, access to drivers may be an issue for some researchers. The limited number of participants that this approach normally results in also affects the statistical power of any quantitative analyses. This is an issue which must be considered very early on in a project, as recruiting train drivers can be a lengthy process. As a compromise, novice participants with rail experience or expertise could be used.

When collecting demographic information about participants, as well as collecting the typical information about age, gender and driving experience, it may also be appropriate to collect additional information about the types of rail technologies that they have used in their driving and their experience of different signalling systems.

The number and type of participants used by previous studies using train simulators have varied greatly. Dunn and Williamson (2012) used 30 professional drivers, and Davies et al. (2007) used 16 professional drivers, whilst Sato et al. (2010) used 7 non-train drivers. Another approach used to increase participant numbers, as used by Marumo et al. (2010), was to recruit equal numbers of two professional drivers and two non-train drivers. This reflects the importance of gaining buy-in from the rail industry prior to designing and building a train driving simulator for research, in order to get a sufficient sample of drivers to participate in the study as well as in terms of the feasibility of the drivers travelling to the research centre.

5.7.2 Training

Here we are referring to issues of training to use the simulator and the simulated route. For instance, if the study is using novice participants, then it is necessary to include extra time for learning the train driving task. Even for experienced train drivers, some time is needed to provide familiarity with the simulator and the simulated route. In the normal train driving task, drivers can usually anticipate the upcoming signal states based on their route knowledge. Over time, drivers will learn points in their routes where they are more likely to be stopped or areas where they are very rarely stopped. The impact of this training on measured driver behaviour in simulators should be considered when designing routes and training. In addition, drivers develop experience of braking points (points in the route where they know they need to start braking to meet the next speed restriction or station stop). If a geotypical route is used, then enough training should be provided so that drivers can develop such basic route knowledge. Davies et al. (2007) compensated for the lack of route knowledge by programming the simulator to automatically provide a verbal announcement prior to a change in line speed. Naweed et al. (2013) argued that the use of a microworld simulator paradigm (described earlier) would mean that very little training would be required as the simple environment would allow for more time to learn the operational tasks. However, this approach would also lack ecological validity.

The duration of training given in previous studies using train simulators has varied greatly, from only five minutes of training with professional train drivers for them to get used to the set-up of the simulator and the controls (Dunn and Williamson 2012), to 15–20 minutes of training with professional drivers (Davies et al. 2007), to one hour of training for a mixture of professional drivers and students (Large et al. 2014). Training should also be used to assess if a novice participant has acquired enough skills to continue with the study. The length of training should be assessed based on the type of participant (novice participants would need longer training) and on the level of fidelity of the train simulator.

5.7.3 Performance Measures

One of the benefits of using a simulator is the opportunity to collect rich performance data. However, the selection of appropriate performance measures needs careful thought. Task time is not necessarily the best performance indicator of a train driving task as the participant may be overspeeding. A more appropriate measure could be driving performance, which can be measured using the number of speed violations (Dunn and Williamson 2012; Large et al. 2014). Speed violations have been measured in the past by both overspeeds and underspeeds. However, this may be a poor representation of performance if drivers do not have any time pressure to meet timetable schedules. Davies et al. (2007) analysed speed with respect to target line speed. Analysing the driver's use of the brake and vehicle deceleration can be used to evaluate braking behaviour and detect unusual driving (Marumo et al. 2010). Errors can also be recorded in terms of signals passed at danger. In addition, the use of other controls such as the throttle, AWS, DSD or screen capture of the use of an in-cab interface could be used as part of the data collected. The data that are captured will depend on the research question. For example Sato et al. (2010) examined the effects of a braking assistance system and were therefore interested only in braking handle position and deceleration data.

Task proficiency could also be measured by calculating fuel efficiency for eco-driving. For new technology, the acknowledgement of alarms may be of interest to the researcher. Driver fault diagnosis may be another performance indicator that could potentially be used as part of a study. This type of data is typically used in training simulators as a form of assessment.

One method used to measure performance is by examining safety and efficiency. Einhorn et al. (2005) investigated safety by monitoring speed control, signal adherence and reaction time and measured efficiency by monitoring stopping accuracy and schedule deviation.

5.7.4 Subjective Measures

Subjective measures can also be collected such as scales for workload (National Aeronautics and Space Administration Task Load Index [Hart and Stavelan 1988]), fatigue, boredom (by using the Boredom Proneness Scale [Farmer and Sundberg 1986]) and task characteristics (Dunn and Williamson 2012). To prevent an intrusion to the driving task, Thomas et al. (1997) administered the Global Vigor Scale (assessing subjective alertness) and the Global Affect Scale (assessing subjective emotional state) during stopping breaks in the simulated route. Usability and presence questionnaires (e.g. Patel et al. 2001) could also be administered at the end of a study to evaluate the simulator. Questionnaires based on motion sickness (e.g. Simulator Sickness Questionnaire [Kennedy et al. 1992], Short Symptoms Checklist [Sharples et al. 2008]) may be of interest, especially if the simulator has tilting motion technology. In addition to questionnaires, subjective data can be collected in the form of concurrent or retrospective verbal protocol. For concurrent verbal protocol, an audio recording device in the simulator is advisable, and for retrospective verbal protocol, video capture is essential.

An alternative method of collecting workload data during the trial is by using the Integrated Workload Scale (Pickup et al. 2004), a nine-point workload scale specifically developed for the rail context and tested with rail signallers and train drivers. Whilst one version of this tool includes a visual presentation of the scale via a tablet device, it would be inadvisable to integrate the collection of these data into the simulator design in the form of presentation on a display as it could interfere with the driving task. Therefore, verbal administration of the tool is more appropriate. An example of the impact of task demand in a rail simulator is provided by Dunn and Williamson (2012), who substituted the number on a speed sign with a basic mathematical calculation. However, caution must be taken that the method of increasing task demand does not interfere with the actual train driving task and affect behavioural validity. Large et al. (2014) increased task demand by increasing the number of stations that a driver had to stop at and by the use of more yellow and red signals.

The Situation Awareness Global Assessment Technique (SAGAT; Endsley 1995) can be used to measure situation awareness and should be considered in the initial design stages. This technique involves pausing a simulation and blanking the display at randomly selected times, then asking participants quick questions about their current perceptions of the situation (Endsley 2000). This technique allows the measurement of all levels of situation awareness, based not only on system functioning and status but also on the simulated environment (Endsley 2000). In the United States, the Department of Transportation has conducted a wide variety of studies using a research-specific human-in-the-loop train simulator to implement SAGAT (Marinakos et al. 2005). They measured situation awareness by asking the drivers to report the block number, speed, milepost, brake pressure, bearing temperature and current through the motors.

5.7.5 Physiological Measures

Train simulators have also been used to collect physiological measures; however, it remains costly in terms of equipment and time to analyse the data, so the value of including physiological data in a study must be clear. Eye movement data can be collected using eye-tracking glasses or by embedding an eye-tracking system into the train simulator. Eye tracking has previously been used in a train simulator to investigate the use of an ERTMS interface and a DAS interface in train driving (Arenius et al. 2012). As well as analysing the eye movement data, recording the scene with or without the overlaid eye movements could be a useful tool for retrospective verbal protocol.

Another form of physiological measure that has been collected in conjunction with a train simulator is cerebral blood flow. Kojima et al. (2005) used functional near-infrared spectroscopy, which measures cerebral blood flow to examine the difference between manual and automatic train operations. There are many other physiological measures that can be captured from data such as heart rate to much more complex measures. The use of physiological measures in train driving research simulators is still at a premature stage.

5.7.6 Data Extraction

The ease of data extraction from a train simulator is very important and is something that needs to be thought through in the early design stages of a new train simulator. A train simulator would not be very useful for the purpose of research if the required data could not be easily extracted. Compatibility of data outputs to pre-existing software analysis tools such as Microsoft Excel and the statistical software package SPSS could be desirable and could potentially reduce analysis time. A post-run visualisation tool of video and audio outputs could be beneficial, especially for retrospective verbal protocols.

5.8 Summary

Train driving simulators for research are underutilised in rail in comparison to other transportation domains. The rail industry needs to support the development and use of research simulators, not only when prompted by an inquiry, but also to support the design and development of the rail system, including the introduction of new automation and control technologies. This chapter provides an overview of the considerations when selecting, designing or commissioning a research simulator, the specific requirements for a rail research simulator and the stages of designing a rail simulator study. A key to effective use of simulation to examine research questions within the rail driving context is to ensure that the capabilities of the simulator, and the study design, fit the requirements of the research question. Therefore, once the approach of using a simulator to conduct rail research has been established, specifying the questions to be considered, and the variables to be measured and manipulated, is a key determinant of simulator complexity and design.

Simulators offer an excellent tool to support analysis of a safety-critical research context such as train driving, providing a safe context in which to understand the impact of changes to current in-cab and external elements and enabling application of methods that would be too intrusive or disruptive in a real-world context. It is therefore important that a requirement-based approach is used to the specification of such a context, to ensure that simulator data inform the research question in sufficient detail and as cost-effectively as possible.

Further considerations when designing future train driving simulators for research include whether it would be beneficial to have a driving simulator linked to a signalling simulator to examine distributed cognition and communication between the driver and the signaller. In addition, the inclusion of new physiological measures should be considered.

The rail industry must follow the example of other transportation domains by investing in the development of research-specific simulators and becoming more proactive in researching train driving, not merely reacting retrospectively to incidents and accidents. Future train driving research simulators would be particularly beneficial prior to the implementation of new technology, new signalling scheme designs and other changes to the rail system.

References

Alexander, A. L., Brunye, T., Sidman, J., and Weil, S. A. (2005) From gaming to training: A review of studies on fidelity, immersion, presence, and buy-in and their effects on transfer in PC-based simulations and games. *The International Journal of Aviation Psychology*, 8, pp. 223–242.

Arenius, M., Metzger, U., Athanassiou, G., and Strater, O. (2012) Planning on track – Introduction of a planning device in the railway domain. In *PSM 11& ESREL 2012, Helsinki: Finland, 25–29 June 2012*.

Buksh, A., Sharples, S., Wilson, J. R., Coplestone, A., and Morrisroe, G. (2013) A comparative cognitive task analysis of the different forms of driving in the UK rail system. In Dadashi, N., Scott, A., Wilson, J. R., and Mills, A. (eds.) *Rail Human Factors: Supporting Reliability, Safety and Cost Reduction*. Taylor & Francis: London, pp. 173–182.

Canas, J., and Waern, Y. (2005) Cognitive research with microworlds. *Theoretical Issues in Ergonomics Science*, 6, pp. 1–3.

Cullen, T. R. H. L. (2001) *The Ladbroke Grove Rail Inquiry: Part 1*. Health and Safety Executive: Liverpool.

Davies, K., Hinton, J., Thorley, P., and Buse, D. (2007) *A Feasibility Study of Head-Up Displays in Driving Cabs*. Rail Safety and Standards Board: London. Report T513.

Dunn, N., and Williamson, A. (2012) Driving monotonous routes in a train simulator: The effect of task demand on driving performance and subjective experience. *Ergonomics*, 55, pp. 997–1008.

Einhorn, J., Sheridan, T. B., and Multer, J. (2005) *Preview Information in Cab Displays for High-Speed Locomotives: Human Factors in Railroad Operations*. John A. Volpe National Transportation Systems Center: Cambridge, MA.

Endsley, M. R. (1995) Measurement of situation awareness in dynamic systems. *Human Factors*, 37, 65–84.

Endsley, M. R. (2000) Direct measurement of situation awareness: Validity and use of SAGAT. In Endsley, M. R., and Garland, D. J. (eds.) *Situation Awareness Analysis and Measurements*. Lawrence Erlbaum Associates: Mahwah, NJ, pp. 131–157.

Farmer, R., and Sundberg, N. D. (1986) Boredom proneness: The development and correlates of a new scale. *Journal of Personality Assessment*, 50, pp. 4–17.

Foerstberg, J. (2000) *Ride comfort and motion sickness in tilting trains: Human responses to motion environments in train and simulator experiment*. PhD thesis, Linköping University: Linköping, Sweden.

Greenberg, J., and Blommer, M. (2011) Physical fidelity of driving simulators. In Fisher, D. L., Rizzo, M., Caird, J. K., and Lee, J. D. (eds.) *Handbook of Driving Simulation for Engineering, Medicine and Psychology* CRC Press: Boca Raton, FL, pp. 7.1–7.24.

Hamilton, W. I., and Clarke, T. (2005) Driver performance modelling and its practical application to railway safety. *Applied Ergonomics*, 36, pp. 661–670.

Hart, S. G., and Staveland, L. E. (1988) Development of the NASA-TLX (Task Load Index): Results of empirical and theoretical research. In Hancock, P. A., and Mashkati, N. (eds.) *Human Mental Workload*. North-Holland: Amsterdam, pp. 139–183.

Hellier, E., Naweed, A., Walker, G., Husband, P., and Edworthy, J. (2011) The influence of auditory feedback on speed choice, violations and comfort in a driving simulation game. *Transport Research Part F*, 14, pp. 591–599.

Hoffman, R. R., Crandall, B., and Shadbolt, N. (1998) Use of the critical decision method to elicit expert knowledge: A case study in the methodology of cognitive task analysis. *Human Factors*, 40, pp. 254–277.

Huang, J. Y., and Gau, C. Y. (2003) Modelling and designing a low-cost high fidelity mobile crane simulator. *International Journal of Human–Computer Studies*, 58, pp. 151–176.

Jentsch, F., and Bowers, C. A. (1998) Evidence for the validity of the PC-based simulations in studying aircrew co-ordination. *International Journal of Aviation Psychology*, 8, pp. 243–260.

Kemeny, A., and Panerai, F. (2003) Evaluating perception in driving simulation experiments. *Trends in Cognitive Sciences*, 7, pp. 31–37.

Kennedy, R. S., Fowlkes, J. E., Berbaum, K. S., and Lilienthal, M. G. (1992) Use of a motion sickness history questionnaire for prediction of simulator sickness. *Aviation, Space, and Environmental Medicine*, 63, pp. 588–593.

Kim, J. S. (2007) A study on a dynamic model of a vehicle simulator with 6 DOF for the Korean tilting train. *Vehicle System Dynamics*, 45, pp. 327–340.

Klein, G. A., Calderwood, R., and MacGregor, D. (1989) Critical decision method for eliciting knowledge. *IEEE Transactions on Systems, Man and Cybernetics*, 19, pp. 462–472.

Kojima, T., Tsunashima, H., Shiozawa, T., Takada, H., and Sakai, T. (2005) Measurement of train driver's brain activity by functional near-infrared spectroscopy (fNIRS). *Optical and Quantum Electronics*, 37, pp. 1319–1338.

Large, D. R., Golightly, D., and Taylor, E. L. (2014) The effect of driver advisory systems on train driver workload and performance. In *Contemporary Ergonomics and Human Factors 2014: Proceedings of the International Conference on Ergonomics & Human Factors 2014, Southampton, UK, 7–10 April 2014*. CRC Press: Boca Raton, FL, p. 335.

Li, G., Hamilton, W. I., and Finch, I. (2003) Evaluation of railway signal designs using virtual reality simulation. In McCabe, P. T. (ed.) *Contemporary Ergonomics 2003*. Taylor & Francis: London, pp. 367–372.

Li, G., Hamilton, W. I., and Finch, I. (2006) Driver detection and recognition of lineside signals and signs at different approach speeds. *Cognition, Technology and Work*, 8, pp. 30–40.

Marinakos, H., Sheridan, T., and Multer, J. (2005) *Effects of Supervisory Train control Technology on Operator Attention*. Federal Railroad Administration, US Department of Transportation: Washington, DC. Report DOT/FRA/ORD-04/10.

Marumo, Y., Tsunashima, H., Kojima, T., and Hasegawa, Y. (2010) Analysis of braking behavior of train drivers to detect unusual driving. *Journal of Mechanical Systems for Transportation and Logistics*, 3, pp. 338–348.

McLeod, R. W., Walker, G. H., and Moray, N. (2005) Analysing and modelling train driver performance. *Applied Ergonomics*, 36, pp. 671–680.

Militello, L. G., and Hutton, R. J. B. (1998) Applied cognitive task analysis (ACTA): A practical toolkit for understanding cognitive task demands. *Ergonomics*, 41, pp. 1618–1641.

Naghiyev, A., Sharples, S., Carey, M., Coplestone, A., and Ryan, B. (2014) ERTMS train driving-incab vs. outside: An explorative eye-tracking field study. In *Contemporary Ergonomics and Human Factors 2014: Proceedings of the International Conference on Ergonomics & Human Factors 2014, Southampton, UK, 7–10 April 2014*. CRC Press: Boca Raton, FL, p. 343.

Naweed, A., Hockey, G. R. J., and Clarke, S. D. (2013) Designing simulator tools for rail research: The case study of a train driving microworld. *Applied Ergonomics*, 44, pp. 445–454.

Naweed, A. (2014) Investigations into the skills of modern and traditional train driving. *Applied Ergonomics*, 43, pp. 462–470.

Nichols, S., Haldane, C., and Wilson, J. R. (2000) Measurements of presence and its consequences in virtual environments. *International Journal of Human–Computer Studies*, 52, pp. 471–491.

Patel, H., Stedmon, A., Nichols, S., D'Cruz, M., Mager, R., Stoemer, R., Schaerli, H., Estoppey, K., and Bullinger, A. (2001) *Usability Test-Battery Manual*. Report on VIEW of the Future Project IST-200-26089 D3.2. University of Nottingham: Nottingham, UK.

Pausch, R., Proffit, D., and Williams, G. (1997) Quantifying immersion in virtual reality. In *Tenth Annual Symposium on User Interface Software and Technology, SIGGRAPH 97, Los Angeles, 3–8 August*. pp. 13–18.

Pickup, L., Wilson, J. R., Norris, B. J., Mitchell, L., and Morrisroe, G. (2004) *The Train Driver Integrated Workload Scale (IWS) Guidance Note*. Rail Safety and Standards Board: London. Report T147.

Roth, E., and Mullter, J. (2007) *Technology Implications of a Cognitive Task Analysis for Locomotive Engineers*. Federal Railroad Administration, US Department of Transportation: Washington, DC. Report DOT/FRA/ORD-09/03. Retrieved online at: http://www.fra.dot.gov/downloads/research/ord0903.pdf.

Sato, H., Marumo, Y., and Tsunashima, H. (2010) Braking assistance system for train drivers by indicating predicted stopping position. In *SICE Annual Conference 2010, Taipei*, 18–21 August 2010. pp. 1353–1357.

Sharples, S., Cobb, S., Moody, A., and Wilson, J. R. (2008) Virtual reality induced symptoms and effects (VRISE): A comparison of head mounted display (HMD), desktop and projection display systems. *Displays*, 29, pp. 58–69.

Sharples, S., Cobb, S., and Burnett, G. (2014) Sickness in virtual reality. In Wiederhold, B., and Bouchard, S. (eds.) *Advances in Virtual Reality and Anxiety Disorders*. Springer: London, pp. 35–64.

Thomas, G. R., Raslear, T. G., and Kuehn, G. L. (1997) *The Effects of Work Schedule on Train Handling Performance and Sleep on Locomotive Engineers: A Simulator Study*. Federal Railroad Administration, US Department of Transportation: Washington, DC. Report DOT/FRA/ORD/-97-09.

Tichon, J. G., Wallis, G., and Mildred, T. (2006) Virtual training environments to improve train driver's crisis decision making. In Bhalla, J. (ed.) *SimTecT 2006 Simulation Conference Abstracts: Challenges and Opportunities for a Complex and Networked World*. Simulation Industry Association of Australia: Melbourne.

Tichon, J. G. (2007) The use of expert knowledge in the development of simulations for train driving training. *Cognition, Technology and Work*, 9, pp. 177–187.

Tichon, J. G., and Wallis, G. M. (2010) Stress training and simulator complexity: Why sometimes more is less. *Behaviour and Information Technology*, 29, pp. 459–466.

Vicente, K. J. (1999) *Cognitive Work Analysis: Towards Safe, Productive and Healthy Computer-Based Work*. Lawrence Erlbaum Associates: Mahwah, NJ.

Waller, D., Hunt, E., and Knapp, D. (1998) The transfer of spatial knowledge in virtual environment training. *Presence: Teleoperators and Virtual Environments*, 7, pp. 129–143.

Wallis, G., Tichon, J., and Mildred, T. (2007) Speed perception as an objective measure of presence in virtual environments. In *Proceedings of SimTect2007*. Simulation Industry Association of Australia: Melbourne, pp. 527–531.

Williams, J. M. (1980) Generalisation in the effects of a mood induction procedure. *Behaviour Research and Therapy*, 18, pp. 565–572.

Wilson, J. R., and Sharples, S. (eds.) (2015) *Evaluation of Human Work* (4th edition). CRC Press: Boca Raton, FL.

Witmer, B. J., and Singer, M. J. (1998) Measuring presence in virtual environments: A presence questionnaire. *Presence: Teleoperators and Virtual Environments*, 7, pp. 225–240.

Yates, T. K., and Sharples, S. (2005) Determining the fidelity requirements for a human factors research train driver simulator. *The Ergonomist*, 5 December.

Yates, T. K., Sharples, S. C., Morrisroe, G., and Clarke, T. (2007) Determining user requirements for a human factors research train driver simulator. In Wilson, J. R., Norris, B., Clarke, T., and Mills, A. (eds.) *People and Rail Systems: Human Factors at the Heart of the Railway*. Ashgate: Farnham, UK, pp. 155–165.

Yates, T. K. (2008) *Design and development of a virtual reality train simulator for human factors research*. PhD thesis, University of Nottingham: Nottingham, UK.

Young, M. (2003) Development of a railway safety research simulator. In McCabe, P. T. (ed.) *Contemporary Ergonomics 2003*. Taylor & Francis: London, pp. 364–369.

Zoer, I., Slutter, J. K., and Frings-Dresen, M. H. W. (2014) Psychological work characteristics, psychological workload and associated psychological and cognitive requirements of train drivers. *Ergonomics*, 57(10), pp. 1–15.

6

Simulators and Train Driver Training

Anjum Naweed

CONTENTS

6.1 Introduction

This chapter discusses how simulators have been designed and used to teach driving skills in rail and how well simulation has lent itself to learning train driving. It draws together the latest applications for simulators in training and provides a helpful resource for researchers and practitioners alike. Whilst many of the themes overlap with other transport modes covered in this book, there are numerous dimensions that are unique to rail. For this reason, the chapter includes sections associated with train driving skills and practice around simulation design. Like most other transport modes, rail has seen incredible advances in technology, and this chapter discusses simulation and train driving in the context of both traditional railway operations and modern high-speed rail.

It is worth setting out some terminology for this chapter. Training can be broadly defined as the act of teaching others a skill, behaviour or set of competencies. Some practitioners dislike the word, suggesting that it has a stale non-specificity and too much emphasis on instruction. In rail, the word *coaching* is the preferred substitute, defined as a 'one-to-one process of helping others to improve, to grow and to get to a higher level of performance' (Pousa and Mathieu 2010). However, *training* is more familiar and will be used in this chapter to describe the act of teaching to others and doing it in a manner conducive to coaching.

The term *simulation* is context driven and can mean different things to different people. The essence of simulation is to recreate an environment or a situation, so the concept is not a technology-bound construct. Given the focus of this book, the word *simulation* will refer to a system that is designed to immerse, interact or engage an individual or user in the experience of driving a train. The word *simulator* will refer to an apparatus or technology designed to interface with an individual or user in the process of a simulation.

There are many classes of trains (electric, diesel, urban etc.) and even more of locomotives. The word *locomotive* has primacy to engine, whereas the word *train* means more in terms of the collective locomotive and carriages, but these terms will be used interchangeably. It would be a crime not to mention the homonymity of *train* and the clumsiness of using it to reference teaching and rolling stock in the same sentence. *Train* and *teach* will be used interchangeably.

6.2 Simulators for Train Driver Training

6.2.1 Brief History

The first rail simulators used for active training arrived in the late 1960s/ early 1970s (Hofsommer 1986). In the United Kingdom, the British Rail commissioned the first train simulator in 1964, which was capable of simulating movement and included a film projection of the route. The cab was suspended in a frame and was tilted by hydraulics, whilst tape recordings provided sound effects (Heyn 1966). Sadly, there are no remnants of it today and, whilst innovative, the reliability of the simulator was dubious at best. The next such simulator in the United Kingdom would appear in the late 1980s, specifically to assist drivers with transitioning from tread to disc brakes (RSSB 2009).

In the United States, investment in simulation to teach train driving was considered in the late 1960s, mainly as a way of getting people 'off the street and into the seat' as quickly as possible. The first simulators were built to address the (then new) practice of creating increasingly long trains and managing the problem of drivers breaking them through poor technique. One such simulator was designed for Southern Pacific (Del Rio News Herald 1970); it cost $1 million and incorporated moving pictures filmed from the railway in an effort to produce 'the first of a new kind of railroad men' (p. 14). Meanwhile, the Canadian National Railway purchased a locomotive simulator in the early 1970s, in an effort to make training realistic and transferrable (Canadian Rail 1972, Congress of the United States 1979).

Originally, the intent of simulator-based training was as a substitute for on-the-job training, but problems became apparent when driving performance failed to improve. At this point, the approach changed to using simulators as an *aid* to learning instead of as a replacement. In the 1990s, the main training philosophy was to provide several weeks of theory involving the simulator before allowing trainees out for practical training. The trainee would then go back into the simulator for an accelerated training programme exploring faults and circumstances that were difficult to replicate in the real world before returning to the field for a final check. Today, this methodology is still prevalent, despite improved technology and advances in training design.

The use of simulators to facilitate driver training found new resolve in the 1990s for maintaining competencies in safety-critical scenarios. This was triggered by the independent inquiry into the rail disaster at Ladbroke Grove; in his inquiry, Lord Cullen said, 'Japanese railway companies make extensive use of technology with a considerable concentration on simulators. These are programmed to help drivers learn about train failures and emergencies. Drivers are expected to demonstrate their capabilities to their supervisors and peers in a two-monthly simulator test' (Cullen 2001, p. 76). The words were powerful, if anecdotal, and simulator developers took note the world over. Today, there are around 40 train simulators being used by rail companies in the United Kingdom alone, costing up to £100 million (RISSB 2007). As far as rail was concerned, Lord Cullen's words justified the existence of simulator-based training.

6.2.2 Evolution of Simulator Design

The designs of early simulators were rudimentary by today's standards, typically comprising desk controls and television screens, but the innovation of the microprocessor meant that they could simulate a range of part-task functions from transmission and engine control systems to control of air-brake systems. More sophisticated designs heightened levels of realism with full cabs, screen projection and motion bases. The scenarios used were typically pre-programmed and aimed at helping drivers to understand the operation, maintenance and diagnostic system of locomotives.

The Santa Fe/Southern Pacific training centre in Kansas City (United States) was one of the world's first fully integrated driver training facilities (see Figure 6.1). Established in 1988, the centre had about 10 training rooms equipped with overhead projectors, televisions and videos and accommodated training for the driver, train dispatcher (controller), conductor (guard), yardmasters and brakeman. The facility had three full-cab train simulators, one with a moving base. The instructor station was remotely located and connected via voice and video. The curriculum was almost exclusively centred on driving technique and appropriate use of the braking and throttle controls. However, the training style was very traditional and the simulators were used more as an aid to learning. Eventually, the route learning part of the task also migrated to the simulator by using video footage of the track synchronised with the simulated train speed. Whilst the speed of the footage appeared normal at line speed, changes in frame rate reduced realism at very high or low velocities.

Modern simulators (see e.g. Figure 6.2) can require up to 50 software or hardware modules to function in a consistent state, including the audio engine and sound model, logging and replay systems, session analysers, scripting engines, controller consoles, track vision centres, the train and world models. Track vision is highly textured, projected digitally, shown on screens or on large LCD monitors. Vision systems incorporate technology that allows drivers to look around the edges of the cab, simulating stereoscopic three dimensions (so, for instance, you can see a signal obscured by the cab itself). The cab contains real-time simulation of all circuit breakers, trip switches and isolators. They can be used for route learning, either through full-task simulation with computer-generated routes or through part-task simulation where drivers only observe routes.

The modern rail simulator is therefore complex with many parts, which invariably increases opportunities for technical fault. Whilst technology has improved their sophistication in design, the training architecture remains

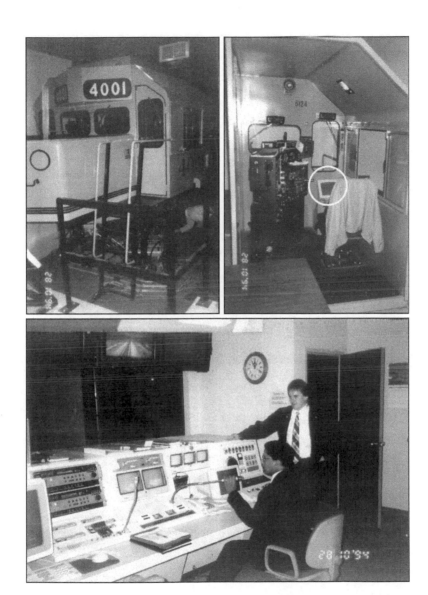

FIGURE 6.1
Santa Fe/Southern Pacific training centre with an external shot of a motion-based simulator (*top left*), the cab (*top right*) and the instructor desk (*bottom*). *Note: white circle* highlights enhanced displays integrated to aid training. (From Fullerton, J. et al., Report on USA Trip, National Rail, Australia, 1994.)

the same – it is a duplication of a cab with a remotely located control centre and used as an aid to training as opposed to a substitute for on-the-job experience. Thus, the biggest change is the focus on competencies and scenario design, and most simulators are designed for recreating and manipulating very complex scenarios.

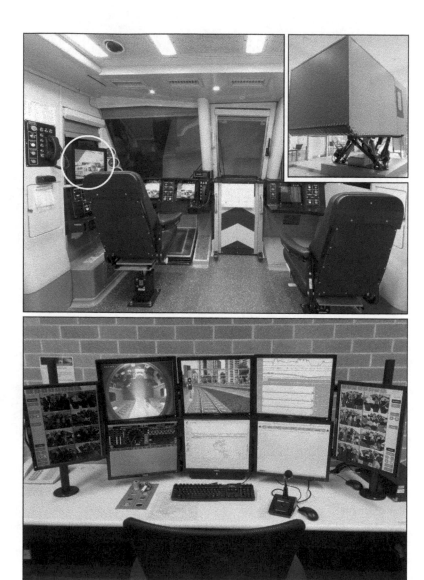

FIGURE 6.2
Petersham training centre with a photo of its state-of-the-art motion-based rail simulator (*top right, inset*), the cab (*top*) and instructor desk (*bottom*). *Note: white circle* highlights enhanced display integrated to aid training. (Courtesy of Transport for New South Wales, Sydney.)

6.2.3 Argument for Using Simulators to Teach Train Driving

A key attraction of using simulators is the ease and control with which degraded conditions, safety-critical scenarios and failure modes can be constructed. Simulators may also be used for trialling technology, testing new processes and evaluating procedures. They also reduce the need to remove trains from service and allow hazardous situations to be recreated without the hardship (or dubious ethics) of real-world consequences.

The argument that simulation is a useful asset for rail is compelling, particularly in the current economic climate. In most countries, the cohort representing the first generation of train drivers is retiring. This generation acquired its experience on the footplate, so there is concern for skill loss, but the industry is also experiencing difficulties training new recruits and *retaining* them once they have been trained.

In Australia, for example, the demand for rail services has increased (Asian Development Bank 2012), but the median age of train drivers remains at 43 years (Transport and Logistics Industry Skills Council 2012) whilst rates of turnover have never been as high (Department of Education Employment and Workplace Relations 2009). The threat of skill shortages had engendered change in workplace learning policies, and the traditional model has adapted to accommodate recruits with no prior rail familiarity. Increases in indoor teaching and assessment have split the focus from on-the-job tuition to the classroom (Naweed and Ambrosetti 2015). The notion of simulator-based training is seen as a panacea, and some training programmes have incorporated principles of social and adult learning into their programmes. Train simulation research generally supports the perspectives underpinning these theories, such that simulator-based learning is typically assumed to afford high consistency, availability and throughput, translating to better knowledge and skill retention.

6.3 Human Dimensions in Rail System Simulation

6.3.1 Overview of the Rail System

One of the most recognisable features of railways is the track itself. This is a key difference from road and other transport analogues. Train drivers are incapable of taking evasive action (other than braking), so – more than any other transport mode – rail human factors relies on movement authority for safe working. The following decomposes the basic layers of the rail system and illustrates some of the dynamic relationships:

- Train: traction, carriages and wagons, weight and speed
- Driver: throttle control, sustained attention/vigilance, shift work and fatigue and reaction time
- Rail infrastructure and terrain: tracks, points and traffic flow; signals and crossings; safe working rules; movement authority; and tunnels, gradients and curvature
- Goals: schedule, time keeping, train control and service delivery
- Weather: visibility, railhead condition, adhesion and temperature

Whilst not comprehensive, the list offers a basic multisystem rail analogy. The dynamism, complexity and lack of transparency present important features that need to be considered when simulating one or more of its parts. The train and driver are restricted to the railway, and the dynamic interactions between rail infrastructure and terrain have implications for control and performance. The infrastructure introduces dependencies on dynamic tasks performed by others (e.g. signallers), and the terrain has its own geographical constraints that define how the system is operated. Train driving is contextualised by performance goals, but everything is strongly influenced (and ultimately determined) by the weather.

The most basic rail systems require drivers to navigate using signals and rudimentary train state indicators (e.g. speedometer). External signals can provide information about the movement authority, the time needed to react and the appropriate speed. Like road, rail signals vary in colour and configuration, but they use *multiaspect* sequencing to provide the driver with early indication of cautionary and stop signals (e.g. clear, caution, stop). In more advanced (and typically faster) rail systems, signalling is supplemented via information displays (i.e. a glass cockpit) giving the driver more time to make decisions. The underpinning technology can coincide with trackside signalling or eliminate the need for it. High-speed rail systems, such as the Train à Grande Vitesse (TGV) in France or the Shinkansen in Japan, use signal communication systems that enforce braking curves on the train, overcoming key human factors limitations of driving at such high speeds. The underlying calculations are based on distances between trains, timetabling and geography. For train simulators, the type of rail system that needs to be simulated is important for defining the user's experience.

The motion dynamics encountered in train driving are unique to rail. Trains are a mass of metal with wheels of steel rolling on steel rails at speed. This means that friction and resistance dramatically influence velocity. Further, the contact point between the wheel and railhead can be as narrow as a few millimetres (around the width of a coin), so principles of lubrication and wear also interact with the surfaces in relative motion. This is why train driving is described as 'driving with the seat of your pants'. Water, dew, light rain and other lubricating agents such as grease, leaves and insects alter railhead characteristics and have implications for train handling (e.g. a heavier-than-required brake application can create an uncontrollable slide), so trains are driven differently based on the conditions. An accurate train simulation must include the means to accelerate and brake realistically in view of the train set and the drag and resistance characteristics of the track.

6.3.2 Train Driving Skill

Over the course of my research, train drivers have referred to themselves as 'crystal ball gazers', and I myself have analogised the task to driving a truck on ice wearing a blindfold. The *truck* and *ice* are about train handling, but the *sight beyond sight* element relates to *route knowledge*. Rail curvature means that movement authorities (signals, speed boards, rail level crossings etc.) are often hidden from view. Route knowledge compensates for the time that it takes to stop and estimate braking distances, just like it did in 1830 on the first intercity railway. In terms of training, route knowledge is considered to require two competencies: (1) an ability to retain vast amounts of knowledge about the route and (2) a demonstrable ability to interrogate it outside a strict sequence to cater for train driving skills (Naweed and Connor 2012).

In its simplest form, train driving is about correcting the discrepancy between the target speed and the current train state (i.e. a standard tracking task). As discussed, train drivers need an accurate understanding of their position at any given moment and it is often said that they must process complex vestibular, kinaesthetic and acoustic information in order to achieve this. In plainer language, these are the motion and sensation characteristics involved with train handling and the auditory and peripheral visual information in a constantly changing environment. In more advanced (typically high-speed) rail networks, the discrepancy between the target and train speeds is enforced by automation.

Figure 6.3 conceptualises the train driving process in terms of the skills that need to be trained and the type of control that the driver performs in different rail modes. The route features provide information about movement authority, and geospatial locations are fundamental for train drivers in traditional networks or in operations where the human has more decision-making authority. This is an example of *real-world target tracking*. In rail modes where there is greater machine agency, a *glass cockpit* usually provides one or more of the following to the human:

1. A preview of the upcoming route and/or in-cab signalling
2. Predictive information about the future train speed profile
3. Ecological display of enforced braking curves

Whilst the first two points can be provided independently of the third, design practice is usually to combine all three, which equates to a *pursuit tracking* task. This means that the human is no longer tracking the speed but an *overlay* or ecological representation of it with its own boundaries. Both traditional and advanced versions of the rail mode have implications for training needs in terms of learning and skill development and provide a range of trainable options for simulation.

6.3.2.1 Training Needs for Different Skills

Advances in technology have done wonders to support the train driver – but the task is not necessarily easier. Today, freight trains are longer, population

External information sources

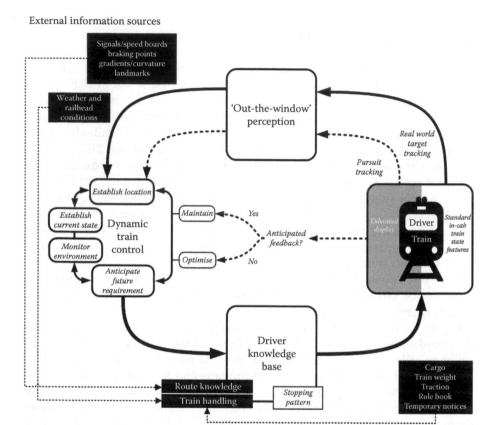

FIGURE 6.3
Model of train driving performance. (From Naweed, A., *Applied Ergonomics*, 45, 462–470, 2014.)

densities are greater, train traffic is higher and service delivery takes precedence. So, whilst the technology has increased, the task demand has shifted from one place to another.

What this means is that train driving requires a number of skills, both technical and non-technical, for which training needs vary and for which simulation has to be appropriately matched. The simulation should encourage the decision-making needed to (1) trade between different driving goals, (2) appreciate the tensions and functions of safety and performance and (3) effectively pursue dynamic train control. This may be achieved in simulators with varying degrees of realism, and different aspects of the task can be simulated to train different skills.

The axes of simulation relating to physical versus psychological needs vary (as do training needs) for different rail modes. For example the skills needed to diagnose a train fault require a good foundation of core driving competencies, but the simulator need not draw on actual driving practice to develop this ability. The simulator would instead be used as a platform to apply different components of the task. As another example, in traditional rail systems, route knowledge retention is a large part of training, but in advanced high-speed networks where the dynamics are different and routes are previewed, this skill would look and behave differently.

6.4 Competence Management and Assessment

6.4.1 Basic Elements of Training and Characteristics of Qualification

As part of the accreditation process needed to satisfy rail regulators, train operators must ensure that their drivers have certificates of competency. Take Australia for example; in this country, competencies are overseen by the Transport and Logistics Industry Skills Council (2014a). The most significant aspect of this particular arrangement is not just that the training is standardised but also that assessments are related to the role. Driving competencies are generally based on three basic elements:

- Safe working system requirements
- Types of train/locomotive
- Rail infrastructure

Safe working requirements vary and are assessed regularly, and train drivers have to be deemed competent for the types of trains that they operate. Competencies associated with the track and rail infrastructure are also reassessed though the frequency and requirement can vary. The competencies needed to qualify depend on the operator's individual business needs. This enables freight/heavy haul and passenger operators to select different competencies appropriate to their system.

The units or modules in train driving qualifications are characterised by retention of certain knowledge and acquisition of various skills. This model places emphasis on assessing the skills and knowledge to achieve competency in the workplace, but it is not related to the amount of time that is spent learning. Thus, a person may be competent to perform a task after relevant training within a few weeks, without necessarily being required to spend a certain length of time at the task before being assessed. In the Australian context, the training framework contains a total of 60 units of competency from which 21 require a successful assessment outcome for a train driver qualification (Transport and Logistics

Industry Skills Council 2014b). Thirteen of these are mandatory and include the following:

- Inspect and prepare a motive power unit
- Apply awareness of railway fundamentals
- Identify, diagnose and rectify minor faults on motive power units and rolling stock
- Identify and respond to signals and trackside signs
- Operate train with due consideration of route conditions
- Operate and monitor a motive power unit
- Use communication systems
- Respond to abnormal situations and emergencies when driving a train

Thus, train driver training programmes require application of a broad range of specialised knowledge and skills. These are characterised by learning, self-management, planning and organising, initiative, enterprise, problem solving and teamwork, and many are built into simulator-based training as key learning outcomes.

6.4.2 Key Learning Stages

The basic elements of training needed to become a train driver are delivered over several stages. Each of these has key learning objectives and assessments with the intention to systematically manage the competencies required of the role. Broadly, there are four stages of learning: (1) company induction and orientation, (2) driving foundations, (3) traction fundamentals and (4) driving tuition. Stages 1 and 3 are typically conducted in the classroom (i.e. off the job) and promote theory development, whilst stages 2 and 4 are part of situated learning, which occur in more authentic workplace settings such as the train cab, shunting yard or maintenance shed – anywhere in the natural environment where classroom-acquired theory is applied. In some cases, each stage can have aspects of the classroom and/or authentic workplace elements (i.e. a blended learning approach).

After the company induction then, the second stage in this process provides trainees with detail about the system of movement applied on the railway. Easily the most galling stage for those unfamiliar with rail, it combines observational and hands-on practical activities including locomotive testing, train preparation and coupling/uncoupling of rolling stock.

The third stage involves learning the fundamentals of train traction and returns the trainee back into a classroom, and this is the most commonly where simulators will be involved in the process. The training includes basic route learning, understanding signals and points, train management work and rosters. Typically coordinated over several weeks, it is the most intensive aspect of a training programme in terms of the amount that has to be learned and the time allocated to this before assessment. It aims to build on the trainee's exposure to the railway by revisiting theory and conceptualising individual driving methodologies and train navigational strategies. Driving activities and safety scenarios may be enacted in teams. The amount of time spent in this stage depends on the route and traction complexity, such as the number of train types and safeworking systems.

Finally, the last stage of learning is tuition with a qualified driver trainer. This stage is almost entirely conducted in the real-world environment, and the trainee learns the route and applies the knowledge and skills from previous stages. It contains the most recognisable apprentice-like components, and trainees are expected to observe and model their knowledge on an expert. It includes activities that try to stimulate intuitive decision-making skills, such as diagnosing and solving faults. This stage is typically modularised to include various tasks, and trainees in urban operations can expect to spend six to nine months under tuition. Trainees in freight/heavy haul can expect to spend anywhere from 12 months to two years in this stage.

6.4.3 Integrating Simulators into Training Programmes

6.4.3.1 Traditional Rail Environments

In traditional rail environments, the use of simulators for training varies greatly. Training frameworks do not suggest how simulators may be incorporated into training, let alone prescribe usage. Consequently, simulator subscription is piecemeal. Some organisations invest in understanding how simulators could sit within their training and competency management frameworks. Unfortunately, many in-house simulators still have a tendency to be bolted on to training programmes, wherever learning and development teams perceive utility, despite recommendations from the Cullen inquiry. Whilst dedicated simulator training and learning centres spend time building simulations with a more formative role, few approach it from a de novo (i.e. bottom–up) perspective. Many simulator managers have confessed to me that despite their intentions, their simulators are often used in a 'here is a cab I prepared earlier' way for new recruits to experience a walkthrough during the induction stage. Simulators also have a tendency to be featured in the traction fundamental stage, back in the classroom. For this reason, there is an emphasis on group learning and observation with simulators rather than on a one-to-one experience.

There is also huge variance in the parts of the task that are simulated. Some trainers use simulators solely for route knowledge, others for developing form and some for evaluating safety-critical decision-making. In any case, evaluation is typically subjective and based on scorecards. In practice, though, simulators can collect and review a large range of training outcomes, such as the driving basics, knowledge of new routes, driving techniques during hazardous conditions, track rules, procedures and effects of changes at the driver level (e.g. mood, experience, fatigue, workload). Similarly, these could be measured in many ways, for example with fuel and brake use, throttle use, speed performance (violations, error, trip time) and train forces (buff and draft, changes in acceleration). However, the emphasis is usually on measures less sensitive to the learning profile, such as a signal passed at danger incident or number of penalty brake applications. For more comprehensive information on data capture, analysis and sense making, the reader is referred to Dorrian and Naweed (2013).

6.4.3.2 Advanced Rail Environments

As a result of the increased technology and speed of operations, advanced rail environments have different driving dynamics, risk profiles and expectations for route learning compared to traditional environments. Automated systems tend to manage speed choices and supervise braking actions, so training is designed around working with these systems and how to work *without* them when they are isolated. Consequently, simulators are generally integrated into basic and refresher training, to monitor and maintain competencies that might otherwise degrade through infrequent use. After acquisition of driving theory, trainees apply what they have learned in the classroom within the simulator to reinforce this knowledge. The focus is on learning to drive, rules and procedures and responses to faults or emergencies. These activities are group-based, and trainees are given opportunities to observe and comment on each other's performance. Driving scenarios are manipulated to elevate workload with role-play.

More sophisticated driving simulators are integrated with simulators of the signalling and controlling tasks and therefore distributed at a system level. This phase is followed by situated learning in the authentic environment, specifically for the section of track and/or depot that the trainee will be joining. Refresher training typically occurs biannually, unless a driver has experienced an incident implicating training issues. This training is usually conducted under the same principles, but scenarios are updated based on current incidents.

6.5 Simulation Design and Skill Transfer

6.5.1 Classes of Simulation for Training

There are a number of difficulties in classifying simulators. First, the literature raises the point that taxonomic classifications do not always capture what simulators can do and thus define them inadequately (Stanton 1996). Second, commercial train simulators are developed and supplied on a competitive basis, so naming conventions and nomenclature indeed tend to vary. This is also apparent in other industries where there is diversity in the scope and depth of simulation (Rehmann, Mitman and Reynolds 1995, Maran and Glavin 2003).

In rail, simulators tend to fall into three broad classes: full cabs, driver desks and desktops (Naweed, Balakrishnan and Dorrian 2013). A full-cab simulator can be considered as a *replica* (cf. Ben Clymer 1980), being an exact duplicate of the cab with a realistic projection of the environment. These are designed with a car body or enclosed capsule and try to replicate the cab to the last sticker. Consequently, they are able to deal with most (if not all) task parameters. Figure 6.4a shows an example of a replica train simulator; note that side windows are also represented. However, some full-cab simulators do not replicate a specific driver interface but incorporate features representative of a broad class of systems, providing a *generic* representation of a train cab. Figure 6.4b shows an example of a full-cab generic simulator, which has an authentic brake stand but a generic console.

Some full-cab simulators are designed with a mixture of authentic equipment but also equipment not found in the real-world counterpart. Ben Clymer (1980) provides a class designation for such simulators and calls them *eclectic*. Thus, a full cab that is identical to its counterpart in every way except for say the inclusion of a training display would be best described as a replica with an *eclectic structure*. This type of blending activity can also happen in driver desk or desktop simulators.

Driver desk simulators forsake the enclosure usually to compromise on space and cost. Whilst they can deal with operational driving, activities that rely on the enclosure (for example operating circuit breakers) are difficult to accommodate, thus limiting the training potential (e.g. for fault detection or relevant procedural training). Unless all of the design elements necessary to simulate a full task are built within the interface of a driver desk, it can simulate only part of the task and its control parameters. Figure 6.4c illustrates a driver desk simulator, whilst Figure 6.4d illustrates an eclectic driver desk simulator that has an additional display for training purposes.

Desktop simulators occupy the lowest rung of physical design. These simulators are usually integrated onto desktops and framed using mock-ups of the driver–cab interface and the functional characteristics of the equipment, but not the actual equipment itself. Thus, they may be very minimalist and a composite of input controls structured to convey as much of the task as is feasible. In virtually all cases, secondary or tertiary aspects of driving are not incorporated. In general, desktop simulators equate with Ben Clymer's (1980) *basic principles* definition of simulators, which omits much in the interest of simplicity but is still able to demonstrate the cognitive work and behaviours. Figure 6.4e illustrates such a simulator. Although the simulator in Figure 6.4f has a more sophisticated desk, it is a mock-up, hence a desktop.

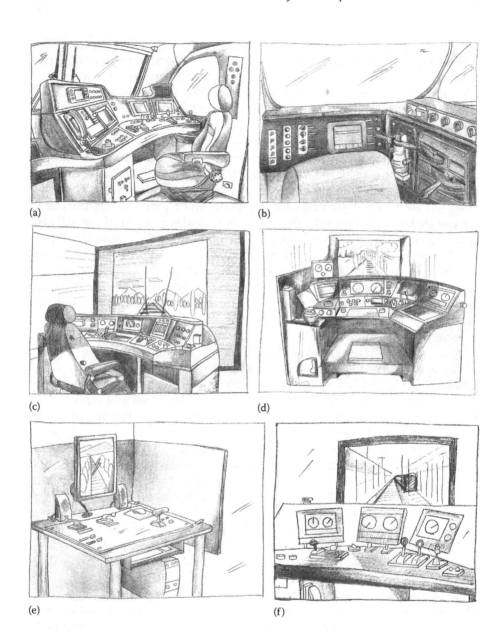

FIGURE 6.4

Different train simulator designs. (a) Replica and (b) generic full-cab simulator, (c) replica and (d) eclectic driver desk simulator and (e) generic and (f) electric desktop simulator. (From Naweed, A. et al., in Bearman, C. et al. (eds), *Evaluation of Rail Technology: A Practical Human Factors Guide*, Ashgate, Surrey, UK, 2013, pp. 171–213.)

6.5.1.1 Advantages and Disadvantages of Simulator Types

Full-cab, driver desk and desktop train simulators present pros and cons for training and competency management. Procuring replicas is an expensive business, and it can be impractical to buy and maintain a whole fleet of them. At the same time, a small number of simulators could cause a bottleneck for throughput and a veritable stranglehold on competency management, particularly as they are more prone to technical failure. Replicas are useful simulators to have at the induction and orientation stage, to both preview the cab and sell the simulator concept. Operationally, they are most useful at the traction fundamental stage and safety scenarios, as well as at driving tuition.

Driver desks can also contribute to most stages of the training process, and if high turnover rates are important, are more likely to be cost effective, simply because they also deliver with respect to signals and points, train management, mainline orientation, distributed power and safety scenarios. Although they might not accommodate fault detection, or some types of procedural training fully, their size does encourage mobility, enabling them to fit within the demands of crew resourcing, rendering them good for inclusion during driving tuition.

Desktops are more limited but are ideal for route learning, signals and points and general familiarisation. A fleet of these would be more practical at both the traction fundamentals and driving tuition stages. I would argue that any simulator should be made available as a revision aid for trainee drivers preparing for assessment, which would also establish more targeted training outcomes for train simulators as educational technologies for e-learning approaches.

6.5.2 Simulator Fidelity

Simulator fidelity is a reasonably simple concept – and extremely contentious. Much of the literature is entrenched in research undertaken over the second half of the twentieth century with some of the old texts undergoing a revival (e.g. Duke 1974). Fidelity is the degree to which the real environment has been faithfully replicated – it is a *measure* of realism in simulators, but the road towards a uniform definition has been rocky. General definitions consider how equipment is reproduced and the degree to which it resembles the simulated situation, whilst more specific definitions also emphasise the role of the environment, task, job and behaviour (Hays 1980).

Interaction, intuition and *immersion* have been highlighted as three components that govern the experience and plausibility of a simulator (Wells 1992). These refer to (1) interaction that is obvious and familiar, (2) the awareness that human actions manipulate the simulation and (3) a virtual environment which feels real enough to provide the experience of being surrounded by the simulation. *Engagement* and *presence* have also been identified as scaffolds for simulator training, where engagement is a deeper form of involvement and presence captures the phenomenon of behaving and feeling as if you are in the real environment, as opposed to a simulation (McMahan 2003).

For train driver training, simulator fidelity is driven by experience and familiarity. Individual perceptions of the same simulation can vary. People come to it with a variety of experiences and knowledge with some more easily immersed than others (Naweed and Balakrishnan 2013). Perceptions of strangeness in design will invariably affect how (and whether) the trainee accepts a simulation. A huge break from familiarity may induce nausea or result in an eeriness that upsets immersion and encourage the experience to be rejected (a concept known as the *uncanny valley*; Mori 1970). Thus, it is more useful to try to deliver skills when using a simulator by constructing familiarity with it as an object in its own right and then try to distribute across both real and synthetic environments.

6.5.3 High Fidelity

The label *high fidelity* is used to provide quality assurance for product design, usually for replica classes of simulator. It has become part of the rhetoric and a badge used to instil hope in the buyer during early phases of the bidding process. But there is still disagreement as to what actually constitutes fidelity, let alone the degrees that it occupies. A simulator perceived to be an exact replication of the real world might create unreasonable expectations and consequently impact buy-in if these expectations are not met. It has even been suggested that fidelity can be *too* high (Wickens and Hollands 2000). Not surprisingly, the subject of high fidelity is polarising.

High fidelity has been praised for maximising training potential (e.g. Bryan and Regan 1972, Roscoe 1980), providing enhanced data capabilities (e.g. Fukazawa et al. 2003) and instilling confidence in trainees (e.g. Stammers and Trusselle 1999). But, in recent years, attention has focused on the psychological aspects, and it has been suggested that whilst physical, visual and equipment fidelity can be scaled, cognitive engagement is the holy vector. A simulator is engaging only to the degree that it is able to involve and occupy its trainees (Gray 2002). Data gathered from *low*-fidelity simulators have been successfully applied in real-world situations, and the argument that they are unsuitable has waned (Stanton 1996). Procuring a high-fidelity simulator is not a predictor of acceptance in trainees, nor a guarantee of improvement in training or successful integration into training (Naweed 2013). Big-budget approaches for designing simulators do not necessarily address fidelity requirements (Lane and Alluisi 1992, Roza, Voogd and van Gool 2000), and buyers of simulators need to direct their procurement initiatives around what they *need* for competence management, not what they believe to be the most realistic.

6.5.4 Simulating Motion

I have come across a train simulator with a moving base designed to accurately simulate motion to the 95th percentile. Yet when speaking to drivers, I learned that they had taken to calling it 'the Vomitron' (with varying degrees of affection). There is debate around the utility of simulating motion via physical cues (yaw, roll, pitch). Although it can be a key reference point for changes in velocity, too often, the benefits are disproportionate to the costs (Young 2003). Simulating physical cues this way is expensive, needs regular maintenance, increases the opportunity for technical failure and requires stringent health and safety measures. Beyond this, it increases the potential for adverse physiological effects, including simulator sickness (Kennedy et al. 1993, Kolasinski 1995). One can argue that incorporating full physical cues of motion is not necessary when features such as sound vibration and a semi-active seat may immerse equally well and enhance the perception of self-motion (RSSB 2007).

For training, a moving base is redundant if the purpose is to teach out-of-course events, equipment failures and situations not requiring physical cues for the perception of self-motion (RSSB 2007). In the absence of movement, alternative cues can be provided, such as richer texture density gradients in the visual environment, changes in the ambient optic array and high-fidelity sound and vibration (Bruce and Green 1990, Padmos and Milders 1992). The main reason for wanting physical motion would be to actually train task aspects that rely on these cues or simply to increase immersion. As far as the former is concerned, there is some speculation around how well this can be achieved. Creating algorithms to realistically replicate rail motion characteristics would involve simulating the effect of the vortices generated when passing other trains and entering tunnels, as well as the vibrations and shockwaves substantive to articulated rolling stock. Then, there is the effect of track adhesion, the centreline profile of the truck, gradients and curvature. Last, but not least, physical cues would incorporate effects of improper train handling. In short, there are many variables that would need to be considered when simulating train driving motion.

6.5.5 Enhanced Information Displays for Training

Most train simulators contain displays that preview route information or show predictive data. For modern rail networks, the display is usually the same as those in the real cabs, but for the more traditional systems, they turn into a peculiarity. Many developers build information displays into simulators. Examples are shown in the simulators in Figures 6.1 and 6.2 (both of which operate traditionally). Information displays are a useful source of data for the instructor, but for real environments that do not feature this information, or indeed the type of display, the simulator class changes to eclectic. This means that any interaction with the simulator (and display) must be considered appropriately not just from the point of training needs, but also skill transfer.

6.5.6 Training and Skill Transfer

Transferring skills attained in simulators is achieved by balancing fidelity. This is the essence of simulation; the cab interface, visuals, acoustics, task and scenario are all designed to complement the learning outcomes. So, whilst levels of realism of the different elements making up a simulation can vary, transfer of training occurs only when the input and output aspects of the simulator (stimulus and response) are matched and compatible with the real environment.

Proactive transfer happens when learned skills carry over to the real environment in a manner that yields a *positive* response and measurably improves performance; transference issues occur when a skill carries over, but the nature of what is learned actually ends up inhibiting performance in the real environment (Wickens and Hollands 2000). Most tasks require skills that transfer positively, and if the stimulus and response characteristics are matched (or compatible), the transfer can be highly positive, but *negative* transfer may happen if the response elements vary from an accurate range of stimuli.

Figure 6.5 illustrates the relationship between stimulus and response by using a train simulator analogy. The stimulus that the trainee is reacting to is abstracted with a series of circles, and the response is represented by a throttle action and a change in speed. A train simulator that displays or characterises the stimulus precisely as in the real environment would transfer highly positively if the response were also the same (for instance, by duplicating the exact throttle, dashboard and response parameters of a cab). However, changing the stimulus so it is different but elicits the same response also has positive transference potential. For example, an individual is likely to perform well if a simulator duplicates the same control and response parameters of the real world, even if the dashboard happens to be different or if say the railway environment is visualised in two dimensions instead of three dimensions. The main distinction is that the trainee's responses need to be the same as they would be in reality, regardless of similarity or difference in the stimulus. However, if the simulator requires the user to push the throttle forward to decrease speed but this action is undertaken with a different control and in a different direction in reality, transference may be negative and induce a longer learning curve. Figure 6.5 shows this in the form of a different-looking throttle lever with a different direction of control. Transfer may be even more negative if the response elements bear no resemblance and are completely incompatible. Say, for example that the controls for decreasing speed are not only different and inverted but the actual notches in the control are also varied, so the train

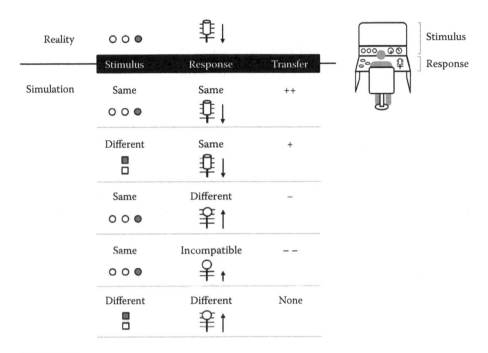

FIGURE 6.5

Relationship between stimulus and response elements with training transfer. (From Naweed, A. et al., in Bearman, C. et al. (eds), *Evaluation of Rail Technology: A Practical Human Factors Guide*, Ashgate, Surrey, UK, 2013, pp. 171–213.)

does not decrease speed in the same way. In this example, skill transfer will be highly negative.

The key point is that the potential for negative skill transfer may be decreased if the intended response (of the trainee) is the same, even if the inputs are different. For example, a driver confronted with a throttle that looks different may tolerate the transfer better if they share an identity (e.g. move in the same direction and possess the same number of notches). Low-cost desktop simulators can therefore offer an effective training platform just as much as a full-cab replica, as long as the simulation elicits the same response as the real world. In this way, whilst the visuals and control stimuli may differ, the trainee's response is essentially the same as in the real world. This also applies to the training potential of very rudimentary simulators, which present very different looking stimuli (e.g. two-dimensional graphs, keyboard buttons) but generate the same response.

Thus, the needs of competency management drive the simulation design and must underpin their procurement. Procedural training and equipment failure scenarios would require the same stimulus as the real world and the same response. On the other hand, basic train driving skills could be learned with different stimuli, as long as the response was closely matched.

6.6 Standardisation and Industry Guidelines

6.6.1 Developing Standards and Guidelines for Rail

In rail, a standard is a statutory instrument that defines mandatory practice, and standardisation strives to reach consensus on a practice or operation for the industry. Developing standards is about creating consistency and cooperation through consultation. Guidelines are developed through the same process but only advocate good practice.

In the United States, rail standards are developed through divisions of the Federal Railroad Administration (FRA) (e.g. Technical Training Standards Division) and organisations comprising accredited standards committees (e.g. Simulation Interoperability Standards Organisation). In the United Kingdom, standards are developed and managed through the Rail Safety and Standards Board (RSSB), which is responsible for developing and managing rules, codes of practice and guidelines. In Australia, the equivalent is the Rail Industry Safety and Standards Board (RISSB). The International Organisation for Standardisation (ISO) and the European Committee for Electrotechnical Standardisation are two international authorities for standards development. Both offer standards for general ergonomics design associated with human–system integration, user interfaces and vehicle design, but nothing that would readily apply to simulator-based training.

6.6.2 Specification of Training Practice in Rail

Table 6.1 shows documents that reference training specifications in rail. Presently, there are few (if any) national standards for training, though there is a code of practice and a number of guidelines. This is partly because consensus is more readily attainable for standards in system design and integrity, and partly because mandatory practice is captured within the curricula and accreditation processes of national training frameworks. In Australia, for example the Australian National Rules and Procedures training and assessment standards are a set of guidelines that extend the competencies outlined by national training frameworks and aim to establish benchmarks for training and assessment.

The guide to developing and maintaining staff competence (Office of Rail Regulation 2007) presents simulators as a way to infer competencies beyond questioning, assessing activities apportioned with risk and reassessing competence to deal with emergencies and/or infrequent events. The guideline to train driver training (RSSB 2008b) mentions simulators in passing as an option for building a higher-fidelity training programme, and whilst it recognises that simulators afford the opportunity to recreate scenarios, it advises against expectations that they reduce training time. Likewise, the guidelines for driver assessment (RSSB 2008a) and competence review and assessment

TABLE 6.1

Sample List of Industry Documents Referencing the Specification of Training Design

Source	Year	Number/Issue	Title
Office of Rail Regulation (UK)	2007	07176 17327	Developing and Maintaining Staff Competence
RSSB (UK)[a]	2007	RS501/2	Good Practice Guide on Simulation as a Tool for Training and Assessment
RSSB (UK) (RSSB 2008a)[a]	2008	RS702/1	Good Practice Guide for Driver Assessment
RSSB (UK) (RSSB 2008b)[a]	2008	RS701/2	Good Practice Guide on Competence Review and Assessment
RSSB (UK) (RSSB 2008c)[a]	2008	RS221/1	Good Practice Guide to Train Driver Training
FRA (US)	2008	Part 240/4	Qualification and Certification of Locomotive Engineers
RISSB (AU) (RISSB 2010)	2010	Version 3	ANRP Training and Assessment Standards

Note: UK, United Kingdom; US, United States; AU, Australia, ANRP, Australian National Rules and Procedures.
[a] In 2013, these RSSB document were superseded by the RS/100 Good Practice Guide on Competence Development (Version 1).

(RSSB 2008b) mention simulators and indicate that as a tool for assessing skill, they can provide evidence of practical handling, competencies in abnormal situations, communication and general human factors issues. The code of practice for the qualification and certification of locomotive engineers (FRA 2008) defines three classes of simulation. These are broadly aligned with increasing levels of graphical, audible and physical fidelity and correspond with criteria for monitoring operational performance and conducting skill testing.

By far the most comprehensive industry document for simulator training is RSSB's guide on simulation (RSSB 2007). It has four modules that explain why simulation should be used as a tool in training and assessment, the methods and types of simulation, guidelines for procuring simulators and future development. It provides information about the specific benefits and drawbacks of simulator training and articulates the value of a comprehensive training need analysis during the initial stages of procurement. Guidance for procurement in terms of the actual process and negotiation has a tendency to be overlooked. This guide offers general advice including lessons learned, information about the structure of the tender and the basic requirements for the various components of a simulator. Presently, it is one of the most useful publicly available guides for simulator-based training in rail.

6.6.3 Standards for Train Simulator Design (or Lack Thereof)

Whilst there are plenty of guidelines for training, simulation and assessment in rail, there is an absence of standards for using and managing these processes. This is significant as the use of simulation is omitted in most accredited training frameworks, which means that there is presently *no established standard for simulator-based training in rail*. Furthermore, there are no standards *or* guidelines at national or international levels for how simulators should be designed. Simulator developers aim to satisfy standards for quality management (e.g. ISO 2008), which generally promote a continuous improvement paradigm, but as far as standards for design are concerned, this information is held at the local level by the developer. At the point of tender, it is a usual practice for the procurer to be given 'customer-facing' documents with high-level information about the design process. Whilst there are national standards that developers may follow for designing their simulator (e.g. track alignment, world generation), the overall process is typically undocumented.

6.7 Emerging Trends and Considerations

6.7.1 Future Developments

RSSB's guidelines for simulation as a tool for training and assessment included a specific module for future developments. In the absence of further guidelines or standards since its publication in 2007, it is prudent to revisit some of these a decade on, to determine if any have been addressed as part of continuous improvement.

6.7.1.1 Linked Training and Integration of Distributed Functions

The practice of coordinating a single scenario with integrated functions is taking off with some rail organisations integrating the driver and guard tasks into their train simulators (see Figure 6.2). Beyond this, a number of organisations are designing and implementing training centres with simulators that integrate several distributed rail functions, such as signallers and controllers, and the capability to emulate them within a single scenario for team training. Once such example is the driving, signalling and traffic management facility in Northern Europe (TRERail 2014).

6.7.1.2 Technology Improvements

The RSSB guidelines identified several areas for technology improvements, including the following:

- Improved visual simulation
- Interactive computer-based training
- Exploitation of low-cost applications
- Easier-to-use control facilities for instructors

Taking these in turn, advances in visual systems technology provide a good example of how the humble train simulator has become a much more immersive environment. Conventionally simulated visuals have no change in parallax with direction of gaze, which makes the view appear static. However, technology can now interact with the driver's viewing point and correct the perspective using real-time tracking. This goes some way towards addressing simulators that have no side windows and assumptions that wrap-around screens compensate for the sense of motion.

Advances in technology have also improved the quality of computer-based training methods, making the desktop simulator a more viable low-cost platform. Interest in using low-cost applications developed for gaming and hobbyist markets has grown, and developers are starting to look more closely at simulator utility outside full-cab or driver desk structures. This invites a change in the perception of fidelity and recognises that functional fidelity is not limited to full-cab simulation.

The point about easier-to-use control facilities still requires attention. Innovations in simulator design have tended to focus more on the experience of the person driving than the person doing the instructing (Balakrishnan and Naweed 2013). The problem is generally steeped in the lack of human factors consideration in the construction and specification of the user interface. More often than not, the logic that informs this comes from software engineers as opposed to usability experts, meaning that control and scenario manipulation can be counter-intuitive and riddled with conceptual leaps. This is partly a legacy issue in that the graphical user interface of an early simulator is often used to model or extend subsequent iterations, so new features are bolted on wherever there is space.

6.7.1.3 Human Factors and Incident Investigations

Some rail simulators are procured with the belief that they can accurately recreate real-life accidents and test theories of causation. Whilst simulator developers may be designing this capability into their technologies, there is little evidence to show that they are being used for this or that procurers are expending resources to make it happen. Part of the problem is that there are variations in the architecture of different event data recorders on board trains, but these are also different from the architecture of simulators. Thus, whilst there may be a general capability for investigating incidents in simulators, the appetite or budget needed to deliver this as part of the procurement is lacking.

6.7.2 Robocop Problem

Problems in simulator design and implementation have been identified as key determinants for rejection by trainees and underutilisation in rail. This has been my experience in Australasia, and I have conceptualised it as the *Robocop problem* (Naweed 2013). This identifies the problem of workers in a highly regulated environment who must integrate, work and train with a new technology designed to improve operations. Considerable time and funds are invested in it, but little consideration is given to determine how workers should interact with it, adapt to it and perceive it. The technology undergoes a number of testing exercises, but the belief that it will work within the existing structure of the organisation is taken at face value. This creates fundamental problems in the perceived reality, relevance and reliability of the training platform with respect to train simulator procurement.

Three 'prime directives' for overcoming the Robocop problem are suggested and include (1) engaging drivers in procurement up until commissioning (serve the end-user trust), (2) linking the simulator with training goals and a well-specified training need analysis (redefine training relevancy) and (3) ensuring that the skills developed through simulation are additive to the learning programme (uphold system integrity).

6.7.3 Design and Specification

Simulator procurement is an asset management process comprising tendering, commercial terms and through-life support. In most cases, specifications developed from a successful bid focus more on highly technical elements of architecture than on usability and design, such that the language can obfuscate meaning. When developing a business case for simulators, it is easy to succumb to the technology and not question its compatibility with the training need. The procurer seldom has adequate opportunity to determine any problems with aspects such as scenario design, so a great deal of unplanned time can be lost post-acceptance, trying to gain this familiarity and develop an understanding of how it all works. Furthermore, there is often little evidence of any empirical exercises that validate levels of simulator fidelity (for function and task), nor is there evidence of rigor in positive skill transfer (Naweed and Balakrishnan 2013). Whilst developers perform verification tests as part of delivery, these are designed to ensure that the simulator meets *contract* specification. The simulation models (e.g. dynamics, functional, visual, audio) usually go through fidelity tests as part of earlier iterations; thus, fidelity revalidation might not be performed on a new simulator, so it is almost entirely taken at face value. To address this issue, an industry standard for the design and specification of simulators for training is needed and should consider the collaborative involvement and integration of human factors and ergonomics design practice.

6.7.4 Looking beyond Train Driver Training and Traditional Learning

Early on in this chapter, I noted the growing demand for rail operators to deliver on service. This has implications for driver training, job satisfaction and staff retention. One of the key threads within this chapter is the ever-changing face of rail and how this is impacting driver training. High-speed rail has altered the way that the train driver works, and automation has changed the error and risk profile.

The fact is that many railways around the world have completely automated operations, a number of which also run 'driverless' trains. Whilst these are typically small, closed systems (i.e. with minimal overlap with other tracks), operators with more complex networks are investigating the possibility of driverless trains for both the passenger and freight/heavy haul environments. The issue has economic and political implications on both sides, and it is difficult to say when (and if) this will happen in earnest and what the implications will be for the remotely operating train driver/controller. However, there is enough automation in the task and enough semi-automated 'hybrid' networks to start evaluating the training needs for the driver caught between varying degrees of human–machine control and explore the application of augmented technology to the range of tasks that it creates.

In the wake of increasing complexity and codependency in tasks, linked training frameworks that integrate different functions are perhaps the only way to train at a *system* level and to review learning on train drivers and other functions alike. At the individual level, research is starting to look more closely at augmented reality to determine how train drivers could harness simulator potential even more. The argument here is to design a training process with learning outcomes that create a consistent and standardised view of the world. A second thread is to consider how coaching and/or mentoring could be delivered with simulators by using *avatars* or *embodied conversational agents* which communicate with the driver on the basis of artificial intelligence. This approach has potential and is largely shepherded by crew resource management concerns. Whether or not it is a viable option in terms of fidelity or skill transfer remains to be seen.

In the face of increasing automation, one must question the role of simulators in a task that is moving towards increased separation between human and machine agency. Simulators have a critical relationship with automation, and for training delivery, they have never been more important. The changing nature of the task and different levels of automation create tensions for workload, cognition, perception and situation awareness. This was evidenced tragically in the Santiago de Compostela high-speed train derailment in 2013. This means that exposure to certain simulated situations is increasingly becoming the *only* way that drivers can learn from, adapt to and prepare for infrequent or novel events before encountering them. Safety and reliability are two different arms of system design, and whilst the driver may operate in an environment where the train does the brunt of the thinking within a safety envelope, there will always be situations of degraded operation calling for the driver to reliably take the reins, placing an imperative on being able to maintain competencies, so they can manage, cope and think through uncertainty.

Train simulators have navigated a long and arduous path to get to where they are – one filled with mismanaged expectations and anecdotal comparisons. Exciting times are ahead, but it has never been more important to stay ahead of the curve.

Acknowledgements

The author gratefully acknowledges the assistance of Ganesh Balakrishnan and Frank J. Hussey in the writing of this chapter.

This chapter is dedicated in the memory of John Wilson.

References

Asian Development Bank. 2012. Asia's shift to greener transport key to global sustainability – ADB forum. Accessed 9 December 2012. http://www.adb.org /news/asias-shift-greener-transport-key-global-sustainability-adb-forum.

Balakrishnan, G., and Naweed, A. 2013. Usability rekindled: In pursuit of easier control station operability. 18th Asia-Pacific Simulation Technology & Training Conference, Brisbane.

Ben Clymer, A. 1980. Simulation for training and decision-making in large-scale control systems. *SIMULATION* 35:39–41.

Bruce, V., and Green, P. R. 1990. *Visual Perception: Physiology, Psychology and Ecology*. Hove, UK: Lawrence Erlbaum Associates.

Bryan, G. L., and Regan, J. J. 1972. Training systems design. In *Human Engineering Guide to Equipment Design*, edited by H. P. Van Cott and R. G. Kinkade, 667–699. Washington, DC: US Department of Defense.

Canadian Rail. 1972. *Waybills*, edited by Canadian Railroad Historical Association. Canadian Rail, Gimli, Canada.

Congress of the United States. 1979. *Railroad Safety – US-Canadian Comparison*. Washington, DC: Office of Technology Assessment.

Cullen, W. D. 2001. *The Ladbroke Grove Inquiry*, edited by Office of Rail Regulation. Suffolk, UK: HSE Books.

Department of Education Employment and Workplace Relations. 2009. Job outlook for train and tram drivers. Accessed 7 April 2010. http://joboutlook.gov.au /pages/occupation.aspx?search=industry&tab=stats&cluster=&code=7313.

Dorrian, J., and Naweed, A. 2013. Evaluating your train simulator part 2: The task environment. In *Evaluation of Rail Technology: A Practical Human Factors Guide*, edited by C. Bearman, A. Naweed, J. Dorrian, J. Rose and D. Dawson, 215–259. Surrey, UK: Ashgate.

Downey, C. 1970. *Train That Goes Nowhere Trains Engineers*. Downey, CA: Del Rio (Texas) News Herald, 14.

Duke, R. D. 1974. *Gaming: The Future's Language*. New York: Sage Publications.

FRA (Federal Railroad Administration). 2008. Qualification and certification of locomotive engineers. In *49 CPR Part 240*, Volume 4, 813–861. Washington, DC: Department of Transport.

Fukazawa, N., Kuramata, T., Satou, K., Sawa, M., Mizukami, N., and Akatsuka, H. 2003. Human errors committed on a train operation simulator. *RTRI Report (Railway Technical Research Institute)* 17:15–18.

Fullerton, J., Hussey, F., and Mann, N. 1994. Report on USA Trip. National Rail, Australia.

Gray, W. D. 2002. Simulated task environments: The role of high-fidelity simulations, scaled worlds, synthetic environments, and laboratory tasks in basic and applied cognitive research. *Cognitive Science Quarterly* 2:205–227.

Hays, R. T. 1980. *Simulator Fidelity: A Concept Paper*. Alexandria, VA: US Army Research Institute for the Behavioral and Social Sciences.

Heyn, E. V. 1966. Simulator trains locomotive engineers. *Popular Science* 188:109.

Hofsommer, D. L. 1986. *The Southern Pacific 1901–1985*. College Station, TX: A&M University Press.

ISO. 2008. *Quality Management Systems – Requirements*. In *ISO 9001*. International Organisation for Standardisation, Australia.

Kennedy, R. S., Lane, N. E., Berbaum, K. S., and Lilienthal, M. G. 1993. Simulator sickness questionnaire: An enhanced method for quantifying simulator sickness. *International Journal of Aviation Psychology* 3:203–220.

Kolasinski, E. M. 1995. *Simulator Sickness in Virtual Environments*. Alexandria, VA: US Army Research Institute for the Behavioral and Social Sciences.

Lane, N. E., and Alluisi, E. A. 1992. *Fidelity and Validity in Distributed Interactive Simulation: Questions and Answers*. Alexandria, VA: Institute for Defense Analyses.

Maran, N. J., and Glavin, R. J. 2003. Low- to high-fidelity simulation – A continuum of medical education? *Medical Education* 37:22–28.

McMahan, A. 2003. Immersion, engagement and presence: A method for analyzing 3-D video games. In *The Video Game Theory Reader*, edited by Mark Wolf and Bernard Perron, 67–86. New York: Routledge.

Mori, M. 1970. The uncanny valley. *Energy* 7:33–35.

Naweed, A., and Connor, A. K. 2012. Understanding train driver route knowledge – Simulator project suite volume 1. Brisbane, Australia: Cooperative Research Centre for Rail Innovation.

Naweed, A. 2013. Simulator integration in the rail industry: The Robocop problem. *Proceedings of the Institution of Mechanical Engineers, Part F: Journal of Rail and Rapid Transit* 227:407–418.

Naweed, A., and Balakrishnan, G. 2013. That train has already left the station! Improving the fidelity of a railway safety research simulator at post-deployment. *Proceedings of the Institution of Mechanical Engineers, Part F: Journal of Rail and Rapid Transit* 227:419–426.

Naweed, A., Balakrishnan, G., and Dorrian, J. 2013. Evaluating your train simulator part 1: The physical environment. In *Evaluation of Rail Technology: A Practical Human Factors Guide*, edited by C. Bearman, A. Naweed, J. Dorrian, J. Rose and D. Dawson, 171–213. Surrey, UK: Ashgate.

Naweed, A. 2014. Investigations into the skills of modern and traditional train driving. *Applied Ergonomics* 45:462–470.

Naweed, A., and Ambrosetti, A. 2015. Mentoring in the rail context: The influence of training, style, and practice. *Journal of Workplace Learning* 27:3–18.

Office of Rail Regulation. 2007. *Developing and Maintaining Staff Competence*. London: Office of Rail Regulation.

Padmos, P., and Milders, M. V. 1992. Quality criteria for simulator images: A literature review. *Human Factors* 34:727–748.

Pousa, C., and Mathieu, A. 2010. Sales managers' motivation to coach salespeople: An exploration using expectancy theory. *International Journal of Evidence Based Coaching & Mentoring* 8:34–50.

RISSB (Rail Industry Safety and Standards Board). 2007. *Review of Simulation Training and Assessment in the GB Rail Industry (T711)*. London: RISSB.

RISSB. 2010. *ANRP Training and Assessment Standards*. Version 3. Canberra: RISSB.

Rehmann, A. J., Mitman, R. D., and Reynolds, M. C. 1995. *A Handbook of Flight Simulation Fidelity Requirements for Human Factors Research*. Wright-Patterson AFB, OH: Crew System Ergonomics Information Analysis Center. Springfield, VI: National Technical Information Service.

Roscoe, S. N. 1980. *Aviation Psychology*. Ames, IA: Iowa State University Press.

Roza, M., Voogd, J., and van Gool, P. 2000. Fidelity considerations for civil aviation distributed simulations. Symposium at the AIAA Modeling and Simulation Technologies Conference, 14–17 August. Denver, CO.

RSSB (Rail Safety and Standards Board). 2007. *Good Practice Guide on Simulation as a Tool for Training and Assessment*. RS/501 Issue 2. London: RSSB.

RSSB. 2008a. *Good Practice Guide for Driver Assessment*. RS/702 Issue 1. London: RSSB.

RSSB. 2008b. *Good Practice Guide on Competence Review and Assessment*. RS/701 Issue 2. London: RSSB.

RSSB. 2008c. *Good Practice Guide to Train Driver Training*. RS/221 Issue 1. London: RSSB.

RSSB. 2009. Review of driver training programmes in Great Britain Railways – Appendix 4. London: RSSB

Stammers, R. B., and Trusselle, T. L. 1999. The effect of reduced simulator fidelity on learning and confidence ratings. In *Contemporary Ergonomics*, edited by M. A. Hanson, E. J. Lovesey and S. A. Robertson, 453–457. London: Taylor & Francis.

Stanton, N. A. 1996. Simulators: A review of research and practice. In *Human Factors in nuclear safety*, edited by N. A. Stanton, 114–137. London: Taylor & Francis.

Transport and Logistics Industry Skills Council. 2012. *2012 Environmental Scan (E-Scan)*. Melbourne, Victoria: Department of Industry Innovation, Science, Research and Tertiary Education.

Transport and Logistics Industry Skills Council. 2014a. *TLI10 – Transport and Logistics Training Package (Release 4.1)*. Canberra, Australia: Department of Education, Employment and Workplace Relations.

Transport and Logistics Industry Skills Council. 2014b. *TLI42613 – Certificate IV in Train Driving (Release 1)*, edited by Employment and Workplace Relations Department of Education. Canberra, Australia: Department of Education, Employment and Workplace Relations.

TRERail. 2014. Norway. Accessed 22 February 2014. http://www.trerail.co.uk/case-studies/norway/.

Wells, M. J. 1992. Virtual reality: Technology, experience, assumptions. *Human Factors Society Bulletin* 35: 1–3.

Wickens, C. D., and Hollands, J. G. 2000. *Engineering Psychology and Human Performance*. 3rd ed. Upper Saddle River, NJ: Prentice Hall.

Young, M. S. 2003. Development of a railway safety research simulator. In *Contemporary Ergonomics*, edited by P. T. McCabe, 364–369. London: Taylor & Francis.

7

Simulators in Rail Signalling

Nora Balfe, David Golightly and Rebecca Charles

CONTENTS

7.1 Introduction

Simulation is a comparatively new phenomenon in the rail signalling domain. At the time of the Ladbroke Grove accident in 1999, only two rail signalling simulators existed on the Great Britain (GB) rail network. Following the accident, Lord Cullen's inquiry report made a recommendation for simulators to take a stronger role in support of signaller training (Cullen, 2000). He argued that the use of simulators facilitated better training of signallers in the wide variety of contingency situations that they may be faced with on the live railway. The ability to rehearse their actions in response to emergency events could only increase their ability to act appropriately in a real emergency. Following this recommendation, the use of simulators in signallers' initial and refresher training was greatly increased in GB.

Along with offering new approaches to signaller training, the installation of simulators has also opened the door to using them in research projects. The aim of this chapter is to set out the applications, benefits and challenges when working with rail signalling simulators. We start with a brief introduction to signalling as a function of railway operations, in order to then describe the different types of simulation that are in use. The role of simulation both in training and in human factors research is then described. Through this description, we offer a number of ways in which signalling simulation has been shown to have practical value or has enhanced theoretical knowledge. This chapter also highlights some of the challenges of designing valid simulator studies and points to opportunities for future work.

7.2 Current Generations of Operational Signalling Systems

7.2.1 Signaller, Task and Environment

The movement of trains through the railway infrastructure is achieved through signalling systems. There are various types of signalling systems, but the primary goal of each is to maintain separation between trains. In the early days of the railway, signalling systems were not necessary as trains had limited routes and ran at slower speeds. As the weight and speed of trains increased, the time taken to slow to a stop also increased. Eventually, this reached the point where a train could not necessarily stop in time upon sighting an obstruction in its path. This meant that if a train was to break down, then a following train was likely to collide with it. A system to instruct train drivers on the safety of proceeding was necessary, and signalling was developed for this purpose.

In essence, signalling divides the railway up into sections and only one train is allowed into a section at a time. This is known as the *block system*. Each of these sections has some form of signal at its entrance to pass information on to the driver about the availability of that section and any points to allow the train to change track. The system is controlled by a signaller with support from the interlocking system. The term *interlocking* refers to the systems developed to ensure that conflicting routes are not set. This is achieved in different ways depending on the signalling system but fundamentally ensures that under normal operations the routes set do not allow two trains to occupy the same section at the same time.

Whilst traffic regulation comprises the core element of signalling, the signaller's role typically also involves many other responsibilities. The signaller is effectively the eyes and ears of the rail network, communicating with drivers, station staff, route controllers and trackside workers to ensure the safe passage of trains and movement of people on the railways. They may also be in charge of level crossings across their area of control. As a result, communication and coordination are a major part of the signaller's work, particularly during periods of disruption.

There are three main types of signalling equipment in use today, all of which can be simulated. The first is the oldest form of signalling, a lever frame, where levers are connected via cables to elements controlled by the signaller (including signals and points). Because the lever had to be directly connected to the signalling equipment, signal boxes had to be located close to the signals, resulting in many signal boxes along a route and a relatively limited control area. As a result, many older signal boxes are physically adjacent to the area that they control. The signaller has a map of his/her control area displayed over the levers and can usually also look out of the window to see the passage of trains or hear or feel the vibration indicating the passage of the train. Even knowing whether it is raining or snowing can tell the signaller about current traction conditions that will influence regulating decisions (for more details, see Roth et al., 2001 or Golightly et al., 2010). The interlocking used in lever frame boxes is mechanical in nature. It works by metal bars attached to the levers in the signal box. When a lever is pulled, the metal bar is positioned so that it blocks other levers which, if pulled or released, would endanger that route. The interlocking is designed using logic tables to describe the releases and locks associated with each lever.

Lever frame signal boxes were the main form of control until the 1950s when entry–exit (NX) panels began to be introduced, a control panel-type technology which used push buttons located on track schematics and solid-state circuitry to set signals. Routes are set by pressing a button at the start and another at the end of the route, and the points and signals are then automatically changed. They represent a significant advancement over lever frames as information on train location and passage is now available on the panel, thus enabling much larger areas of the railway to be controlled by each signaller. The move to electrical control of signals meant that fewer boxes could control more signals, from a greater distance. NX panels use route relay interlocking instead of mechanical interlocking. The principles of logic and the purpose are the same, but the data are held in electrical circuitry rather than mechanical bars.

The next major step in the evolution of railway control systems was in the 1980s with the development of visual display unit (VDU)-based control systems, which are similar in nature to the NX panel, but the signaller interface is now via computer screens rather than a control panel. These facilitate more advanced forms of automation and highly centralised control. The ultimate vision for signalling is that all signalling in the future will be through the VDU-based control, particularly as the railways move to digital, in-cab signalling in the form of the European Train Control System and European Rail Traffic Management System. With the advent of modern, centralised control systems, however, signallers are increasingly removed from the area of control and more dependent on the information provided through their interface.

From the brief description earlier, some key requirements for signalling simulators are clear. Primarily, they have to simulate the movement of trains through an area of infrastructure as represented to the signaller and simulate the means by which the signaller sets signals, points or routes to control the movement of trains. Whilst lever frames require a smaller area of control to be modelled, the flexibility of the VDU-type display (similar workstation hardware can be reconfigured to simulate many different signalling areas), the fact that much of the information that they use is represented through the display and communication systems (rather than through looking out of the window) and their growing prominence as the major form of rail signalling mean that VDU-type workstations are most amenable to simulation. In the following section, we expand on the different types of simulation and the components required.

7.2.2 Elements of Simulation

Before considering the different types of signalling simulation, it is worth considering the technological elements required in order to offer sufficient realism given the systems described earlier. Typically, a simulator is built upon an infrastructure model that defines the track, signals, junctions etc. under the area of control and a traffic model which mimics the behaviour of trains. This infrastructure model may need to include aspects of the world not explicitly represented in the signalling interface. One example is topography, as gradients will substantially affect the speed of trains. The traffic model varies depending on the type of traction, top speed, load and so on. A heavy-freight train will behave in a completely different manner from a two-carriage commuter train. This traffic model is also controlled by a timetable, which triggers when vehicles will enter the workstation, call at stations and so on. Recreating these two elements – the infrastructure and traffic model – is far from a trivial task. Signallers are hugely dependent on both local knowledge (Pickup et al., 2013) and learned expectations about how traffic will behave (Golightly et al., 2010), so the accuracy of these models is critical to validity, even more so than the physical fidelity of the signalling environment (see Chapter 1 for a discussion of validity and fidelity).

Sitting on top of the traffic model is the physical set-up of the simulator. This is the human–machine interface (HMI) aspect of the simulation that recreates the input/output characteristics of the workstation. In theory, different simulated HMIs can be used with the same infrastructure/traffic model. Finally, simulators typically try to recreate some aspect of the communication environment, for example by having a simulation of incoming radio and telephone calls from drivers and other control centres. Typically, this aspect of simulation can be recreated only in a simplistic manner, although there are examples where the full-communication environment is mimicked offline from simulators, such as through role-playing exercises. It is, however, difficult to simulate all of the cues of active overhearing (i.e. overhearing relevant information from adjacent signallers), weather etc. that might influence signaller decision-making.

Usually, the simulated workstation is attached to a PC that runs the simulation and allows a trainer or researcher, usually on a desk nearby, to trigger events, make calls to the signaller and capture performance metrics generated by the simulated environment. Typical measures include scores for a number of performance metrics – number of trains running late (subdivided into categories such as zero to three minutes late and three to five minutes late), number of trains held at red signals, number of trains signalled into correct platforms and so on. This may then be used to calculate an overall performance score for the session.

With this broad definition of the elements of signalling and signalling simulation, it is possible to understand the major types of simulation available.

7.2.3 Lever Frames

There are a limited number of lever frame simulators in existence, and most of the signaller's training is achieved on the job. The space constraints in lever frame boxes together with the relative simplicity of lever frames mean that it is impractical to develop specific high-fidelity simulators for individual boxes, although there may be an opportunity to utilise lower-fidelity computer-based part-task trainers to train basic principles. To the authors' knowledge, such simulators are not in widespread usage; however, lever frame simulators are used in a limited capacity in GB for initial training (see Figure 7.1 for an example).

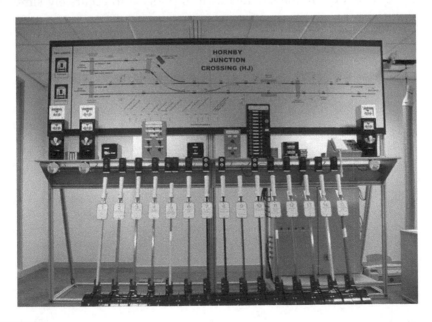

FIGURE 7.1
Lever frame simulator.

Where these simulators exist, the level of physical fidelity is high, with weights being attached to the levers to simulate the actual force required to set a route and to provide an accurate replication of the interlocking system. However, the operation of lever frame boxes relies extensively on manual recognition of trains passing the signal box. These simulators do not replicate the visual and audio cues that signallers would use in real operations, although the information passed via the 'block bell' (a telegraph system) from the adjacent signal boxes can be simulated. This limitation, coupled with their geographical specificity, might explain why their use has not been more widespread. They are used in basic training to explain the concepts of lever frame signalling and can be used for basic lever pulling skills. The accurate replication of the interlocking means that they are a good tool for basic training and explanation of key railway signalling concepts.

7.2.4 NX Panel

These panels have been successfully simulated, although again, they are not in widespread use, perhaps because of the cost of simulating what can be very large panels of several metres in length. Small panels can be faithfully replicated, as with the Network Rail training simulator shown in Figure 7.2, and the level of fidelity is high with the look and feel of the signalling interface faithfully replicated as well as the reaction of the simulator to control inputs. Unlike lever frame simulators, it is possible to accurately simulate

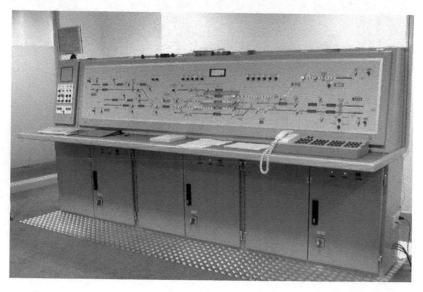

FIGURE 7.2
NX panel simulator.

the approach and passage of trains on an NX panel since the information is provided through the signalling interface itself rather than the external cues. However, the simulators replicate only a single area of control and larger signal boxes may have upwards of 10 separately operated areas. Therefore, some of the operational realism associated with signallers working and communicating in close physical proximity is not replicated in an NX panel simulator. External communications, however, can be provided by a trainer or simulator operator via a simulated telephone system.

It has also been possible to simulate NX panels by using a bank of touch screens. These can be more easily simulated using modern technology but represent a lower level of physical fidelity due to the difference in the interface and the lack of physical buttons. They have not been widely adopted, and only one is in existence on the GB network.

The NX simulators are primarily located in training centres and are used as part of the basic training. More detailed training specific to a particular location is usually completed on the real panel, with a qualified signaller. Simulators are therefore predominantly used to explain the concepts of signalling on this type of interface and how it differs from lever frames.

7.2.5 VDU Based

The most recent, and most widely simulated, form of signalling system is the VDU-based signalling system (Figure 7.3). Most large VDU-based signal boxes in GB now have a simulator available in the control centre, which can be configured to represent any of the workstations. This is a more cost-effective solution because one simulator can replicate all the workstations in the signal box.

These simulators are typically based on the same screen views as the actual signalling workstation and, assuming the use of an accurate infrastructure and traffic model, can be run with an operational automation component

FIGURE 7.3
VDU-based signalling simulator.

that recreates the automation support experienced on the real workstation. As with NX panel simulators, the approach and passage of trains can be easily simulated, but the lack of realism around the interaction with adjacent workstations remains an issue. In contrast to both lever frame and NX panel simulators, VDU-based simulators are deployed in individual signal boxes providing a training environment specific to that location.

7.3 Use of Simulators in Training

Typically, signaller training begins with an eight-week classroom-based course in which trainees cover the rules and regulations governing railway operations. Simulation may be used at this stage to illustrate concepts presented in the classroom, and there is a growing recognition that simulation can bring additional benefits to signaller training, particularly in terms of allowing signallers to practice their application of the knowledge gained in classrooms through properly simulated scenarios. The whole process and nomenclature of signalling are steeped in legacy, and therefore, the lever frame can be used to introduce the trainee signaller to the first principles of rail traffic control. The NX simulator then develops this knowledge with more complex control scenarios typically encountered on the network. It should also be noted that whilst both the lever frame and NX simulator illustrated in this chapter (and used in the Network Rail central training facilities) behave in a theoretically realistic manner, they use fictional locations and infrastructure/traffic models, which afford the trainer a high degree of flexibility.

After this initial training, signallers are trained on the job at their assigned signal box, usually shadowing a more experienced signaller. Again, there is increasing recognition of the role that simulation can play at this stage, particularly in terms of allowing the trainers, mentors and assessors to present the trainees with bespoke signalling scenarios tailored to the environment in which they are working. Here, the value of the authentic traffic and infrastructure model becomes paramount, as the signaller moves from learning the principles of signalling, to the specific patterns that they need to learn for the workstations that they will operate.

One area where simulators are currently widely used for training is in VDU-based boxes, where they are used to train signallers on the interface, particularly the menu structures required to control the automation or change the information displayed. This is not required on the older technologies as the interface is simpler and more intuitive. The VDU interface is significantly more complicated than the older technologies, and simulation also serves a purpose in allowing the trainee to become comfortable with the technology before using it on the live railway.

There is also a role for signalling simulation for more experienced signallers. These signallers benefit from refresher training on emergency procedures which may not be used frequently in practice or new procedures that might be introduced (such as new track worker protection procedures). Alongside the benefits to the trainee, the simulations provide auditable training and assessment records, which can be used to provide assurance of improved signaller competence. Additionally, the reconfiguration of a workstation or the introduction of a new workstation requires training for signallers of all levels of experience to familiarise themselves with the working of the infrastructure, traffic patterns and any automation that may be included in the workstation. Signallers may also use the simulator to familiarise themselves with new versions of the timetable.

It is an observation, however, that the use of simulation is not as widespread as it is in sectors such as aviation, where use of simulation is also a voluntary, informal part of maintaining and developing skills. This may be in part due to simulation scenarios still lacking some of the crucial fidelity associated with communications and additional information picked up in the real signal box environment.

7.4 Use of Simulators in Research

The use of simulation for research purposes in this sector is still reasonably low, but the projects undertaken so far include the use of eye-tracking equipment to study the signallers' eye movements, identification and analysis of signaller strategies, evaluation of the benefits of graphical interface tools, comparison of technologies and exploration of potential infrastructure changes. The predicted growth of rail transport and the continuous drive to improve both safety and performance mean that there is a demand for knowledge that can be used to support the required improvements; the studies described here are examples of some of the ways in which rail signalling simulators offer the opportunity to obtain much needed data to support enhancement of the signalling system. Often, it is not feasible to obtain these data by using the live systems due to safety concerns or the lack of flexibility of the live system. In general, the objective of research using simulators in rail signalling can be split between understanding current methods of working, testing new designs and concepts within the current operational framework and prototyping new operational tools.

7.4.1 Understanding Current Operations

A key area of interest in research on railway signalling is investigating the strategies that signallers use to prioritise train movements. Qualitative data may be collected from interviews and observations on live signalling work-stations (e.g. Roth et al., 2001, Haavisto et al., 2010, Patrick et al., 2013), but the scenarios cannot be controlled and it is difficult to compare the strategies of different signallers. Simulation is ideal in this area of research since it is repeatable and controllable (see Chapter 1 for a summary of relative merits of simulator research).

Network Rail has used simulators in different ways to attempt to elicit further understanding of the strategies that signallers use (Cheung, 2011, Balfe et al., 2016). Traditional, dynamic simulations have been employed on a high-fidelity VDU-based simulator in a signalling centre in which conflicts (i.e. trains wishing to use the same piece of track at the same time) between trains are programmed for participants to resolve. A 40-minute simulation was recorded with each participant and a debrief method used whilst viewing the recording to gather qualitative data on the choices made by the participants. The variation in strategies employed by the six participants in this experiment was relatively low but showed high-quality responses, with almost 75% of strategies employed being graded as good and no strategies were graded as poor.

In another piece of research, Balfe et al. (2012) used a similar VDU simulator to compare levels of automation in rail signalling. Three levels of automation were defined on the simulator: a high level in which an automated agent set routes for trains by using timetable information, a medium level in which trains were routed along pre-defined paths and a low level where the operator (signaller) was responsible for the movement of all trains. Performance, subjective workload and signaller activity were measured for each level of automation running under both normal operating conditions and abnormal, or disrupted, conditions. Choice of disruption was a key part of the experimental design as a noticeable effect on workload was required to understand whether the impact of automation was different for different levels of demand. Many disrupted conditions on the railway involve a high degree of communication, which is more difficult to simulate accurately. It was therefore necessary to control communications as far as possible to ensure that they did not affect the results as communications can contribute to increased workload and may have masked any workload effects of the automation. Therefore, the selected form of disruption was closure of a section of track – a platform at a busy station. The local simulator was used, which, although not an exact physical replica, functions in an identical manner to a real workstation and has the same number of screens and identical input devices (i.e. tracker ball and keyboard). The simulator gives a percentage score based on performance compared to the timetable, but it is not always clear on what this score is based. The experiment produced some interesting results indicating that the perceived workload increased and the performance was most consistent with the highest level of automation. It would be impossible to undertake this kind of research outside a simulated environment, since the running of the real railway is inherently variable.

A similar study employed simulators in the study of situation awareness for signalling. Signalling simulators lend themselves well to this area as the availability of a communication system allows the researcher to apply the situation present assessment method (SPAM; Durso and Dattel, 2004). In SPAM, not only the accuracy but also the time taken to respond to queries about events currently occurring on the workstation is an indicator of the signaller's awareness of the current state of events (Golightly et al., 2012). Simulation is particularly appropriate for this research as the queries presented to participants need to be carefully choreographed to match the events occurring on the workstation. Thus, it is possible for the investigator to spend time in piloting and modifying queries as well as, potentially, the simulation scenario to develop a study protocol that is realistic, challenging and consistent between participants. This kind of planning is completely impractical when running studies on a live network.

A more novel use of simulators was undertaken by Cheung (2011) (see Balfe et al., 2016), also at Network Rail, who used the simulators to generate 'static scenarios'. These scenarios were printed from a paused scenario on a high-fidelity VDU simulator and presented to signallers who were asked what strategy they would employ for the trains in the area at that time. Six different scenarios were created, and supporting information in terms of additional train running information was also mocked up. Information was slowly added, including the time of the scenario, timetable information and delay information, and changes to the signaller's original strategy were noted in order to understand the effect of the information on their strategies. The scenarios were based on real situations that had occurred and caused significant delays. Each participant was given 60 seconds to generate or update their strategy after each piece of information was presented and was then asked to give their answer verbally to the experimenter and to draw it on the blank schematics provided. The experiment found that as more information was added, the variation in the strategies decreased. The quality of the strategies (as measured by the total resulting delay minutes) also improved as more information was added but was considerably lower than the quality of strategies found in the dynamic simulation discussed earlier. The static scenario method of simulation was found to be less realistic, and this may have resulted in the lower quality of strategies, but it did offer benefits over the dynamic simulation in terms of controllability, ease of administration and repeatability. It would not be possible to investigate the relative importance of individual pieces of information on the strategies generated by signallers in a full simulation.

Besides direct measures or observations in simulations, eye-tracking equipment can be used to understand where gaze is directed on the interface in order to determine the information used by participants during their decision-making. Eye tracking is a common research method in train driver studies (see Chapter 5), where point of gaze is a critical research issue in better understanding new driving interfaces or incidents where signals are passed at danger (e.g. Itoh et al., 2001, Merat et al., 2002, Luke et al., 2006, Naghiyev et al., 2014). Whilst this research approach is less commonly used in railway signalling, it can reveal useful information on the cues used by operators and the key points of interest on a particular workstation, thus potentially facilitating future interface design. Balfe (2010) used eye-tracking equipment to understand the differences in signaller gaze patterns under the different levels of automation described earlier. A head-mounted eye tracker was used which, once calibrated to a specific screen, superimposed the point of gaze on a video of the environment. The data were analysed by coding the point of gaze at one-second intervals against the areas of interest on the screen. The results revealed that for all three conditions, the signaller's gaze typically followed the line of route – that is, their scanning was not randomly distributed across the screens but followed a logical progression along the primary path. The distribution of time spent scanning different areas on the workstation was also largely similar, but significantly more time was spent when monitoring the most complex area under automated conditions than when signalling manually.

Further eye-tracking research was undertaken by Mellis (2010) in an unpublished master's thesis. This work aimed to derive conclusions for future training of signallers by analysing eye movements during a simulation. It was undertaken on an NX panel simulator and compared results from a novice group with those from an expert group. Areas of interest on the simulator, including a clock, a level crossing, timetable information, train identifiers, signals and platforms, were all identified in advance and coded against the recorded eye movements. The experiment identified eight main strategies employed during the simulations:

1. Checking the time and then the timetable information to note approaching trains
2. Setting a route for a train and then checking back over the route to ensure that it is set
3. Monitoring train progress by looking at the rear of the train indication on the panel
4. Looking ahead to the next signal when pressing a signal button
5. Planning a train route by looking along the route whilst the train is in the platform
6. Reading the timetable information and then locating the entry point for approaching trains
7. Monitoring other train movements whilst operating the level crossing
8. Reading the timetable information when there is a natural break in train movements

The novices were more likely to engage in the first three strategies, and the experts were more likely to engage in the final three. These results indicated that the experts were more confident in the actions that they undertook and were less likely to check their success whilst instead they focused their attention on their future tasks. The recommendations from this work included highlighting to novices the importance of locating entry and exit points for trains on their area of control, using periods of low workload to plan ahead for approaching trains and setting routes as far ahead as possible.

It is not easy to undertake research using additional equipment such as eye tracking outside of the simulated environment, due to the potential for distraction on the live railway. Simulators therefore offer a valuable opportunity to gather data on signalling operations, which would be difficult or impossible in the live environment. However, the limitations of the simulators, particularly in terms of the closed nature of the simulated environment, which limits communications, mean that this research must be carried out in parallel with suitable methods based on the real environment. The work to date in this area has only scratched the surface of the knowledge that can be gained through the use of simulation, particularly in conjunction with technologies such as eye trackers. Improvements in eye-tracking technology mean that the data can now be collected and analysed more easily, making it a more realistically achievable research method. Only through advanced research using simulation can sufficient knowledge be gained to support future signalling system design.

7.4.2 Testing New Concepts

Incremental changes to the signalling environment are often proposed in response to emerging issues and challenges in rail operations. Simulating these changes in advance of their live deployment is one way to understand the risks and put measures in place to mitigate them.

One such proposed change in the GB rail network is the introduction of a new format for giving signallers information on the planned platforms for trains in large stations. A form of part-task simulation was undertaken by Network Rail (Charles, 2014) to investigate differences in performance due to the format of the information provided. The overall aim of the experiment was to compare the use of list-based, graphical and electronic tools used when re-platforming trains in order to understand the relative merits and disadvantages of each tool. Whilst lists and basic graphical representations are being used in operation and the findings for the use of graphical tools are encouraging (Charles et al., 2013), electronic tools of this kind are not currently in use, so simulation was the natural choice for further investigation. Scenarios were presented to non-expert participants by using a simplified version of railway rules, and the participants had to decide how best to address the scenario. The results of the experiment found that the participants were able to complete all 10 scenarios in the graphical conditions, with 100% completion rate for the electronic graphic-based tool, 97% for the paper graphic-based tool and 88% for the paper list. The electronic graphic-based tool also facilitated the most efficient solutions to the scenarios, again followed by the paper graph and the list, with delay minutes to trains considerably lower in the electronic graph condition. The experiment concluded that all three of the display types assisted the participants when solving the scenarios. However, all three behaved very differently, with electronic graph users displaying lower understanding of the underlying task. The use of low-fidelity simulation in this case allowed different representations to be directly compared and produced data on the potential usefulness and possible disadvantages of developing new technology to support signalling operations. However, more detailed and high-fidelity studies using expert participants are still required to develop the concepts further.

A less cognitive and more practical issue emerged during the move from NX panels to VDU technologies, and that was the use of a tracker ball to interface with the signalling system, rather than directly operating the panel buttons. This indirect method of control input led to anecdotal reports that signalling was slower and more error prone when using a tracker ball, and a desire to use a more standard mouse was raised. Internal unpublished research at Network Rail used an experiment on a VDU simulator to compare the speed and accuracy of mouse usage compared with tracker ball usage. Participants were timed and observed whilst setting a series of routes of varying lengths and complexities using both technologies. In total, seven different types of routes were measured and the mouse technology was significantly faster for all the seven routes. In contrast, the mouse was less accurate than the tracker ball, with an average of four miss–hits (i.e. missing the target signal) per signaller when using the mouse but only two miss–hits when using the tracker ball. However, the majority of these were quickly noticed and rectified, but there were more unnoticed errors with the tracker ball than the mouse – an average of more than one unnoticed errors for the tracker ball but only 0.25 for the mouse. This experiment used a high-fidelity simulator to test the different routes and provide concrete data on the relative speed and accuracy of the different choices available.

A more novel approach to testing new designs and concepts is gaming simulation. Gaming simulation is a method of involving stakeholders in simulations to solve or further understand complex real-world problems. This technique has typically been used for educational purposes, frequently using computer games (Greitzer et al., 2007). Meijer (2012a) described the application of gaming simulation in Dutch railways, whilst van Luipen and Meijer (2009) suggested that serious games are suitable for exploring complex problems when the full details of the real-world state are not available.

The method relies on gathering qualitative data from participants to better understand the practical issues within the scenario and to start to identify potential resolutions. These data may be collected during the scenario but more often are collected during debriefs. The distributed, complex nature of the rail environment (and the cost of changes) makes low-fidelity gaming a useful method for testing out new concepts in early design. In the applications on the Dutch railways, the simulations used both computer and tabletop simulations featuring representations of the railway, which varied depending on the problem under consideration. van Luipen and Meijer (2009) described four different case studies ranging from a policy game involving senior management examining the allocation of capacity to the freight market, to testing the differences between two mechanisms for handling disruption on a busy rail corridor. This latter simulation involved a wide range of operational roles including traffic controllers, passenger information, driver resourcing and rolling stock resourcing. The participants assumed the role that they were employed in and ran through a scenario by using the current method of working and the proposed new method of working. The new method was expected to deliver resilience benefits, allowing the railway to recover from disruption more quickly, but the simulation showed that in practice, the new solution would not deliver the expected benefits.

van Luipen and Meijer (2009) described another gaming simulation used in Dutch railways to evaluate the effects of new rail infrastructure or timetables. The simulation was based as much as possible on real timetables, track and working principle data, but the interfaces were not realistic, as befits a gaming simulation. Ten operational staff (train drivers and controllers) participated in the experiment which examined new control concepts for proposed high-frequency passenger trains. In particular, they were interested in how well the concepts would be supported by the different levels of operational control in place in the Dutch railways. The simulation was successful in identifying infrastructure features necessary for the success of the new control concepts, but the operational staff found it difficult to relate to the detailed control of the infrastructure through interfaces that did not closely match those that they used daily.

The use of gaming simulation is not widely adopted in rail signalling and is not an obvious form of simulation, but there are clear benefits to this approach, including the ability to include a wide variety of roles in an integrated simulation, the ability to model complex problems for which there are insufficient data to create a full simulation and the relatively low cost associated with this type of simulation. Meijer (2012b) described the necessary level of simulation to ensure that participants become immersed in the game and suggested that the infrastructure should be presented in an abstract format, whilst the timetable and train running information should be presented in a format matching that of their daily use. These simulations, however, can be complex and time consuming to run, and engagement of the participants is crucial to their success.

Simulation, both in advanced high-fidelity forms and in lower-fidelity representations of the signalling environment, provides a valuable environment in which new ideas can be tested away from the safety concerns of the live environment. It is a key step in developing new approaches to existing issues.

7.4.3 Prototyping

Simulation and prototyping are key evaluation techniques for new rail traffic control technologies. Solutions can be evaluated against the key concerns of a particular network in order to better understand any mismatches between the technology and the infrastructure ahead of deployment. This use of simulation is similar to testing new concepts but uses higher-fidelity simulations to tease out issues prior to deployment, rather than testing whether basic concepts may be useful.

Prototyping of new control systems is increasingly possible due to the computer-based nature of modern signalling control systems. Research in the Swedish rail system has identified that current traffic control systems provide limited support to operators (Kauppi, 2006), and in order to address the issues raised in the course of researching the limitations of current operations, a prototype system was developed to simulate the proposed future solution. Users were involved in evaluating the prototypes and providing feedback for the design in an iterative process. The solution envisaged a system that allows controllers to re-plan the timetable as disruption occurs, instead of the current situation where decisions are made in real time. Kauppi (2006) described the use of both high- and low-fidelity prototypes, with low-fidelity prototypes (such as drawings of scenarios) used to facilitate discussion, whilst high-fidelity prototypes were found to be of more use in examining how the future system will work and in communicating the proposed ideas. Meanwhile, Sandblad et al. (2000) described the development of the simulator, which was based on an existing modelling tool used for the development of timetables. The design allowed different presentation systems to be connected to the core simulator, providing an ideal prototyping tool. A test environment was developed using a projection of the graphical user interface and a camera to record the operator as they interacted with the system. The operator can be presented with pre-defined scenarios, and numerous variables of interest can be recorded for future analysis (Wikström et al., 2006). This was later converted into a portable test environment (Kauppi et al., 2006). The results of interactive demonstrations using these prototypes provide proof of concept for new interfaces as well as important feedback for the system designers. However, the need to structure and time constrain sessions on the prototype may limit the novel findings from such sessions. A solution to this might be an Internet-based version of the prototype, which allows the participants to access the simulation more regularly.

Further along the development life cycle, prototypes are an excellent way of testing the 'softer' aspects such as workload and role allocation. Network Rail recently used three signalling system prototypes to conduct an unpublished ergonomics evaluation in order to assess their potential as future signalling systems. These fully functioning prototypes, located in dedicated model offices, were subjected to a three-month trial, using trained operators. The evaluation consisted of three strands – workload assessment, roles and processes assessment and usability assessment. The data collected for the three strands varied, but the overall approach was the same. The operators involved in the assessment rotated around the model offices, spending a month in each one. Each prototype was modelled on the same signalling area, and 12 test scripts were devised to test different scenarios, from normal, routine operations to extreme disruption – such as theft of vital cables on track, resulting in failed signalling equipment or a fatality. One script was run per day, and the operators were assigned different roles. Each trial was treated as a live environment, and the operators were expected to apply current rules and regulations to the operation, unless otherwise stated.

Several methods were used to actually measure the required elements, from physical data logging by the system to observations and application of existing Network Rail ergonomics tools. Workload was assessed during the trial using a subjective workload scale (Integrated Workload Scale [IWS]; Pickup et al., 2005), and afterwards, different specially developed questionnaires were administered based on the IWS scores. Usability was measured after each script by using an online questionnaire based on the Software Usability Measurement Inventory scale (Kirakowski and Corbett, 1993). Role and process issues were also evaluated using an online questionnaire to assess which role allocations worked or did not work and how the operators approached different problems. The results not only provided a clear idea of workload and usability issues to be addressed in the next stage of the project but also offered an excellent way to clearly see how each supplier approached different problems and compare the merits of different designs.

Even though they were all given the same basic requirements, all three systems performed very differently in terms of functionality and usability. This may have been due to the fact that the suppliers were given relatively high-level requirements, which did not constrain them and left them free to develop their designs in different ways. This is a vital point when considering new technology, especially at a prototype level; although it may seem tempting to set strict requirements and a substantial wish list, in doing so, you may constrain creativity and miss out on intuitive, well-designed solutions as a result. The prototyping used in this case enabled an evaluation of potential designs at a relatively early project stage, and thus, new design ideas could be trialled without the risk of compromising operations. By utilising the results of the evaluation, along with gaining a clear understanding of each of the suppliers' approaches, detailed requirements were drawn from the prototypes to be used by the suppliers going forward. Simulation and prototyping can therefore be key evaluation technologies for new rail traffic control solutions, allowing new ideas and solutions to be tested against the real concerns of a particular network.

7.4.4 Observations from Using Simulators in Research

It is worth adding a concluding note here on how to use simulators effectively in research, with an emphasis on using simulators with experienced operational staff rather than, say, student volunteers. First, despite their use in training, signallers typically do not use simulation in the very routine way that, say, pilots might. Therefore, appropriate briefing is essential for participants to understand the aims of the study and particularly to address any concerns around whether the outputs of the simulation are being used as a form of competence assessment. Ethics and clear consent are crucial.

Second, as noted earlier, signalling simulation is all about the behaviour and realism of the traffic scenario, and so, the fidelity of the traffic and infrastructure models and scenarios is crucial. Therefore, it is useful for the researcher who is unfamiliar with a particular signal box to take some time to understand the mechanics and peculiarity of a particular workstation and ensure that the scenario is as realistic as possible. For example in the Golightly et al. (2012) study of situation awareness, a minor disruption was introduced to measure how long a signaller took to react. It took some time to find an event that was realistic enough to not stand out as something unnatural but was noticeable to the signallers. Timing is a crucial element of this. Similarly, Balfe et al. (2016) took several observation and simulation design sessions to create a scenario that featured an appropriate and measureable increase in signaller workload. Piloting scenarios is a vital stage of study preparation and should not be overlooked.

Finally, signallers need appropriate debriefing on the aims of the study and the value of their input. This is again not only to allay concerns, but also because signallers are usually forthcoming with information and insight on the general signalling task that can be very valuable context, for example in giving background as to why the workstation or timetable has evolved in the way that it has.

7.5 Conclusions and Future Developments

Whilst not as widespread and mature as simulation in some other industries, rail signalling utilises simulation to a degree that is appropriate for the domain. The technology is relatively expensive, meaning that there is little point in widespread use of simulation for training when on-the-job training is available and appropriate. Training in simulators is therefore largely limited to early training stages. The simulators are highly useful for research purposes, however, since it is not always appropriate to carry out research on a live workstation in a safety-critical environment. In particular, three areas of application for simulation research are evident: understanding current operations, testing new ideas for the operational environment and prototyping new control systems.

Each of these has its own requirements, but rail signalling simulators are typically not developed for research purposes. To better meet the needs of research, signalling simulators need to be more flexible with easily modified timetables and even modification of the underlying infrastructure. Communications are not well supported currently, and the research presented in this chapter has had to be designed around this limitation. Future research simulators could address communications through automated interfaces simulating communications with key actors, rather than relying on the human operator.

Whilst the existence of simulation is well established, particularly in early training, there are many opportunities that lie ahead with relevance to human factors in signalling. First, the work on prototyping is likely to escalate as rail networks worldwide look to integrate greater levels of automation and newer forms of data or representation into the control environment. One particular innovation will be to further close the loop between control and driver by increased use of speed-based (rather than route-based) movement authority. Although this exists already on some lines and in some countries, a major development will be adapting that in-vehicle speed-based control information for capacity and eco-driving. As a result, the driver's behaviour and response to control information will have a bearing on the suitability of actions issued in the control centre. If a driver fails to follow capacity-based speed instructions for example the signaller's ability to predict performance on high-capacity lines will be hampered.

Therefore, there is the interesting possibility of linking together driving simulators with signalling simulators to give a more authentic training and research experience. This would allow signallers to experience a traffic model that is not just generated as a simulation but is based on authentic driving behaviour and for drivers to experience authentic control decisions. One of the drawbacks of this approach is that both the driving simulator and signalling simulator need highly accurate and equivalent traffic models. A solution is to use descriptions of infrastructure and timetable based on railway markup language (railML). railML is an extensible markup language-based standard that allows diverse rail applications to communicate including timetable and infrastructure descriptions required for homogenous environments across rail simulators.

Another opportunity is to be able to introduce workload and demand estimation in the signalling environment. Whilst signalling demand estimation tools such as operational demand evaluation checklist (Pickup and Wilson, 2007) and signaller workload exploration and assessment tool (Shanahan et al., 2012) require manual computation of demand parameters, some of these parameters can be captured directly from the infrastructure and traffic models and tested in different configurations – for example with reduced infrastructure or to compare demand when two or more workstations have the potential to be combined through the introduction of automation. In future, actual user strategy and performance, taken directly from interaction with the signalling HMI, can be factored in as a weighting parameter. For example a signaller with a very reactive style of regulation may be more prone to certain demand factors than someone with a more proactive style.

However, despite these possible improvements, simulation is already a key tool in understanding the potential issues of new rail signalling technologies and for providing information to support their further development.

References

Balfe, N. 2010. *Appropriate Automation of Rail Signalling Systems: A Human Factors Study*. Nottingham, UK: University of Nottingham.

Balfe, N., Wilson, J.R., Sharples, S., and Clarke, T. 2012. Effects of level of signalling automation on workload and performance. In *Rail Human Factors around the World: Impacts on and of People for Successful Rail Operations*, ed. J.R. Wilson, A. Mills, T. Clarke, J. Rajan and N. Dadashi, 404–411. London: Taylor & Francis.

Balfe, N., Houghton, R., Cheung, J., and Sharples, S. 2016. Investigating strategies in rail signalling: Comparison of simulation methods. Paper presented at EHF 2016, Daventry, UK.

Charles, R. 2014. *Using Cognitive Artefacts to Aid Decision Making in Railway Signalling Operations*. Nottingham, UK: University of Nottingham.

Charles, R., Balfe, N., Wilson, J.R., Sharples, S., and Carey, M. 2013. Using graphical support tools to encourage active planning at stations. In *Rail Human Factors: Supporting Reliability, Safety and Cost Reduction*, ed. N. Dadashi, A. Scott, J.R. Wilson and A. Mills, 427–432. London: Taylor & Francis.

Cheung, J. 2011. *Routing strategies in rail traffic control*. Master's thesis. Nottingham, UK: University of Nottingham.

Cullen, W.D. 2000. *The Ladbroke Grove Rail Enquiry: Part 1 Report*. Norwich, UK: HSE Books.

Durso, F.T., and Dattel A. 2004. SPAM: The real-time assessment of SA. In *A Cognitive Approach to Situation Awareness: Theory, Measures and Application*, ed. S. Banbury and S. Trembley, 137–154. Aldershot, UK: Ashgate.

Golightly, D., Wilson, J.R., Lowe, E., and Sharples, S. 2010. The role of situation awareness for understanding signalling and control in rail operations. *Theoretical Issues in Ergonomics Science* 11: 84–98.

Golightly, D., Wilson, J.R., Sharples, S., and Lowe, E. 2012. Developing a method for measuring Situation Awareness in rail signalling. In *Human Factors of Systems and Technology*, ed. D. de Waard, N. Merat, A.H. Jamson, Y. Barnard and O.M.J. Carsten. Maastricht, Netherlands: Shaker Publishing.

Greitzer, F.L., Kuchar, O.A., and Huston, K. 2007. Cognitive science implications for enhancing training effectiveness in a serious gaming context. *Journal on Educational Resources in Computing* 7.

Haavisto, M., Ruuhilehto, K., and Oedewald, P. 2010. Train traffic controller's task demands during major railroad construction work. Paper presented at European Conference on Human Centred Design for Intelligent Transport Systems, Berlin, Germany.

Itoh, K, Arimoto, M., and Akachi, Y. 2001. Eye-tracking applications to design of a new train interface for the Japanese high-speed railway. In *Usability Evaluation and Interface Design: Cognitive Engineering Intelligent Agents and Virtual Reality*, ed. M.J. Smith, G. Salvendy, D. Harris and R.J. Koubek, 1328–1332. Mahwah, NJ: Lawrence Erlbaum Associates.

Kauppi, A. 2006. *A Human–Computer Interaction Approach to Train Traffic Control*. Uppsala, Sweden: Uppsala University.

Kauppi, A., Wikström, J., Sandblad, B., and Andersson, A.W. 2006. Future train traffic control: Control by re-planning. *Cognition, Technology and Work* 8: 50–56.

Kirakowski, J., and Corbett, M. 1993. SUMI – The Software Usability Measurement Inventory. *British Journal of Educational Technology* 24: 210–212.

Luke, T., Brook-Carter, N., Parkes, A.M., Grimes, E., and Mills, A. 2006. An investigation of driver visual strategies. *Cognition, Technology and Work* 8: 15–29.

Meijer, S.A. 2012a. Introducing gaming simulation in the Dutch railways. *Procedia – Social and Behavioural Sciences* 48: 41–51.

Meijer, S.A. 2012b. Gaming simulations for railways: Lessons learned from modeling six games for the Dutch infrastructure management. In *Infrastructure Design, Signalling and Security in Railway*, ed. X. Perpinya, 275–294. Rijeka, Croatia: InTech.

Mellis, S. 2010. *Monitoring strategies in rail traffic control*. Master's thesis. Nottingham, UK: University of Nottingham.

Merat, N., Mills, A., Bradshaw, M., Everatt, J., and Groeger, J. 2002. Allocation of attention among train drivers. In *Contemporary Ergonomics 2002*, ed. P.T. McCabe, 185–190. London: Taylor & Francis.

Naghiyev, A., Sharples, S., Carey, M., Coplestone, A., and Ryan, B. 2014. ERTMS train driving in cab vs. outside: An explorative eye-tracking field study. In *Contemporary Ergonomics and Human Factors 2014*, ed. S. Sharples and S. Shorrock, 343–350. London: Taylor & Francis.

Patrick, C., Balfe, N., Wilson, J.R., and Houghton, R. 2013. Signaller information use in traffic regulation decisions. In *Rail Human Factors: Supporting Reliability, Safety and Cost Reduction*, ed. N. Dadashi, A. Scott, J.R. Wilson and A. Mills, 409–418. London: Taylor & Francis.

Pickup, L., Wilson, J.R., Norris, B.J., Mitchell, L., and Morrisroe, G. 2005. The Integrated Workload Scale (IWS): A new self-report tool to assess railway signaller workload. *Applied Ergonomics* 36: 681–693.

Pickup, L., and Wilson, J.R. 2007. Mental workload assessment and the development of the operational demand evaluation checklist (ODEC) for signallers. In *People and Rail Systems: Human Factors at the Heart of the Railway*, ed. J.R. Wilson, B. Norris, T. Clarke and A. Mills, 215–224. Aldershot, UK: Ashgate.

Pickup, L., Balfe, N., Lowe, E., and Wilson, J.R. 2013. "He's not from around here!": The significance of local knowledge. In *Rail Human Factors: Supporting Reliability, Safety and Cost Reduction*, ed. N. Dadashi, A. Scott, J.R. Wilson and A. Mills, 357–366. London: Taylor & Francis.

Roth, E.M., Malsch, N.F., and Multer, J. 2001. *Understanding how railroad dispatchers manage and control trains: Results of a cognitive task analysis.* Report No. DOT/FRA/ORD-01-02. Washington, DC: US Department of Transportation, Federal Railroad Administration.

Sandblad, B., Andersson, A., Jonsson, K.E., Hellström, P., Lindström, P., Rudolf, J., Storck, J., and Wahlborg, M. 2000. A train traffic operation and planning simulator, 241–248. In *Proceedings of COMPRAIL 2000*, Wessex Institute of Technology, Southampton, UK.

Shanahan, P., Gregory, D., and Lowe, E. 2012. Signaller workload exploration and assessment tool (SWEAT). In *Rail Human Factors around the World: Impacts on and of People for Successful Rail Operations*, ed. J.R. Wilson, A. Mills, T. Clarke, J. Rajan and N. Dadashi, 434–443. London: Taylor & Francis.

van Luipen, J.J.W., and Meijer, S.A. 2009. Uploading to the MATRICS: Combining simulation and serious gaming in railway simulators. Paper presented at the 3rd International Conference on Rail Human Factors, Lille, France.

Wikström, J., Kauppi, A., Andersson, A.W., and Sandblad, B. 2006. *Designing a graphical user interface for train traffic control.* Technical report 2006-25. Uppsala, Sweden: Department of Information Technology, Uppsala University.

Section IV

Air

8

Flight Simulator Research and Technologies

Barbara G. Kanki, Peter M. T. Zaal and Mary K. Kaiser

CONTENTS

8.1 Introduction

The use of flight simulators for research may traditionally be conceived of as a laboratory-based, empirical endeavour. But in reality, the research-oriented flight simulator is but one methodology along the continuum from laboratory to field environments. This continuum also encompasses both pure and applied research. Research outcomes, such as human factors insights and recommendations, are typically published and handed off to customers (e.g. manufacturers, regulators), who have their own goals and use these deliverables as a foundation on which to develop their own products (e.g. building aircraft components, developing rules and advisories, designing training modules). It is therefore sometimes hard to distinguish between the research and development activities performed by manufacturers, regulators, operators and others. The line that we try to draw in this chapter is based on the primary goal of the simulator user, which tends to dictate the simulation priorities.

In this context, the chapter covers three main areas. Section 8.2 contains broad descriptions and examples of several key areas in which flight simulator research has been used to answer human factors questions and has furthered the use of flight simulation. Section 8.3 discusses technical aspects of simulators that can limit research design and the questions that they can meaningfully answer. Section 8.4 focuses on a few recent simulator technologies and how flight simulator research can be redirected in the future to expand its applicability to complex airspace systems.

8.2 Flight Simulator Research

Three topic areas – namely pilot proficiency, technology evaluation and accident investigation – have been well served by flight simulator research. What these areas have in common is the goal of understanding how humans perform in flight operations: what skills are involved, how they will interface with new technologies and procedures and why they performed in a particular way during an emergency.

8.2.1 Pilot Proficiency

It is well known that flight simulators are commonly used for pilot training and evaluation (see Chapter 9). What is less well known is how research helps to determine which pilot skills are essential to ensure pilot proficiency, particularly since aircraft and airspace operations are continually evolving. Flight simulators have helped to address this issue by providing a controlled experimental setting to isolate key factors and determinants of behaviour. At the same time, innovative techniques for capturing behavioural data – critical for conducting human-in-the-loop (HITL) research – have been developed. In turn, both the findings and the techniques feed into training and assessment. An early example of how flight simulators played a major role in such research is in the development of non-technical skill training.

Ruffell Smith (1979) conducted what was to become a classic study of flight crew performance in a Boeing 747 simulator at National Aeronautics and Space Administration (NASA) Ames Research Center. The experiment originally investigated pilot vigilance, workload and response to stress, but the focus shifted and became a dramatic demonstration of how crew interaction, communication and resource management could impact overall performance. In addition to demonstrating the yet untapped potential of high-fidelity, full-mission simulation for research and training purposes, it confirmed that technical skills alone were not enough to guarantee effective crew performance. Although crew resource management (CRM) was not yet a known concept, this study showed that specific team behaviours, such as crew communication and coordination, could be realistically elicited, clearly identified and characterised. These developments provided a needed proof of concept that could be modified or refined for a variety of both research and training uses.

Ruffell Smith (and many researchers to follow) shifted the focus of crew performance from technical performance alone to additional, essential non-technical skills, thus opening the door to researching human factors and performance in the simulator and to a useful methodology for reliably assessing crew performance. The early studies pointed to the importance of non-technical skills, but researchers needed to characterise group process behaviours associated with performance differences in order to provide useful information to trainers. It was clear that data collection methods would need to be devised for the simulator environment that would capture specific behavioural (and trainable) indicators of CRM and other non-technical skills.

In the interest of capturing targeted behaviours, flight simulators were sometimes fitted with more realistic external radio communication channels (e.g. with air traffic control [ATC] or airline dispatchers). Within-cockpit interactions and communications were typically recorded by time-synchronised video of pilots (sometimes with multiple views), which could be viewed in a split-screen display alongside flight displays and parameters. Researchers also experimented with using eye-tracking cameras as a means of seeing what the pilot was seeing. However, as with any behavioural methodology, care had to be taken to avoid subjective interpretation of observable data and many researchers worked on developing reliable systems of coding. Although such coding could be performed, in theory, in real time in actual operations, the simulator ensured that crews were all performing under the same flight conditions.

An early example was a study that explored whether crew speech patterns could differentiate higher- from lower-performing crews (Kanki and Foushee 1989). Full verbal transcripts were made and analysed in terms of speech act sequences from the videos of each flight in the simulator. Such detailed analysis, including the timestamp for pilot and aircraft actions, would not be possible outside high-fidelity, full-mission simulation. Among the results, the study showed that higher-performing crews had greater *information exchange and validation and greater participation from the First Officer in task-related topics*. Because the research was exploratory in nature, other researchers could use the same data transcripts to look for other kinds of communication sequences, such as those pertaining to decision-making and information management (Orasanu 1993, p. 157).

At the same time that researchers were developing new data collection methods, some findings were handed off to trainers who adapted these innovations to fit their training and evaluation purposes. For example instructor/evaluators could tag video footage of the simulator performance in order to show pilots the specific areas where performance was above or below average. Developing reliable performance metrics was a challenge for researchers and trainers alike, and many approaches to improving assessment reliability were studied and developed for the benefit of both (Seamster et al. 1995, Goldsmith and Johnson 2002, O'Connor et al. 2002).

Although CRM skills were a needed addition to technical skills alone, it was not enough to consider them in isolation from the technical skills that together comprise a complete performance. The following example focusing on pilot recovery skills demonstrates how both types of skills must be fully integrated to be an indicator of an effective crew.

In their simulator study, Casner et al. (2014) sought to understand how prolonged use of cockpit automation could affect the retention of manual flying skills. The results showed that although manual control skills were relatively unaffected by automation, it was the associated cognitive skills in dealing with system failures that appeared to create the more serious problems. These issues reveal a mix of technical and non-technical skills in the appropriate management of aircraft automation.

Designing a simulated flight scenario that includes such a mix of skills is not difficult in a research setting as long as full-mission realism is maintained. However, determining the appropriate behavioural indicators may be challenging. Skills such as monitoring (i.e. early recognition of upset conditions), appropriate crew and automation management and recovery skills (aircraft control and energy management) are all relevant and must be elicited by appropriate and realistic environmental and aircraft states throughout all phases of the flight. The researcher must be able to assess (or control) any sequence of crew actions. For example if crews are allowed to anticipate and prevent upset conditions, the performance measures must be able to be meaningfully compared to crews that enter upset conditions but respond successfully. The researcher simply has to decide whether to control for these kinds of variations or devise performance measures that account for them.

8.2.2 Technology Evaluation

Whilst some environmental conditions and operational elements are typically controlled as a part of the flight scenario, aspects of the task and new technologies may be defined as independent variables in themselves, so that they can be systematically evaluated. In these cases, human performance measures provide indicators of where the technologies and/or procedures work well or need improvement.

Aircraft are continually updated and retrofitted with new technologies. Many aspects of the hardware and software may be evaluated in the laboratory by applying human factors standards and lower-fidelity tests of usability. But the use of high-fidelity full-mission HITL simulation becomes more critical as the evaluation becomes more operationally focused. Thus, the resource-intensive high end of flight simulation may be reserved for the most integrated testing, that is, when you need a realistic context for evaluating crew performance under critical operational conditions.

An example was a simulator study in which an electronic flight bag (EFB; a tablet-sized device holding electronic versions of the traditional paper flight deck documents) was operationally evaluated (Seamster and Kanki 2007). A flight scenario was developed to fully exercise the use of flight deck documents with an emphasis on evaluating how the EFB was used compared with traditional paper documents. The scenario encompassed a three-hour flight made up of six event sets that required document usage during preflight, engine start, cruise with several route changes, a divert and a low-visibility taxi-in. The combined results demonstrated an advantage for EFB in terms of reduced head-down time and workload, particularly in some higher workload phases of flight (route changes and diversions). Because the study yielded a useful preview of the best ways to use EFBs during every phase of flight, it served as a foundation for developing policies, procedures and best practices.

Recently, Next Generation (NextGen) air transportation system improvements (e.g. the transition from radar to satellite-based systems and digital data communication) in the US National Airspace System have been moving from the research and development phase to partial implementation. Described as an integration of new systems rather than a single system, the implementation of new ATC technologies and aircraft is complex at a socio-technical level and requires a coordinated implementation across pilots, air traffic controllers etc. Details of these changes are described each year in a Federal Aviation Administration (FAA) update (FAA 2014).

Given that the large-scale incremental improvements require changes by all users of the system, operational testing and evaluation are crucial. However, despite the use of realistic pseudo-controllers, flight simulators have been limited in focus to the pilot role. In more recent years, ATC simulators have been developed to focus on the controller task with the participation of pseudo-pilots. For example simulators have recently been used to study controller coordination procedures and decision support tools (Prevot et al. 2010, Chevalley et al. 2014, Parke et al. 2014).

As early as 2005, the concept of integrating multiple simulation facilities was demonstrated in the Virtual Airspace Simulation Technology-Real Time project at NASA Ames Research Center. Part of this project tested a decision support tool that could deliver automated arrival, departure and timed taxi clearances to aircraft via datalink. In this test, the Future Flight Central Tower simulator was networked to the Crew Vehicle Systems Research Facility (flight simulators). Although a specific datalink study was not conducted at that time, it proved the concept that system integration issues (i.e. between pilots and controllers) could be investigated (a topic that we return to towards the end of this chapter).

In support of anticipated NextGen changes that will affect both pilots and controllers, an example of coordinated research was conducted by Wing et al. (2013). They conducted two HITL simulation experiments to investigate the allocation of separation assurance functions between ground and air and between humans and automation under mixed-operation conditions (i.e. conditions in which ground-based separation and self-separation operations coexist in shared airspace). This clearly offers another approach to addressing the complexity of NextGen research questions and how they may be studied using simulation technologies.

8.2.3 Accident Investigation

When an aircraft accident occurs, a concerted effort is made to determine both the underlying causes of the accident and what actions the aviation community should undertake to avoid similar accidents in the future. In almost every accident, pilot error is cited as a cause or a contributing factor. However, most accidents result not from a single error, but rather from a series of compounding events. The aim of the investigation is to understand the *cascade of failures* that led to the accident.

Because it is so critical to examine each causal factor in the context of the accident scenario, HITL simulation can prove to be a highly valuable investigative tool. By recreating the accident scenario, investigators can examine critical factors and their interactions in real time, thus supporting or negating the plausibility of hypothesised causal chains.

The utility of simulators in this context is limited by two primary factors. First, there is the degree to which the simulator can recreate the operational environment in which the accident occurred (i.e. the fidelity of the flight simulator). Second, there is the quality of the data that the investigators possess about the events preceding the accident (e.g. aircraft state, pilot input commands, communications between crew and ground control).

As it happened, both of these factors achieved sufficient levels of maturity in the mid-1990s. By that time, flight simulators reached a reasonable level of operational fidelity in their motion and visual display technologies. On the accident data side, cockpit voice recorders (CVRs) were augmented with aircraft telemetry data from both cockpit and ground-based sources. Thus, the US National Transportation Safety Board (NTSB) started adding simulation studies to its suite of investigative tools in the mid-1990s.

We have selected two accident investigations in which HITL simulations provided useful insights into the probable cause of an accident. The cases were chosen to illustrate the range of factors that can be investigated via HITL simulations and, by virtue of drawing on cases from the mid-1990s and 2010, to demonstrate how simulator studies have matured as an investigative tool. Interestingly, these studies have also motivated the aviation community to acquire better event data, thus further increasing the ability of HITL simulations to illuminate the probable causes, and possible mitigations, of severe accidents.

8.2.3.1 Aircraft Performance

Just after 1900 hours (Eastern Daylight Time) on 8 September 1994, USAir Flight 427 crashed whilst manoeuvring to land at Pittsburgh International Airport. The aircraft, a Boeing 737-300, had left Chicago O'Hare Airport at approximately 1800 hours (Central Daylight Time); the entire flight occurred in daylight, visual meteorological conditions. At 1900 hours, the flight crew was in communication with Pittsburgh approach control and executing a normal approach pattern. Three minutes later, the aircraft impacted hilly, wooded terrain near Aliquippa, Pennsylvania, approximately 6 miles northwest of the airport. The five crewmembers (two pilots, three flight attendants) and all 127 passengers were killed; the airplane was destroyed by ground impact and post-crash fire.

Reconstructing those final three minutes of flight, wherein a commercial airliner transitioned from a stable approach to a 300 mph nose-first dive into terrain, required an almost five-year effort on the part of the NTSB (NTSB 1999) and resulted in a primary report of almost 300 pages (over twice the average length of a hull-loss report). The paucity of surviving physical evidence (much less any human survivors) meant that the NTSB would have to draw heavily on their best existing analytical tools. It also meant that they would need to create new ones, one of which was HITL flight simulation.

Several critical items did survive the devastating crash: both of the flight recorders (CVR and the flight data recorder [FDR]) were recovered; furthermore, crucial components of the flight control system were reconstructed and examined. Physical evidence was coupled with prior evidence of rudder control issues in the Boeing 737 to derive a working hypothesis of probable cause: loss of control due to rudder malfunction.

Whilst much of the engineering analysis focused on mechanisms for the rudder malfunction, HITL simulation examined the plausibility of suspected pilot actions and aircraft performance issues. Initial tests were conducted using the multipurpose cab engineering simulator at Boeing. FAA test pilots flew a variety of failure or malfunction scenarios. Of all the simulations flown, 'only the rudder hardover simulation produced results that were generally consistent with the data from USAir flight 427's FDR' (NTSB 1999, p. 59). This gave further credibility to rudder failure being a probable cause.

However, there was still the question of whether pilot inputs caused or exacerbated the rudder hardover. Of particular concern was whether the initial extreme movements of the aircraft caused spatial disorientation, leading the pilot(s) to input improper commands (rather than control reversal by the rudder system). Examining this question in a simulator was challenging because most transport simulators are designed primarily to reproduce vehicle motions typical of normal flight. The extreme yaw and roll rates recovered from the FDR of Flight 427 could not be recreated on a Boeing 737 simulator with a hexapod motion platform.

Thus, the NTSB team elected to use the Vertical Motion Simulator (VMS) at NASA Ames Research Center. This simulator (shown in Figure 8.1) was designed to recreate the motion characteristics of a wide range of vehicles, ranging from transport aircraft to high-performance jets and from rotor-craft to the Space Shuttle. Given the extensive velocity/acceleration capabilities of VMS (see Table 8.1), it was possible to produce the extreme motions that the crew of Flight 427 experienced, as well as recreate the visual cues (e.g. horizon line) that would have been available to help the crew maintain orientation.

Based on the results of these simulated flights, the investigation was able to conclude that pilot disorientation was unlikely to have been a major contributor to the accident. Of course, the VMS studies relied on the subjective impression of subject matter experts and test pilots because the investigators did not have data regarding all of the pilots' control inputs during flight, nor did they have data regarding the control surface positions of the aircraft. In fact, the data recorder onboard the aircraft recorded only the parameters required by regulations in place at the time. Whilst this list of parameters

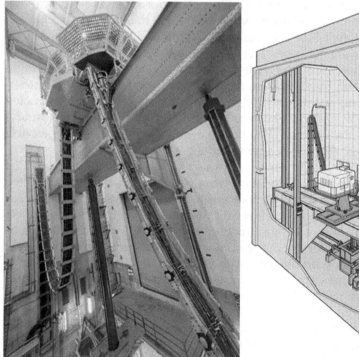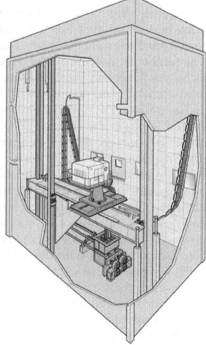

FIGURE 8.1
VMS at NASA Ames Research Center, with cutaway diagram (*right*).

TABLE 8.1

System Limits of the VMS Motion Base

Axis	Displacement	Velocity	Acceleration
Roll	±18 degrees	±40 degrees/second	±115 degrees/square second
Pitch	±18 degrees	±40 degrees/second	±115 degrees/square second
Yaw	±24 degrees	±40 degrees/second	±115 degrees/square second
Long.[a]	±4 feet	±4 feet/second	±10 feet/square second
Lat.[a]	±20 feet	+8 feet/second	±16 feet/square second
Vertical	±30 feet	±16 feet/second	±24 feet/square second

Note: Long., longitude; Lat., latitude.
[a] Test cab can be oriented to maximise either longitudinal or latitudinal limits.

was extensive, once one attempts to create a HITL simulation to compare test pilots' performance to that of the crew, the paucity of the database becomes glaringly apparent. A critical lesson learned from the Flight 427 investigation was that far more data should be collected by flight recorders. Based on the recommendation of NTSB, the FAA required that additional flight parameters be recorded.

Thus, the NTSB investigation of USAir Flight 427 not only provided critical insights into the probable cause of the accident (and recommended corrective actions), it also ensured that future investigations would have the benefit of the additional data needed to recreate crucial flight events.

8.2.3.2 Extreme Flight Environments and Pilot Training

Just after 1800 hours (Standard Mountain Time) on 20 December 2008, Continental Airlines Flight 1404 began its take-off from Denver International Airport. The weather was clear, but the wind was blowing from the west at 27 knots – strong, but still under the 33-knot limit for the conditions. As the aircraft accelerated, the captain shifted his attention from the thrust levers to outside visual reference, keeping the plane tracking the runway centreline. The first officer focused on the engine instruments (as per company policy); once the power was set, he monitored the airspeed so he could provide the standard airspeed callouts (which start at 100 knots).

As the airplane rolled down the runway, ambient winds increased, gusting up to 48 knots. The captain attempted to compensate with rudder pedal and then control wheel inputs. The captain's inputs became more extreme (peaking at nearly 90% maximum rudder). Less than six seconds later, the control wheel input transitioned from left to right. The airplane left the runway about 2600 feet from the approach end as the captain called to reject the take-off attempt. The aircraft crossed a taxiway and an airport service road before grinding to a stop near an air rescue and firefighting station, its nose pointing midway between the intended take-off direction and the crosswind.

Despite the rough ride and post-accident fire, there were no fatalities. The captain and five of the 110 passengers were seriously injured; the pilot and 38 passengers received minor injuries, whilst the remaining crewmember and passengers escaped unscathed. The aircraft, although substantially damaged, emerged fairly intact.

Following the recommendation from the USAir investigation and other NTSB efforts, the investigators of the Flight 1404 incident had a wealth of additional flight data to aid their efforts. Because of these rich data sources, investigators were able to recreate the exact time-correlated event sequence of the journey of the aircraft down and off the runway. Of particular interest were the comparisons among wind speed, rudder pedal input, aircraft heading and lateral load. This timeline indicated that the crew encountered unexpectedly high and gusty winds, and their control inputs were inadequate to compensate for these.

HITL simulations proved highly informative in determining the probable cause. Test pilots were taken through the standard crosswind training scenarios of the Continental Airlines Flight 1404, and all were able to perform take-offs in crosswinds of 0, 25 and 35 knots (rating the difficulty of these conditions as 'very easy', 'neither difficult nor easy', and 'slightly difficult', respectively; NTSB 2010). However, it is important to note that in these simulations, the crosswind was constant, and pilots were accurately informed of the wind intensity in advance. The accident crew had neither of these advantages. Moreover, the weather briefings that the crew had received gave them no prior warning of the intensity and variability of the crosswind.

To further elucidate the causal sequence of the accident, a second, 'back-drive' simulation was performed. In a back-drive simulation, data from the FDR (both the crew's control inputs and the resulting aircraft motions) are used to 'drive' the simulator through the runway roll – participants are simply along for the ride, but it can be a highly informative ride. In this case, the participants generally agreed that the captain's large right rudder inputs were understandable given the strong and gusty left crosswind for which he was compensating. They considered it likely that the captain would have been able to keep the aircraft on the runway if he had continued making significant right rudder inputs when the airplane veered left. However, in the light of findings from a UK Air Accident Investigation Branch HITL simulation study on pilot's response time for unexpected events (AAIB 1981), it was also understandable that the captain failed to do so, especially considering that the conditions were well outside his training window.

Based largely on these simulation studies, the NTSB recommended that the FAA develop procedures to provide crews with more accurate and timely wind conditions and to develop 'realistic, gusty crosswind profiles for use in pilot simulator training programs' (NTSB 2010, p. 62).

8.3 Limits of Flight Simulation in Research

This chapter has so far demonstrated the value of flight simulation in research. However, there are limitations that need to be considered. Even though today's flight simulators are capable of producing realistic and immersive scenarios, it is by no means possible to fully replicate all aspects of the real aircraft and its operational environment. Some of these aspects are simply beyond the limits of the current simulator technology, whilst others may be technically possible but the costs involved make them unfeasible.

However, these limitations do not necessarily constrain the usefulness of a simulator as a tool to conduct effective pilot research. It is not always required to simulate all aspects of aircraft behaviour with high fidelity, depending on variables such as the type of flight task, the phase of flight or the research focus. For example research focusing on pilots' integration of visual and motion cues when performing manual flight control tasks requires high-fidelity motion and visual stimuli, but the fidelity of the cockpit and operational environment is less important. Meanwhile, research investigating the effects of new cockpit technology on pilot performance might require a high operational fidelity, whilst the fidelity of the motion stimuli is less important. It is the task of the researcher to determine the level of fidelity required in each area of the simulation.

The fidelity of a flight simulation is difficult to define, as it can be determined at many different levels, such as the dynamic model used to simulate the aircraft response or the cuing systems that generate visual, motion and auditory stimuli. Furthermore, there is not a single method to measure fidelity. It can be defined using subjective pilot ratings or with various objective metrics, such as flight task performance.

Taking a system perspective, the simulator offers a closed-loop control system. A pilot makes control inputs to the aircraft dynamic model with the flight controls. The aircraft dynamic model in turn provides inputs for the different simulator cueing systems, such as cockpit displays, audio systems, the motion system and the out-the-window (OTW) visual system. These different systems then provide stimuli that may be perceived by pilots and aid them in performing their task. Using this information, the pilot makes a new control input, effectively closing the control loop. Flight simulation fidelity can be defined at different stages in this control loop, and past research has focused on each of these stages. The remainder of this section is structured to discuss fidelity issues at each stage.

8.3.1 Aircraft Model Fidelity

Aircraft dynamic model fidelity is the degree to which the dynamic model is able to replicate the real aircraft response. In a flight simulation, this is the most basic form of fidelity. It can be measured relatively easily by comparing the model response with the real aircraft response by using time histories. The dynamic model provides the data for the different simulator cueing systems. Obviously, if the dynamic model is not able to reproduce the real aircraft response accurately, the different simulator cueing systems – such as the motion system – will not be able to provide the pilot with the correct stimuli. For this reason, aircraft model fidelity in itself is not a practical measure of fidelity for the overall simulation. Rather, high dynamic model fidelity is a necessity for achieving a high fidelity at other levels in the simulation.

In some regions of the flight envelope, a model can have a higher fidelity than in others. Usually, the highest fidelity is obtained in the region of normal flight operations. However, depending on the focus of a particular research study, high fidelity might be required in less common flight conditions. In the current flight simulators, aircraft dynamic model fidelity is particularly low in abnormal situations that take the aircraft out of its usual flight envelope and in unstable flight conditions, such as an aerodynamic stall. In the last decade, researchers have examined modelling issues associated with civil transport stalls and have identified aircraft model deficiencies (Foster et al. 2005, Gingras and Ralston 2008, Schroeder 2012). These include the lack of reduced control effectiveness and angular stability near the stall point and the lack of randomisation in the model response. In addition, the effects of environmental phenomena on the aircraft dynamics, such as icing, are often not modelled accurately. Obviously, when a research study focuses on the influence of such phenomena on pilot performance, accurate models need to be developed and validated.

Typically, an aircraft simulation model is built to match the actual aircraft response within certain tolerances. However, flight data in abnormal conditions are difficult to obtain for commercial aircraft. Alternatively, models for certain areas of the flight envelope, such as an aircraft stall model, could be built from computational aerodynamics, wind tunnel data and input from test pilots who have experienced these scenarios. Recent research investigated if a stall model developed using these alternative methods would satisfy civil transport training objectives (Schroeder et al. 2014). It was found that when flight data are not available, using a representative model not based on flight data appears satisfactory for training stall recoveries. This suggests that a high-fidelity type-specific aircraft model is not always needed to effectively train pilots or to conduct research.

8.3.2 Cueing Fidelity

8.3.2.1 Visual Cueing

A pilot viewing the outside scene from the cockpit of the real aircraft is experiencing a much richer visual environment than what can be provided in a simulator OTW visual display. Pilots routinely operate in visual extremes:

from bright sunlight whilst flying above a solid cloud deck to the darkness of flying under cloud cover on a moonless night, from unlimited visibility to zero visibility, from rich visual environments to featureless. Vision is a required element to conduct take-offs and landings; although guidance systems, instrumentation and airfield markings and lighting assist the pilot operating in very low-visibility conditions, some level of visibility is almost always required to perform take-offs and landings.

In flight simulation, the goal is to provide the necessary and sufficient visual cues to allow the pilot to accomplish a task (or multiple tasks) in a manner that produces pilot behaviour that is representative of how the pilot would act in the real aircraft. Additionally, we want to avoid influencing pilot behaviour with visual artefacts (cues/conditions that would not be present in the naturally viewed scene). Not surprisingly, the necessary and sufficient cues vary as a function of task and environment. It is up to the researcher to determine the visual cueing requirements for a particular domain.

There are fundamental limitations present in any OTW visual display. Some limitations are associated with the physics of the light production compared with natural scenes. Other limitations are associated with the way in which the pilot's eyes and body position interact with the display, related to the optical qualities of the light. Another is related to the temporal and spatial sampling of the display (related to update rate and resolution). Lastly, the actual display content typically does not exactly match the perceptual experience. The relative importance of these limitations to a particular study or type of research varies. Thus, human factors researchers must consider carefully what characteristics of the visual environment need to be veridically simulated.

8.3.2.1.1 Physics/Light Production

OTW displays have fundamental limits in the levels of luminance that they can produce, as well as the range of luminance that the display can depict. Luminance is a measure of the amount of light that passes through a solid angle. A number of units are used to express luminance, but the measure commonly used in flight simulation is the foot-lambert (fL). Full sunlight on a cloud deck has a luminance level of 1×10^6 log fL, whilst a starless, moonless night has a luminance of 1×10^{-5} fL. Within these extreme luminance conditions, humans are sensitive to luminances as low as 0.0003 fL and can perceive comfortably up to approximately 300 fL (when luminances exceed this, we will reduce what reaches our eye with a protective eyewear, or failing that, by squinting our eyes).

Research simulators frequently incorporate visual system technologies developed for training purposes. These visual systems typically operate in the range of 6–50 fL or beyond, but low luminance levels in simulators rarely approach the limits of human perception. The FAA requires its simulators to demonstrate a contrast ratio (high luminance/low luminance) of 5.0; for our 6 fL display, the low luminance would be achieved at 1.2 fL, well above the human vision limits of 0.0003 fL. Note that the highest contrast levels are achieved when the minimum luminance is truly black, that is, not emitting any light. As visual displays typically cannot achieve a true level of black, they cannot present the levels of contrast that would be visible in a naturally viewed scene.

Fortunately, most studies and investigations do not require this level of visual fidelity. Studies in which the visibility of features or lighting conditions are thought to be significant factors should not be conducted with a simulated visual scene. Examples include operating in visual glare and visibility of cockpit instruments due to environmental lighting.

8.3.2.1.2 Sampling Artefacts

We live in a continuous, analogue world. Our vision appears to us to be completely continuous; we do not detect any display lines or sharp edges at the limit of our vision, and objects moving continuously appear to move smoothly.

A displayed visual image, whether generated by a camera or a computer, is actually sampled in time and in space. The spatial sampling is seen in the individual pixels; temporal sampling refers to the update rate. When the pixels are sufficiently fine and the update rate is fast enough, the pixels, updated at time intervals, simulate the experience of viewing a natural scene.

Typically, the resolution levels of simulator visual systems are not close to human visual acuity limits. Normal human vision is considered to be 20/20; this means that high-contrast (e.g. black-on-white) features as small as 1.0 amin can be discerned. Additionally, the FAA and military both require corrected vision of 20/20 or better; many pilots possess visual acuities higher than 20/20. Extremely high-resolution displays are typically not feasible due to cost constraints. Taking FAA requirements as an example, the spatial resolution requirement is 3.0 amin. In order to achieve 1.0 amin, the number of pixels would need to be increased by a factor of 9: at a minimum, it would require nine times the projectors and nine times the image generators.

Currently, the highest-resolution simulator existing is the Operational Based Vision Assessment simulator located at Wright–Patterson Air Force Base. This fixed-base simulator was developed to achieve a resolution of 0.5 amin, equivalent to 20/10 vision. This was designed around a specific research requirement to study the effects of visual acuity (and other characteristics of vision) on pilot operational performance within their pilot population, possessing an average visual acuity of 20/13 (Sweet and Giovannetti 2008). However, this represents a significant outlier among research simulators.

It should be noted that pixel resolution does not determine actual visual system resolution. Other factors such as display contrast, image generator characteristics, scene content, blending and ambient light (including scattering at the display surface) reduce the effective resolution. Winterbottom (2004) described a technique for measuring the simulator visual system resolution from a perceptual standpoint.

Temporal sampling can produce artefacts. Real-time, computer-generated images can appear blurry when the content contains motion (such as a moving horizon in a simulator). This is called motion-induced blur, and it is due to an interaction between the pilot's eye movements and the display (Sweet and Hebert 2007). It can be quite noticeable in flight simulation applications; multiple methods exist to reduce the magnitude of this artefact. These methods reduce the amount of time the display is emitting light during the frame. Specific methods include mechanical or electronic shuttering, black frame insertion and direct light modulation; typically, the blur reduction methods are fully integrated in the display and are based on the characteristics of the display technology. Whilst effective at preventing blur, these methods also reduce the overall brightness of the display. Blur is also reduced by increasing the update rate, without any loss of display brightness.

The last artefact to be discussed is related to the number of levels of brightness or colours that are available in a visual display, termed a *quantisation artefact* rather than a *sampling artefact*. Colours in typical OTW displays are specified at 256 levels; although this leads theoretically to a very large colour space, from white to black, there are only 256 levels. For lower luminance conditions (night scenes), quantisation can create very visible artefacts.

Similar to physics limitations, sampling artefacts are not an issue in a broad array of research applications. Resolution limits would need to be considered carefully in designing research related to visual identification tasks, such as reading and interpreting airport signage. Blurring artefacts are directly related to image motion, and in transport category aircraft, the greatest contributors are likely to be pitch and roll. The impact of blur on closed-loop manoeuvring (such as in a loss of control) should be considered in research applications.

As stated earlier, typical simulator visual systems are not ideal for low-light condition scenarios due to the limitations in producing low luminances. The quantisation artefacts described here are also most visible in low-light conditions. Experimental design of low-light condition scenarios should be done with careful consideration of visual artefacts.

8.3.2.1.3 Scene Perception

In viewing a natural scene, we are functionally constructing a three-dimensional world from a two-dimensional image. There are a number of features in a scene that we exploit to compile the layout and depth map of our perceived world (a summary and a discussion of these cues are available in Gibson 1950). Image generators are capable of generating most of these cues veridically, but two cues that are not universally available (or correct) are shadows and texture gradient.

Shadows are sometimes an important cue to determine the relative distance or depth of objects (such as determining wingtip clearance with other aircraft and structures in the gate environment). Not all image generators are capable of generating shadows, and the database rendering the characteristics of capable systems needs to be specifically programmed to provide shadows.

Another cue that is typically not veridical is texture gradients. Image generators employ a number of methods to prevent spatial sampling artefacts (called anti-aliasing). Using the most sophisticated, realistic anti-aliasing techniques also reduces the speed of rendering. Aliasing artefacts can reduce the realism of the visual scene, potentially reducing the immersion of the subject. Researchers sometimes need to consider the trade-offs between scene or database detail and artefacts.

8.3.2.2 Motion Cueing

One of the main difficulties in flight simulation is realistically simulating the motion of the aircraft. A real aircraft has virtually unlimited motion capabilities, whilst the simulator cabin is confined to a very limited motion envelope. The fidelity of simulator motion stimuli is dictated not just by the capabilities of the motion system hardware (size, configuration, actuator dynamics etc.), but also by the software algorithms that transform the accelerations from the aircraft dynamic model to simulator accelerations. The purpose of these motion algorithms is to keep the motion of the simulator within the limits of its motion envelope. Motion algorithms contain gains to reduce the overall magnitude of the aircraft motion and washout filters to attenuate translational and rotational motions at lower frequencies. The behaviour of the motion algorithms can be adjusted by changing the values of different parameters, such as the gains and filter cut-off frequencies.

8.3.2.2.1 Motion System Configurations

Many different motion system configurations exist. The most common configuration is the six-degree-of-freedom hexapod or steward platform (Figure 8.2). An advantage of this configuration is the compact design, but a disadvantage is the coupling between the degrees of freedom. This means that, for example, when the motion platform is pitching, it will not have its full roll capability. A common way to increase motion fidelity within the capabilities of the motion system is by exploiting the limits of human perception. Tilt coordination is used to simulate sustained accelerations in the longitudinal degree of freedom (for example during a take-off). By tilting the simulator forward or backward, the impression of longitudinal acceleration is given, as the human sensory system cannot distinguish between the gravity vector and specific force resulting from linear acceleration. In this case, it is important to rotate the simulator at a rate below the pilot's sensory threshold, as perceived rotations break down the illusion.

To facilitate studies that require larger motion that is more similar to real flight, research simulators have been designed in the last decade with novel motion configurations. For specific manoeuvres (e.g. upset recovery manoeuvres), the simulation of certain aircraft (e.g. military fighter jets)

FIGURE 8.2
SIMONA Research Simulator at Delft University of Technology.

or certain types of research (e.g. spatial disorientation research), different motion configurations can provide much higher levels of fidelity by being able to simulate sustained accelerations more effectively. One solution is to use centrifugal force to generate the perception of sustained acceleration, by combining the motion system capabilities from a human centrifuge with the capabilities from a hexapod motion base. If a subject is rotated continuously, the human semicircular canal system adapts and the centrifugal force generates the perception of a sustained acceleration.

The Desdemona research simulator (Figure 8.3) features a unique hybrid centrifuge motion system. Its geometrical dimensions give the motion system a large cylindrical motion space and a broad range of dynamic performance capabilities. The simulator has a gimballed cockpit mounted on a framework, which adds vertical motion. The framework is mounted on a lateral track attached to a rotating platform. The track allows the simulator cab to be positioned at different radii from the centre of rotation, allowing for sustained forces up to 3.5*g*. Desdemona was specifically designed for spatial disorientation training, advanced military flight simulation and research on human motion perception (Wentink et al. 2005).

Figure 8.4 depicts the CyberMotion Simulator at the Max Planck Institute (MPI) for Biological Cybernetics in Tuebingen, Germany (Teufel et al. 2007). This simulator is based on an industrial robot arm manufactured by KUKA GmbH. The robot arm has similar degrees of freedom to those of a human arm and has been modified for use as a real-time motion simulator. Rotation

FIGURE 8.3
Desdemona simulator developed by the Netherlands Organisation for Applied Scientific Research and AMST Systemtechnik GmbH.

FIGURE 8.4
CyberMotion Simulator at the MPI for Biological Cybernetics.

around the base axis is continuous through the use of slip rings. In addition, the robot arm is positioned on a linear track with a range of 9.88 m. All axes of the MPI CyberMotion Simulator are electrically driven. The robot arm can be equipped with a seat or an actuated cabin.

The eight degrees of freedom of the simulator are not coupled, and therefore, the motion envelope is extended compared to the hexapod motion platforms. Hence, the simulator can be used to move participants into positions that cannot be attained by such platforms. For example, the enclosed cabin can be positioned above the vertically extended robot arm, such that participants can be rotated along the vertical axis indefinitely, and differential thresholds of human subjects in yaw can be determined. In a different configuration of the simulator, participants can be positioned in any orientation in the enclosed cabin or the seat, even upside down, to investigate the influence of gravity on perception.

Other solutions include the VMS, the largest flight simulator in the world, which has large vertical and lateral motion tracks (discussed earlier and shown in Figure 8.1). The highest motion fidelity can be achieved in in-flight simulators, such as the Convair NC-131H Total In-Flight Simulator. These research aircraft are usually equipped with additional flight control surfaces to be able to simulate the motion dynamics of different types of aircraft.

The novel motion system designs discussed earlier might allow for better research into pilot training, behaviour and performance in manoeuvres that induce high-sustained accelerations, such as aircraft upsets. Pilots are significantly affected by these sustained accelerations and are often startled when first encountering them. In order to produce meaningful experimental results, the conditions of these high-sustained acceleration manoeuvres need to be replicated as closely as possible. In addition, these motion system designs allow for new opportunities in human perception and spatial orientation research in active and operationally relevant control tasks previously not possible. These novel motion configurations might not necessarily be adopted for future training simulators, as they are often very costly. However, certain aspects of their designs might be used in future research simulators or flight simulators for specific types of training, such as upset recovery training.

8.3.2.2.2 Motion Algorithms

As with motion system configurations, there are also many different types of motion algorithms – indeed, the type of motion system has a big influence on the selection of the motion algorithm. A commonly used motion algorithm in hexapod simulators is the classical washout algorithm developed by Reid and Nahon (1985). In most training simulators, adaptive motion algorithms are used where the value of certain algorithm parameters is changed based on the current simulator state. Most commonly, the motion filter gains are reduced as the simulator approaches the limits of the motion envelope.

Although such filters may increase the fidelity of the motion cues in certain respects – for example by better replicating the onset of the cue – it becomes more difficult to assess the overall motion fidelity, as these filters do not express constant behaviour over time.

Motion fidelity is not determined only by how closely the aircraft motion is replicated by the motion algorithm; the algorithms themselves may also introduce false cues, such as a motion cue in the simulator that is in the opposite direction to that in the aircraft or when none was expected in the aircraft. No objective motion criteria exist for flight simulators, and the parameters of motion algorithms are mainly tuned using subjective pilot evaluations. This often leads to inconsistent results, false motion cues and variations in motion fidelity between simulators of the same make and model. Recently, objective motion cueing criteria have been proposed to bring more standardisation of motion cueing in the training community and to get more insight into the variations in motion fidelity between different training simulators (Hosman and Advani 2013).

8.3.2.2.3 Research in Motion

For research, the motion of a simulator is often specifically tuned for the purpose of the experiment. If the research focuses on a specific manoeuvre with a dominant degree of freedom, the fidelity of motion cues in this degree of freedom can be optimised at the expense of motion fidelity in other degrees of freedom. For example if the manoeuvre is a straight-in approach and landing, motion fidelity in the vertical and pitch degrees of freedom can be optimised to simulate the sink rate and landing flare with higher fidelity.

The effects of flight simulator motion on pilot training and performance have been a hot topic for flight simulator research. Many studies have investigated the effects of variations in motion fidelity on pilot behaviour and performance (Schroeder and Grant 2010). These studies show that motion cues strongly affect pilot skill-based behaviour and performance, depending on the controlled aircraft dynamics and the structure of the task. Motion is more important in instances that require human lead equalisation (i.e. a response to rate information), for example when controlling unstable aircraft dynamics. Furthermore, motion effects are the biggest in compensatory tracking tasks, where the only visual information is on the error between the desired and actual state of the aircraft. When the compensatory task is more of a disturbance-rejection task, motion helps to increase performance, whilst in a target-following task, motion helps to increase stability margins. Motion also has been shown to improve performance in conditions with higher workload, such as when multiple axes (e.g. pitch and roll) are controlled at the same time. In addition, several studies directly compared control behaviour from real flight to behaviour in a simulator with reduced motion fidelity (Zaal et al. 2012). These studies showed that when simulator motion more closely resembles the real aircraft motion, simulator pilot behaviour and performance more closely resembles that of real flight. Studies also suggest that some components of the total aircraft motion are redundant and that others have a more negative effect on pilot performance (Zaal et al. 2009).

To date, there is no consensus on the benefits of motion for pilot training. Whilst the positive effects of motion on pilot performance in skill-based control tasks are well understood, it is far more difficult to determine the effects on training and transfer of training (de Winter et al. 2012).

8.3.2.3 Auditory Cueing

When flying an aircraft, pilots hear a wide range of sounds in the cockpit. In addition to radio communications between pilots and ATC, sound originates from aircraft systems and the interaction of the aircraft with the environment. Sound originating from systems in the cockpit, such as warnings and alarms, is usually relatively easy to replicate. However, ambient noise from aircraft systems such as the engines and the interaction of the aircraft with the environment is more difficult to replicate as it depends on the flight condition.

Sounds may be replicated in a flight simulator to increase the fidelity of the simulation. Depending on the research being performed, though, sound can also be an important cue that influences pilot performance. Discrete sounds, such as alarms or the sudden wind-down of an engine, prompt pilots to take appropriate actions. Changes in ambient noise from the engines or the slipstream can indicate changes in aircraft attitude and may increase performance when manually controlling the aircraft. Usually, a separate sound system is used to generate these sounds, taking inputs from other modules in the simulation, such as engine RPM from the engine module and slipstream magnitude from the dynamic model.

Relatively little research has been performed on the effects of auditory cues on pilot performance in a simulator. However, it has been shown that human operators can control equally well with auditory cues as with visual cues in a compensatory tracking task (Vinje and Pitkin 1971). This might suggest that continuous auditory cues (e.g. resulting from a changing pitch in engine noise or slipstream) can be helpful when manually controlling the aircraft in certain tasks.

8.3.3 Operational and Environmental Fidelity

Whilst there are endless combinations of operational and environmental variables, the researcher has the freedom to identify and define the factors most relevant to the research question. Often there are too many variables to incorporate into a single experiment and choices must be made. Once the key independent variables are identified, the appropriate level of simulator fidelity must be achieved so that a flight scenario effectively elicits a valid crew response. The following study is an example of how a lack of fidelity with respect to radio communication simulation may lead to false expectations of what a crew can achieve in actual operations.

8.3.3.1 Radio Communication Simulation

Lee (2001) conducted a study to determine whether the fidelity of radio communication simulation (RCS) would impact crew performance. Twelve current Boeing 747 aircrews flew a relatively challenging line operational scenario in which severe weather activity and nose gear malfunction would require crews to miss the approach, change runways and re-programme the flight management system amidst increased ATC communications. Presumably, the increase in workload during the approach and landing phases would present an opportunity to observe crew performance differences across conditions in a low-fidelity RCS flight deck environment compared to a high-fidelity RCS flight deck. In this study, low-fidelity RCS meant that radio communications would be spoken by the instructor/evaluator (as is often done in training) and high-fidelity RCS meant that radio communications would be simulated realistically by a pseudo-controller playing the role of ATC outside the cockpit.

As expected, pilot communications increased for the high-fidelity RCS crews. Whilst the overall frequency of radio communications did not differ significantly between groups, high-fidelity RCS crews communicated more than twice as often with ATC, primarily coordinating the missed approach procedures. Associated with these increased ATC communications was the increased time required to execute the missed approach, an average of 82.3 seconds compared to an average of 37.0 seconds for the low-fidelity RCS crews. Since the management of abnormal conditions and changes during approach and landing is critical, it would be misleading to think that the average missed approach time was only 37.0 seconds. In short, aspects of the simulation fidelity may have a systematic influence on performance metrics (in this case, time to execute a missed approach). Whilst it is known that crew process can be costly in terms of flight critical time, communication and coordination across teams can cost in terms of process, so the context for these processes must be simulated as realistically as feasible.

8.3.3.2 Designing Abnormal Operation Scenarios

Many simulation studies are interested in how crew performance is affected when normal operations are disrupted by some kind of anomaly (weather, traffic, aircraft malfunction, navigation changes etc.). Researchers design scenarios to be challenging for different reasons. Sometimes, it is merely to induce differences in crew performance. If scenarios are too easy, the manipulated variable might show no effect on overall performance because a significant workload threshold has not been reached. Other times, researchers design scenarios with malfunctions that are of specific interest (such as how a crew will respond with a loss of automation or multiple failures; e.g. Sherry and Mauro 2014). Obviously, the way in which malfunctions occur in the flight scenario must be realistic; the concomitant simulator displays and actions need to make sense so that performance differences can be induced and measured reliably. It is challenging to introduce a functional complexity failure into a simulator scenario and equally challenging to introduce realistic (yet surprising) combinations of subtle sensory cues, system alerts and outside communications in order to study how pilots monitor system states, detect and recognise when they have a problem.

8.4 Conclusions and Future Directions

Flight simulators are useful and versatile research tools, but because research objectives vary so greatly, it is not possible to develop a single set of requirements that will ensure that simulators will be effectively used. As the complexity and volume of traffic in our airspace increase, the need to simulate the presence and interactions of its component systems increases. In complex system studies, the coordination of the flight simulator with other simulator systems (such as ATC) may be necessary.

In the last decade, technology has developed to enhance flight simulation by integrating it with other system elements, thereby allowing the pilot to interact within the airspace system more realistically. Over a decade ago, one of the authors of this chapter predicted that one of three likely future trends in simulator systems would be the following:

> An emergence of networked simulator facilities, which allow pilots to engage in shared simulation environments. These networks should become increasingly tolerant of hosting a variety of simulator types and platforms. The military is already conducting networked simulations, but this technology will extend to civil transport aircraft, and involve modelling large segments of the national (and, ultimately, international) airspace. (Kaiser and Schroeder 2002, p. 469)

Yet even she did not anticipate how far networked simulations of integrated aviation systems would advance in such a short time. As stated earlier, the military was an early adopter of networked simulation. In fact, military research labs were already developing such systems. However, the information-technology communication infrastructure (or the lack thereof) proved a major obstacle towards development. Given the state of network communications, it was difficult to achieve reliable, high-volume transmission of data in real time. Further, graphical workstations that could support reasonably high-fidelity simulations were expensive, and much of existing graphical software was manufacturer specific. Zyda et al. (1992), who developed NPSNET for the US Navy, provided an insightful view of the challenge facing networked simulation in the 1990s.

In the civil aviation world, NASA Ames worked (under FAA funding) to network their existing high-fidelity flight and ground simulation facilities. The goal of this Virtual Airspace Simulation Technologies project was to demonstrate the feasibility of creating an integrated test bed that included the Boeing 747 and Advanced Concepts Flight Simulator cabs, the ATC simulators and Future Flight Central (a high-fidelity control tower simulator). Ultimately, the project was expanded to include the Ames Research Center VMS (which could add simulated rotorcraft to the airspace), as well as off-site simulation facilities (such as the FAA Technical Research Center). By the end of the project, the researchers had demonstrated the feasibility and utility of networked, distributed aviation simulation for evaluating HITL performance with current and next-generation ops concepts (Lehmer and Malsom 2004).

In parallel to these formal projects conducted by the Department of Defense, NASA, FAA and other large aviation organisations, several impressive grassroots networks began to emerge in the early years of the twenty-first century. Advances in computer technologies (and connectivity) made it feasible for individuals to acquire the hardware and software necessary to configure flight and ATC simulators of reasonable fidelity. Whilst individuals initially operated their home systems in an isolated mode (or occasionally on a one-to-one basis between pilot and controller), most realised that their experience was diminished by not being part of a larger airspace system. Artificially intelligent agents could only do so much in terms of adding additional traffic and controllers.

At the same time, the computer gaming industry was developing the software architecture and communication protocols to enable massively multiplayer online games. It was an obvious marriage of technologies, and soon online communities of aviation gamers developed. However, unlike most gaming communities, those interested in aviation simulation took their 'play' quite seriously. Two dominant online flight simulation networks emerged: the Virtual Air Traffic Simulation Network (VATSIM) and the International Virtual Aviation Organization (IVAO). Both VATSIM and IVAO have over 100,000 registered users worldwide. IVAO claims nearly 39,000 active users, consisting of both pilots and air traffic controllers, many of whom are active professionals. Users of VATSIM likewise include many active aviation professionals, as well as students wishing to acquire additional exposure to a realistic flight environment at minimal cost. A random visit to the VATSIM site found around 800 engaged users, including over 650 pilots, nearly 100 controllers and 10 observers.

Both the airspaces of IVAO and VATSIM are divided into regions, with protocols and regulations in each region reflective of its real-world counterpart. Impressively, these simulation networks have gained recognition from the FAA and other regulatory agencies as providing a useful environment for training and informal testing and observation. However, the very same aspects that lend strength to these networks as an organically grown simulation environment limit their utility as a controlled simulation system for research and development.

Ultimately, the aviation community benefits from having both institutionally controlled integrated simulation facilities and the more ad hoc (although still monitored) online simulation environments. Both perform the critical function of recreating the complexities and interactions of the modern airspace, evoking the emergent behaviour of both humans and the systems that serve them.

References

AAIB (Air Accident Investigation Branch) (1981). *An Experiment Designed to Measure Response Times of Pilots to a Locked Elevator Condition at Rotation Speed*. Report on the Accident to Bae HS 748 G-BEKF at Sumburgh Airport, Shetland Islands, on 31 July 1979, Appendix 5, Aircraft Accident Report 1/81. London: Department of Transport.

Casner, S. M., Geven, R. W., Recker, M., and Schooler, J. W. (2014). The retention of manual flying skills in the automated cockpit. *Human Factors*, 56, 1506–1516.

Chevalley, E., Parke, B., Lee, P., Omar, F., Yoo, H-S., Kraut, J., Rein-Weston, D., Bienert, N., Gonter, K., and Palmer, E. (2014). Decision support tools for climbing departure aircraft through arrival airspace, *33rd Digital Avionics System Conference*, Colorado Springs, CO, Oct. 2014.

de Winter, J. C. F., Dodou, D., and Mulder, M. (2012). Training effectiveness of whole body flight simulator motion: A comprehensive meta-analysis. *International Journal of Aviation Psychology*, 22, 164–183.

FAA (Federal Aviation Administration) (2014). *NextGen Implementation Plan*. Washington, DC: FAA.

Foster, J. V., Cunningham, K., Fremaux, C. M., Shah, G. H., Stewart, E. C., Rivers, R. A., Wilborn, J. E., and Gato, W. (2005). Dynamics modeling and simulation of large transport airplanes in upset conditions, *AIAA Guidance Navigation and Control Conference*, San Francisco, CA, Aug. 2005.

Gibson, J. J. (1950). *The Perception of the Visual World*. Boston, MA: Houghton Mifflin.

Gingras, D., and Ralston, J. (2008). Aerodynamics modeling for upset training, *AIAA Modeling and Simulation Technologies Conference and Exhibit*, Honolulu, HI, Aug. 2008.

Goldsmith, T. E., and Johnson, P. J. (2002). Assessing and improving evaluation of aircrew performance. *International Journal of Aviation Psychology*, 12, 223–240.

Hosman, R. J. A. W., and Advani, S. K. (2013). Are criteria for motion cueing and time delays possible? Part 2, *AIAA Modeling and Simulation Technologies Conference*, AIAA Paper 2013-4833, Aug. 2013.

Kaiser, M. K., and Schroeder, J. A. (2002). Flights of fancy: The art and science of flight simulation. In P. S. Tsang and M. A. Vidulich (Eds.), *Principles and Practice of Aviation Psychology*. Mahwah, NJ: Lawrence Erlbaum Associates.

Kanki, B. G., and Foushee, H. C. (1989). Communication as group process mediator of aircrew performance. *Aviation, Space and Environmental Medicine*, 60, 402–410.

Lee, A. T. (2001). *Radio Communications Simulation and Aircrew Training.* BRI-TR-130901, September 2001. http://www.betaresearch.com/BRITR130901ATCCommSim747 .pdf. Los Gatos, CA: Beta Research.

Lehmer, R., and Malsom, S. (2004). Distributed system architecture in VAST-RT for real-time airspace simulation. *AIAA Modeling and Simulation Technologies Conference and Exhibit*, Providence, RI, 16–19 Aug. 2004.

NTSB (1999). *Uncontrolled Descent and Collision with Terrain, USAir Flight 427, Boeing 737-300, N513AU, Near Aliquippa, Pennsylvania, September 8, 1994.* Aircraft Accident Report NTSB/AAR-99/01. Washington, DC: NTSB.

NTSB (2010). *Runway Side Excursion During Attempted Takeoff in Strong and Gusty Crosswind Conditions, Continental Airlines Flight 1404,* Boeing 737-500, N18611, Denver, Colorado, December 20, 2008. Aviation Accident Report/AAR-10/04. Washington, DC: NTSB.

O'Connor, P., Hormann, H-J., Flin, R., Lodge, M., Goeters, K-M., and the JARTEL Group. (2002). Developing a method for evaluating crew resource management skills: A European perspective. *International Journal of Aviation Psychology*, 12, 263–285.

Orasanu, J. (1993). Decision-making in the cockpit. In E. L. Wiener, B. G. Kanki and R. L. Helmreich (Eds.), *Cockpit Resource Management.* San Diego, CA: Academic Press.

Parke, B., Chevalley, E., Lee, P., Omar, F., Kraut, J. M., Gonter, K., Borade, A. et al. (2014). Coordination between sectors in shared airspace operations. *33rd Digital Avionics System Conference*, Colorado Springs, CO, Oct. 2014.

Prevot, T., Lee, P., Callentine, T., Mercer, J., Homola, J., Smith, N., and Palmer, E. (2010). Human-in-the-loop evaluation of NextGen concepts in the Airspace Operations Laboratory. In *Proceedings of the AIAA Modeling and Simulation Technologies Conference.* Toronto, Canada, Aug. 2010.

Reid, L. D., and Nahon, M. A. (1985). *Flight Simulation Motion-Base Drive Algorithms: Part 1: Developing and Testing the Equations.* Toronto, Canada: University of Toronto Institute for Aerospace Studies.

Ruffell Smith, H. P. (1979). *A Simulator Study of the Interaction of Pilot Workload with Errors, Vigilance, and Decisions.* NASA TM-78483. Moffett Field, CA: NASA Ames Research Center.

Schroeder, J. A., and Grant, P. R. (2010). Pilot behavioral observations in motion flight simulation, *AIAA Guidance, Navigation, and Control Conference and Exhibit*, Toronto, Canada, 2010.

Schroeder, J. A. (2012). Research and technology in support of upset prevention and recovery training, *AIAA Modeling and Simulation Technologies Conference*, Minneapolis, MI, Aug. 2012.

Schroeder, J. A., Burki-Cohen, J., Shikany, D. A., Gingras, D. R., and Desrochers, P. (2014). An evaluation of several stall models for commercial transport training, *AIAA Modeling and Simulation Technologies Conference*, National Harbor, MA, Jan. 2014.

Seamster, T. L., Edens, E. S., and Holt, R. W. (1995). Scenario event sets and the reliability of CRM assessment. In *Proceedings of the 8th International Symposium on Aviation Psychology*, 613–618. Columbus, OH.

Seamster, T. L., and Kanki, B. G. (2007). Beyond electronic flight bag approval: Improving crew performance. In *Proceedings of the Fourteenth International Symposium on Aviation Psychology.* Dayton, OH: Wright State University.

Sherry, L., and Mauro, R. (2014). Controlled Flight Into Stall (CFIS): Functional complexity failures and automation surprises. *IEEE Proceedings of the Integrated Communications Navigation and Surveillance (ICNS) Conference*, Herndon, VA, D1-1–D1-11.

Sweet, B. T., and Hebert, T. M. (2007). The impact of motion-induced blur on out-the-window visual system performance, *IMAGE Conference*, Scottsdale, AZ, Jul. 2007.

Sweet, B. T., and Giovannetti, D. P. (2008). Design of an eye limiting resolution visual system using commercial-off-the-shelf equipment, *AIAA Modeling and Simulation Technologies Conference and Exhibit*, Honolulu, HI, Aug. 2008.

Teufel, H. J., Nusseck, H.-G., Beykirch, K. A., Bulter, J. S., Kerger, M., and Bülthoff, H. H. (2007). MPI motion simulator: Development and analysis of a novel motion simulator, *AIAA Modeling and Simulation Technologies Conference and Exhibit*, Hilton Head, SC, 2007.

Vinje, E. W., and Pitkin, E. T. (1971). Human operator dynamics for aural compensatory tracking, *Seventh Annual Conference on Manual Control*, 339–348. University of Southern California, Los Angeles, CA, 2–4 June 1971.

Wentink, M., Bles, W., and Hosman, R. J. A. W. (2005). Design and evaluation of spherical washout algorithm for Desdemona simulator, *AIAA Modeling and Simulation Technologies Conference and Exhibit*, San Francisco, CA, 2005.

Wing, D., Prevot, T., Lewis, T., Martin, L., Johnson, S., Cabrall, C., Commo, S. et al. (2013). Pilot and controller evaluations of separation function allocation in air traffic management. In *Proceedings of the Tenth USA/Europe Air Traffic Management Research and Development Seminar*, Chicago, IL.

Winterbottom, M. D. (2004). An integrated procedure for measuring the spatial and temporal resolution of visual displays, *Interservice/Industry Training, Simulation, and Education Conference*. Orlando, FL, 6–9 Dec. 2004.

Zaal, P. M. T., Pool, D. M., de Bruin, J., Mulder, M., and van Paassen, M. M. (2009). Use of pitch and heave motion cues in a pitch control task. *Journal of Guidance, Control, and Dynamics*, 32, 366–377.

Zaal, P. M. T., Pool, D. M., van Paassen, M. M., and Mulder, M. (2012). Comparing multimodal pilot pitch control behavior between simulated and real flight. *Journal of Guidance, Control, and Dynamics*, 35, 1456–1471.

Zyda, M. J., Pratt, D. R., Monahan, J. G., and Wilson, K. P. (1992). NPSNET: Constructing a 3D virtual world. In *Computer Graphics, Special Issue on the 1992 Symposium on Interactive 3D Graphics*, MIT Media Laboratory, New York, 29 March–1 April 1992, 147–156.

9

Flight Training

John Huddlestone and Don Harris

CONTENTS

9.1 Simulation and Flight Training

This chapter takes a look at the history of pilot training in simulators of all types, followed by a review of current industry standards and guidelines. Finally, it discusses the effectiveness of simulator training and concludes with an overview of the methods for assessing the transfer of training from simulators to flight.

From the outset, it has to be emphasised that flight simulation *does not* provide training. Put an untrained person in a simulator and no matter how long they spend in there, they will not emerge as a competent pilot. Flight simulation *supports* the training process, but simulation facilities can only be specified after the training objectives have been set, not vice versa.

In commercial aviation, the training process, design and operational requirements for flight simulators are heavily regulated.* Military requirements are not so highly regulated; nevertheless, training is still tightly specified and simulators are subject to detailed specification and oversight. The range of flight training devices (FTDs) used in military aviation is also greater, stemming from the wider variety of missions undertaken by pilots in the armed services. The focus in this book is on civil simulation, but it is worth noting one more key difference between civil and military aviations: with the exception of operations, military crews spend most of their life training in one form or another. In civil operations, training costs time and money: it is a large financial overhead which takes pilots away from revenue-producing operations. Crews in simulators are not generating money; hence, training and simulation also need to be cost-effective.

However, flight simulators are more than just simply about providing initial training. For a professional pilot, recurrent training is required as is the demonstration of competence through mandated regular checks.

9.2 Early History of Flight Simulation

From the start of manned flight, there has been a requirement for training devices. At the beginning of the twentieth century, learning to fly was

---

* See the regulations in *Code of Federal Regulations*, Title 14, Part 60 ('Flight Simulation Training Device Initial and Continuing Qualification and Use') and Part 61 ('Certification: Pilots, Flight Instructors, and Ground Instructors') and JAR-FCL 1 (*Joint Airworthiness Requirement – Flight Crew Licensing – Aeroplanes*) (JAA 1999).

dangerous, especially as there were no training aircraft. Haward (1910) pointed out the following:

> The invention therefore, of a device which will enable the novice to obtain a clear conception of the workings of the controls of an aeroplane, and of conditions existent in the air, without the risk personally or otherwise, is to be welcomed without a doubt. (p. 1006)

The first training devices produced were essentially de-rated aircraft, with low-powered engines so that rudder control could be practised without becoming airborne. A Bleriot monoplane with reduced wingspan was used at the French École de Combat enabling pilots to get used to the feel of the controls whilst taxiing. The Sander's Teacher was an aircraft mounted on a universal joint. When it was pointed into a wind of sufficient strength, the controls acted in the same way as when airborne, but the vagaries of the weather limited its utility. However, these devices could better be described as ground-based aeroplanes, rather than flight simulators.

One of the first devices that can be called a truly synthetic FTD was the 1910 Antoinette trainer. Simulator training at this time focussed on the development of psychomotor skills for aircraft handling. This device comprised two half-barrels mounted on top of each other. The trainee used representative flight controls to line up a reference bar with the horizon in response to disturbances caused by instructor inputs. The significance was that there was now a marked difference between an aircraft and the training apparatus: the device bore little resemblance to an aircraft except for features being replicated for a specific training goal (i.e. the use of controls and their effects on aircraft motion). It had limited but essential realism.

It was the development of instrument flying that saw the widespread adoption of flight simulation. Early instrument flying was dangerous, and 'controlled flight into the ground' was common. In 1934, when the US Army Air Corps was ordered to fly the airmail, five pilots were killed in the first few days of providing the service from a lack of experience of flying at night and in poor weather. Some years earlier, Edwin Link had developed a trainer using the pneumatic components from organ mechanisms to provide control actuation. His initial attempts to sell the trainer to the military failed as the US military considered them to be unaffordable. In response to these accidents, Edwin Link added functional flight instruments and a magnetic compass to his initial design, and after further demonstrations, the Air Corps duly purchased six Link Trainers.

The requirement to teach instrument flying had an important effect on simulator design. The instructor now needed a display of the student's position relative to the track that they were required to fly. Consequently, instructors' stations also had to be enhanced. In the Link Trainer, a course plotter was added, which depicted the student's track on a chart using an inked wheel. The instructor could then also control the transmission of simulated radio beacon signals to the Link Trainer. This chart enabled the instructor to provide high-quality performance feedback to the student, thereby aiding their learning. The inclusion of navigational aids also represented another important development, with the integration of elements of the environment into the simulation. What was also notable about this initial use of simulators was that the training delivered was concerned with the procedural and cognitive skills of radio navigation, rather than inculcating psychomotor flight skills per se.

World War II brought about an unprecedented expansion in aviation along with significant advances in technology. This had a significant impact on the development of flying training devices. The volume of training led to an upsurge in the sales of training devices with over 7000 Link Trainers ('Blue Boxes') being sold to the US military alone. The Blue Box became a part of most military pilots' training. Furthermore, during WWII, a diverse range of simulation requirements began to emerge encompassing the need for both procedural and multicrew training.

Silloth Trainers, developed by the Royal Air Force, were developed to meet the requirement for procedural training. They were constructed from an actual Halifax bomber cockpit, using simulated engine, electrical and hydraulic systems with controls and indicators driven by pneumatic systems operating in the normal manner. An instructor station was provided for crew monitoring and the insertion of system malfunctions by an instructor. These devices partially simulated the aircraft, containing only the elements required for the tasks to be trained. However, the provision of such devices removed the requirement for aircraft to be used for procedural training. Devices similar in concept are still heavily used today (usually described as cockpit procedure trainers), albeit using modern technology.

One early device developed for multicrew training was the Type 19 trainer used for training the crews of radar-equipped night fighters. It was a fixed-base device with positions for the pilot, air intercept controller and instructor. It had computers for the simulation of attacking aircraft and the calculation of aircraft positions, a visual projection unit and course plotters. An image displayed on a hemispherical cyclorama mounted in front of the pilot showed the night sky and ground and a tail silhouette of a target which moved in the correct manner relative to the fighter.

9.3 Post-War Developments

The development of simulators during WWII was clearly driven by the training requirement; however, subsequent developments were largely led by technological advancements, in particular advancements in digital computing and display technology. These advances facilitated the development of ever more complex training solutions such as the full-flight simulator with full visuals and a six-axis motion platform. Until relatively recently, the training requirement per se was almost secondary.

One of the most hotly debated topics has been the requirement for motion systems in flight simulation. Adorian et al. (1979) noted that early motion systems were rejected by users in the post-war period because they did not fly like the real aircraft. In addition, the common customer perception was that pilots should fly by their instruments rather than the 'seat of their pants'. It was not until the late 1950s that airlines started to purchase motion systems again, when analogue computer technology had developed to the point where motion could be modelled more accurately. Nowadays, six-axis motion bases are the norm for airline simulators; indeed, the Civil Aviation Authority (CAA), European Aviation Safety Agency (EASA) and FAA mandate them for zero-flight time simulators. However, for military fast jet use, the range and speed of motion bases are inadequate. In modern fast jet simulators, motion cues are often provided using an 'active seat', where techniques such as the use of inflatable panels to apply pressure to the pilot's body are employed to give the impression of acceleration and movement.

Visual systems depicting the outside world have developed considerably. The first step was the introduction of television technology. Closed-circuit TV cameras, used in conjunction with scale models of the ground, provided visual scenes that were displayed through the simulator cockpit windows. Increasingly sophisticated computer graphics have now superseded model boards and cameras. Furthermore, commercial pilots spend the bulk of their time flying straight and level on instruments. It is only when they approach and land at an airport that they fly using visual references. Consequently, a commercial aircraft simulator will have detailed models of just a selection of airports, often with highly accurate models of their layout, but little else. To get appropriate cues, the level of scene detail is particularly significant, as is a wide field of view, both vertically and horizontally.

Figure 9.1, adapted from a model developed by Rolfe (1985), shows the major components of a modern flight simulator. All modern simulators use digital computers, with software programmes controlling the behaviour of the system. Figure 9.1 is split into two parts showing the key hardware and software components. The hardware typically consists of a student station, an instructor station and the computer(s). The student station can be as simple as a computer terminal on a desk or may be as complex as a fully enclosed simulacrum of a flight deck.

The visual system provides a view of the outside world. It may take the form of a wrap-around display system offering the same field of view as in the aircraft, or it may be restricted to a monitor offering a more limited field of view. There may be no visual system at all in the case of a procedure trainer. There will certainly be displays and controls, which the student

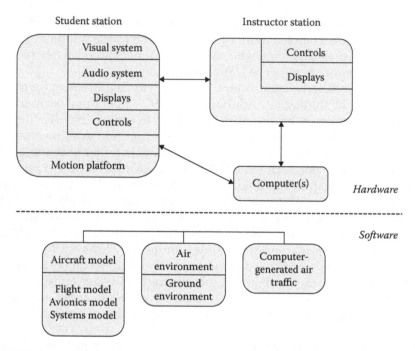

FIGURE 9.1
Structure of a modern flight simulator. (Adapted from Rolfe, J. M., *Fidelity: An evaluation of the concept and the implications for the design, procurement, operation and evaluation of flight simulators,* Unpublished Defence Fellowship thesis, Wolfson College, Cambridge, UK, 1985.)

uses to operate the simulation. However, these can range from using actual aircraft-style components to simple virtual representations on a computer.

An audio system provides the necessary auditory environment that may include such elements as warnings, engine noise, ground roll 'rumble' and intercom. Finally, the whole simulation system (including the instructor station providing the facilities to control the simulation and view the scenario) may be mounted on a motion platform to give the sensation of motion. Instructor displays may include replications of the flight deck displays, as well as information concerning simulator settings and alternative views such as a 'God's-eye' view of the aircraft and its environment.

The software system structure typically consists of a number of models, implemented as programme components. The central model is that of the aircraft itself handling the performance of the aircraft as well as the various aircraft subsystems. Separate software models provide the environment, such as the visual scene, weather effects and other external entities such as navigation beacons. The position, altitude and heading of the aircraft in its simulated world are calculated and are reflected on the navigation instrumentation. There may even be interactions with other simulated aircraft or ground traffics.

9.4 Advantages of Using Simulators

The advantages of using simulators as a training device are manifold (see also Chapter 1). Rolfe and Staples (1986) listed several broad areas of advantage of this training medium:

- *Increased efficiency* – Training is not affected by availability constraints; it is also possible to repeat manoeuvres without the time and cost of re-positioning. If there is the requirement to train in a wide range of geographical areas using different airports, a simulator provides all these environments at the one location. Night-time conditions can be made available at any time of day.

- *Increased safety* – Manoeuvres may be undertaken that would be unwise to perform in flight unnecessarily; non-normal and emergency procedures, including system failures, may also be trained. Airline pilots are required to practise their response to emergencies as part of both initial and annual refresher trainings.

- *Lower training costs* – Flight simulators are cheaper to purchase and operate than large commercial aircraft. Such a requirement for cost minimisation has led directly to the development of zero flight time training for converting qualified pilots onto new aircraft types. Annual instrument rating tests are also typically conducted in simulators for the same reason.

- *Reduction in environmental impact* – Simulators provide a reduction in emissions of greenhouse gases. Other environmental pressures also come into play; for example night flying around airfields can be unacceptable due to the noise pollution.

- *Training effectiveness* – Simulation allows the trainer and trainee to log, replay and review performance, providing almost instantaneous feedback. Simulators also permit the employment of a richer variety of instructional strategies. The facility to freeze a simulation allows the instructor to intervene at critical points in an exercise and review the situation with the students rather than having to wait until the end, resulting in a significant reduction in non-productive time. Student progression may also be logged.

9.5 Current Issues in Flight Simulation

The original driver for the development of flight simulators was safety. This is still significant today. Initial use of flight simulation concentrated upon the development of the technical flight skills associated with the control and management of the aircraft and its systems. However, cost saving has become an ever-increasing factor with the desire to move as much training as possible to simulators for this reason.

Civil flight simulation, especially in the area of conversion training, has matured. Albett (1989) stated the following:

> The technical fidelity of simulators for the large transport aircraft has in general gone as far as is necessary to fulfil their role as a replacement for training on the real aircraft. The task for manufacturers is to produce the same product at a reduced cost. (p. 21.7)

Andrews, Carroll and Bell (1995) observed that in the past, simulators with limited or 'selective' fidelity were specified partly as a result of the technological difficulties of achieving high fidelity. They argued that analytical methods based on cognitive theory needed to be developed to facilitate the identification of important cues required for a given training task. Furthermore, they suggested that the appropriate fidelity will be based on 'matching the level of fidelity of the simulator to the ability of the trainee to absorb stimuli' (p. 31) rather than on technological characteristics of the device. This can be regarded as a 'human centric', training-based approach to flight simulation rather than one driven by the available technology.

9.5.1 Training Requirements

A great many of the competencies required by the airworthiness authorities in professional pilot licensing relate to flight skills and technical knowledge. These requirements have to be matched to specific capabilities if the flight simulator is to be accredited for training. However, there are changes in the content of professional pilot training. Airlines are being given the opportunity to specify and deliver aspects of training better matched to the specific requirements of their operation. In the United States, under the Advance Qualification Program (AQP*), the process of designing and delivering training is as important as the competence of the pilots that it produces. The emphasis in the AQP is away from time-based training requirements to fleet-specific, proficiency-based requirements. Training needs are identified and developed directly from line operational requirements. The applicant (not the regulatory authority) develops a set of proficiency objectives based upon that airline's particular requirements. The AQP is based upon a rigorous task analysis of operations with emphasis firmly placed upon the cognitive aspects of the flight task, such as decision-making or the management of automation (see Seamster et al. 1998).

---

* See Advisory Circular AC 120-54 (FAA 1991a).

The first step in the AQP requires a task analysis covering the full range of conditions internal and external to the aircraft. From this breakdown, the required skills, knowledge and abilities are elicited, including the applicable performance standards. These may be generic skills or flight phase independent. The task analysis output can then be developed into a curriculum based upon a hierarchical series of enabling objectives to prepare crews for subsequent training in an operational flight deck environment. Individual training objectives are subsequently allocated to various curriculum components to construct the syllabus.

The development of all training proceeds through a series of generic stages. The Interservice Procedures for Instructional Systems Development (IPISD) has become a benchmark approach for training design, although it was originally intended for training design in larger systems than those on the flight deck (Branson et al. 1975). The IPISD has five major functional phases:

- *Analyse* commences with a training needs analysis (TNA), which can identify performance gaps which can be addressed via training. All tasks that need to be undertaken are identified and described along with their required performance standards. They should also be categorised in terms of their difficulty, importance and frequency. Tasks need to be categorised to determine their most suitable instructional setting. An option analysis should review potential alternatives for training methods and the media required by these different options (i.e. in this case, the potential for training to utilise a flight simulator).

- *Design* develops the output from the TNA into an instructional design. Each task selected for training needs a learning objective (i.e. what the trainee should be able to accomplish at the end of this instructional step), which is then subject to further decomposition and analysis to identify the steps necessary for their mastery. Tests are developed to examine each learning objective.

- *Develop* classifies the learning objectives to identify and develop guidelines for optimum learning. The media selection process is also undertaken during this phase, which involves identification of the type of simulation that should be used. However, it should also be noted that simulation is not always the best answer. The best answer should also take into account factors such as the trainee characteristics (e.g. what stage of training are they at?), how many trainees there are, what are the required/available simulation facilities, whether there are other criteria relating to the setting of the training (e.g. distance) and whether cost is a major factor. On the basis of these considerations, the instructional materials should be developed and tested.

- *Implement* trains the personnel delivering the course in its content, methods for evaluation and the instructional management system.

Opinions concerning facilities and future revisions to the syllabus and/or its method of delivery should be collected. Finally, the training should actually be delivered.

- *Control* evaluates the training syllabus against targets and, if required, revises it. Internal evaluation involves the analysis of trainee performance; external evaluation requires an assessment of performance on the job to determine if the training is producing graduates of the required standard. Evaluation data are used as a quality control measure to assess and revise the training.

The AQP encourages applicants to integrate the use of advanced flight training equipment, including full-flight simulators, into pilot training. It suggests that airlines should use appropriate technology for curriculum delivery at a given stage of the process based upon the TNA, not necessarily the highest level of technology (or fidelity). It has been demonstrated that CRM (non-technical) skills can be taught successfully using a desktop PC running a copy of Microsoft Flight Simulator. The physical fidelity of the simulation did not matter: what mattered was the complexity of the crew task and the cognitive and communication demands placed upon the trainees (Johnston et al. 1997). By adopting such an approach, airlines can significantly reduce the need for use of a full simulator in favour of lower-cost training facilities.

9.5.2 Regulatory Categories of Civil Flight Simulation Training Devices

In an airline, full-flight simulators are supplemented by cockpit procedure trainers, part-task trainers, computer-based training facilities for aircraft systems and procedures and 'regular' classroom facilities (see Farmer et al. 1999; Moroney and Moroney 1999; or Kaiser and Schroeder 2003). A cost-effective and successful training programme uses the appropriate training aids in the right way.

In Europe, flight simulation training devices (FSTDs) fall into four major regulatory categories:

- Full-flight simulator (FFS)
- FTD
- Flight navigation procedure trainer (FNPT)
- Basic instrument training device (BITD)

The FAA and EASA categorise FFSs into four further subcategories based upon their capabilities (FAA 1991b [AC 120-40B, *Airplane Simulator Qualification*]; JAA 2008 [*JAR-FSTD A: Aeroplane Flight Simulation Training Devices*]). All FFSs *must* simulate a specific type and variant of an aircraft.

Level A FFSs have the lowest level of complexity. They must have an enclosed, full-scale replica of the flight deck including simulation of all systems, instruments, navigational equipment, communications, caution and warning systems. The control forces and displacement must replicate the aircraft simulated and should respond in the same manner. However, only basic motion, visual and sound systems are required. The visual system must provide a field of view of at least 45 degrees in the horizontal and 30 degrees vertical (per pilot). Control lag must not be greater than 300 milliseconds more than that experienced on the aircraft simulated. In addition to the all requirements for level A, a level B device must use validated test data for flight performance and system characteristics. It is also required to include representative ground handling and aerodynamic effects.

A level C FFS must have a visual system providing each pilot with a collimated field of view of at least 180 degrees in the horizontal and 40 degrees vertical, capable of displaying daylight, twilight and a night visual scene. It also requires a six-axis motion platform and wind shear simulation. A sound simulation is required that can represent environmental and significant aircraft noises. The response to control inputs must not be more than 150 milliseconds greater than that in the aircraft simulated. The requirements for a level D simulator are almost identical to those of a level C device; however, it must also have complete fidelity of all sounds and incorporate motion buffeting.

The capability of the simulator determines the training and licensing functions that may be undertaken on it. A level A FFS JAR-FSTD A (JAA 2008) permits only procedure training; instrument flight training; transition/conversion training; testing and checking (excluding take-off and landing manoeuvres); and recurrent training, testing and checking (type and instrument rating renewal/revalidation). A level B simulator also permits pilots to maintain recency of experience (three take-offs and landings in 90 days) and to undertake type conversion training for take-off and landing manoeuvres. Aircraft-type conversion testing and checking may be undertaken in a level B FFS, except for take-off and landing manoeuvres. Level C simulators allow for testing and checking take-offs and landings for experienced flight crew. In addition, a level D simulator may also be used for zero flight time conversions from one aircraft type to another for experienced pilots.

FTDs are full-sized replicas of a specific type of aircraft flight deck. For EASA level 1 approval, such a device may be either open or closed. It does not require either a force cueing system for the controls or a visual system, but it must have the flight deck equipment and simulation capability to represent fully at least one system. Level 1 FTDs can only be used as part of an approved training course for the systems represented. A level 2 FTD has all flight deck systems fully represented in a closed cockpit and includes a navigation database to support the aeroplane systems and the training requirement. The aircraft flight dynamics and primary flight controls need only be representative of that class of aircraft.

A type I FNPT has a generic representation of the flight deck of a certain class of aircraft. Type II FNPTs have more capability. They also incorporate a limited visual system providing an out-of-the-window view. Both simulate the flight deck systems of a generic multiengined aircraft. A BITD represents only part of the pilot's position representative of a certain class of aeroplane. It may use facsimiles of aircraft instruments on a computer screen and spring-loaded flight controls as it provides a training platform for only the procedural aspects of instrument flight. The UK CAA suggests that five-hour BITD credit may be allowed towards initial private pilot licence training and up to five hours of credit against the instrument flying portion for a commercial licence. However, BITD cannot provide credit towards the requirements for an instrument rating (*CAA Standards Document 18*; CAA 2003).

The US FAA uses a slightly different system of designation and categorisation for FTDs that are not FFSs based upon *ICAO Document 9625 'Manual of Criteria for the Qualification of Flight Simulation Training – 3rd Edition'* (International Civil Aviation Organization 2009). The FAA categorises all lower-level flight simulators as FTDs. There are seven categories (although at the moment level 1 is not used). Level 2 and 3 FTDs are generic devices (they do not simulate a specific aircraft type), whereas levels 4–7 are type specific. However, the basic principles to those used in Europe are identical.

9.6 Evaluation of Simulator Training Effectiveness: Transfer of Training and Cost-Effectiveness

Training in any simulator should transfer positively to the real environment. Positive transfer occurs when performance in the aircraft improves as a result of training. Negative transfer occurs when the training delivered actually degrades performance in the live environment. The reason why the issue of simulator fidelity is of interest is as much one of cost as it is of training effectiveness. Lower-level training devices are much cheaper to purchase and operate than a FFS. As a result, many arguments relate to using the *appropriate* technology for curriculum delivery based upon particular training requirements; this may not always be a FFS.

Thorndike (1903) provided one of the earliest theoretical accounts of transfer. He proposed that transfer would occur between two tasks if they contained identical elements. His position was extreme in that he stated that transfer would only occur if the match was exact. The implication of this is that the number of identical elements between the two tasks should be maximised; consequently, the simulation should be as realistic as possible if transfer of training is to occur. This has often been used as the justification for the development of highly realistic, 'high-fidelity' simulators. Osgood (1949) suggested that the greater the correspondence between the stimuli presented in the training environment and in the live environment, the greater the transfer of training will be (the 'transfer surface' concept). This model allows for a continuum of similarity between the stimuli presented in the synthetic environment and those in the live environment. It additionally looks at the corresponding response demands of the real and virtual environments. The underlying principle is that more fidelity equates to more transfer: maximum positive transfer should occur when the stimuli in the two environments are identical, and hence, the required responses are identical. However, the significance of Osgood's model was that it suggested that transfer could still occur even if the training environment and live environment were *similar* but not identical. Nevertheless, there has been a push for greater and greater fidelity in flight simulation, which some argue has led to the development of ground-based aeroplanes rather than training devices.

The training effectiveness is determined by the degree to which conducting training in a simulator has the desired effect on the performance of trainees in the real flight environment. AGARD (1980) identified two essential criteria for such an evaluation. The first criterion is training transfer, established by evaluating the degree to which performance is improved by training in the simulator. The second is cost-effectiveness, which is concerned with determining if desired performance can be achieved more quickly or by utilising a less expensive training resource than the actual aircraft. The same training effectiveness may also be achieved with greater cost-effectiveness if more hours of training are conducted in a cheaper training resource than the aircraft.

There are a number of methods for evaluating simulators. Caro (1977) provided a comprehensive list of these models, composed of two distinct sets. Analytical models are based on a static analysis of the simulator properties. The second set is based upon performance evaluations conducted in the simulator.

9.6.1 Analytical Models

Three main types of analytical models may be used for evaluating simulators (see Table 9.1). These are relatively easy to implement as they do not involve complex trials, but they do have a number of weaknesses.

TABLE 9.1

Analytical Models for Simulator Evaluation

Model	Method
Simulator fidelity	Evaluation of the fidelity of the simulator compared with the real aircraft
Simulator programme analysis	Evaluation of the suitability of the training programme
Opinion surveys	Evaluation of the opinions of operators, instructors, training specialists and students of the effectiveness of the simulator

Source: AGARD, *Fidelity of Simulation for Pilot Training*, NATO/OTAN, Neuilly-sur-Seine, France, 1980.

9.6.1.1 Simulator Fidelity

This model is based on the notion that the greater the fidelity of the simulator, the greater the effectiveness of the training conducted in it (see Osgood 1949). It may identify if some critical aspect of fidelity is missing. However, on their own, the results obtained do not provide evidence that training in the simulator will be effective. Furthermore, fidelity is not a unidimensional construct so it is difficult to describe any simulator as being of either low or high fidelity. Liu et al. (2009b) described two major dimensions: physical fidelity and psychological–cognitive fidelity. These dimensions themselves have subdimensions to them.

- *Physical fidelity* – The degree to which a simulator looks, sounds and feels like the environment that it replicates (Allen 1986). This comprises the following:
  - Visual–audio fidelity
  - Equipment fidelity
  - Motion fidelity
- *Psychological–cognitive fidelity* – The degree that a simulator replicates factors such as workload, communication and situation awareness and demands the same degree of engagement as when flying the real aircraft. This dimension comprises the following:
  - Task fidelity
  - Functional fidelity (how well the simulator reacts to inputs [Allen 1986])

To these dimensions, Brooks and Arthur (1997) added that of 'motivational fidelity', the degree to which a training device engages the user and motivates them when using it.

Gross (1999), in a report from the Simulation Interoperability Standards Organization Fidelity Implementation Study Group, outlines the parameters by which fidelity can be described and quantified. These may be applied judiciously to some of the parameters described earlier:

- *Accuracy* – the degree to which a parameter within a simulation conforms to reality
- *Capacity* – the number of instances of an object represented by the simulator
- *Error* – the difference between simulated and real-life values
- *Fitness* – the capabilities required for a function or application
- *Precision* – a measure of the rigour with which the computational processes are described and/or performed in the simulation

- *Resolution* – the degree of detail used to represent aspects of the real world
- *Tolerance* – the maximum permissible error between the maximum and minimum allowable values in any component in the simulation
- *Validity* – the quality that determines if an aspect of the simulation is rigorous enough for acceptance for a specific use

9.6.1.2 Simulator Programme Analysis

In contrast to the dimensions described previously, other authors suggested that focussing on the physical aspects of a simulator was not an appropriate way of assessing fidelity. Steurs et al. (2004) argued that greater emphasis should be placed upon the psychological–cognitive aspects. They suggested that a better way to measure fidelity was to quantify the extent to which a simulator induced pilot control behaviours comparable to those observed in the real aircraft. This aspect is considered in the simulator programme analysis model described by AGARD (1980), which was concerned principally with the instructional overlay. Whilst this aspect merits consideration as it may identify weaknesses in the instructional strategy utilised or in the provision of facilities for instructor control of the simulation, it cannot give a direct indication of the degree to which the simulator will be effective.

9.6.1.3 Opinion Surveys

The different groups of user are asked for their opinions of the suitability of the simulator. One of the difficulties with this approach is the validity of the responses elicited. In the context of trainees' opinions, Salas et al. (1998) pointed out the following:

> This provides a favourable evaluation of high-fidelity simulation because trainees like the 'bells and whistles' of high-fidelity simulation; consequently, high-fidelity training appears to be highly effective for training. (p. 203)

Notwithstanding these weaknesses, though, user opinion data are worthwhile, as underlined by Bell and Waag (1998):

> Although hardly sufficient to establish the value of simulator-based training, we believe that positive user opinion is, nonetheless, a necessary condition for the acceptance of a simulator. User acceptance is a necessary first step in obtaining the support necessary to conduct more rigorous evaluations of simulator-based training. (p. 225)

These analytical models can provide some indication of potential weaknesses in a simulator and may be useful as diagnostic tools if it does not appear to deliver training as effectively as anticipated. However, as a result of the subjective nature of the judgements made and the related issues of validity and reliability, they are of limited use in determining the effectiveness of a simulator in actually providing pilot training. Furthermore, they do not provide clear evidence as to the degree to which a simulator meets the criteria of training transfer and cost-effectiveness. Ultimately, trainee performance has to be evaluated.

9.6.2 Performance Evaluation Models

Within a civil aviation context, there are really no issues concerning the ultimate training effectiveness of the highest-fidelity full-flight simulators. It is almost inconceivable that airlines will even consider recommencing pilot training in the aircraft itself. The main concerns usually revolve around establishing if the required level of trainee performance can be achieved more quickly or with use of a less expensive training resource. There is evidence that many cognitive tasks (as opposed to psychomotor tasks) can be trained very successfully using very low-fidelity simulation (training) systems. For example Pfeiffer et al. (1991) found that instrument training on a low-fidelity training device transferred positively to performance in instrument flight. Schiewe (1995) and Johnston et al. (1997) both observed that low-level simulation using a PC-based flight simulator could be used to develop CRM skills.

Figure 9.2 provides a summary of the various performance evaluation models. The first model is the classic transfer-of-training model. All the remaining models (except the simulator performance improvement model) are variations of the transfer-of-training model, designed to overcome various resource constraints.

9.6.2.1 Transfer-of-Training Model

In the pure transfer-of-training model, matched control and experimental groups are trained simultaneously in different environments and their performance is assessed in the aircraft itself. Ideally, the control group is trained on the aircraft and the experimental group is trained in the simulator under evaluation. The relative performance of the two groups gives a comparative measure of the effectiveness of the simulator compared with the aircraft. The difference in performance is attributed to the degree of transfer from the training environment to the aircraft.

True transfer-of-training studies are very rare in aviation human factors. This is principally associated with the cost of undertaking such work, its complexity and a lack of perceived necessity (airlines are not about to give up using their expensive FFSs). In a variation on the full transfer-of-training design, the control group is trained on an existing simulator system rather than the aircraft. This yields data on the effectiveness of the new simulator compared with the old simulator rather than the aircraft. Nevertheless, implementation of this latter variation on the transfer model provides a

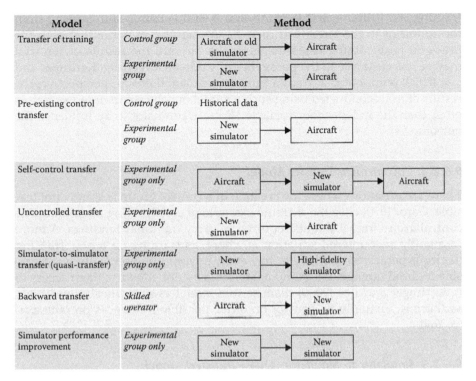

Model		Method

Model		Method
Transfer of training	*Control group*	Aircraft or old simulator → Aircraft
	Experimental group	New simulator → Aircraft
Pre-existing control transfer	*Control group*	Historical data
	Experimental group	New simulator → Aircraft
Self-control transfer	*Experimental group only*	Aircraft → New simulator → Aircraft
Uncontrolled transfer	*Experimental group only*	New simulator → Aircraft
Simulator-to-simulator transfer (quasi-transfer)	*Experimental group only*	New simulator → High-fidelity simulator
Backward transfer	*Skilled operator*	Aircraft → New simulator
Simulator performance improvement	*Experimental group only*	New simulator → New simulator

FIGURE 9.2
Performance evaluation models for simulator evaluation.

significant challenge, both in terms of providing matched control and experimental groups and in running two different training patterns in parallel, which may have licensing implications if undertaken as part of a training programme approved by the airworthiness authorities.

9.6.2.2 Pre-Existing Control Model

This uses historical data to provide the experimental control. The validity of this approach is predicated upon two conditions. Firstly, the control group is matched with the experimental group (for example there has been no change in input standards to the training system over time). Secondly, those performance data were collected under the same conditions that would apply to a control group trained in parallel with the experimental group. If these conditions hold true, there is a little difference between this and the classic transfer model.

9.6.2.3 Self-Control Model

In this model, student performance in the aircraft immediately preceding simulator training is compared with performance in the aircraft after training. Any difference in performance is again attributed to transfer from the simulator to the aircraft. If there is any significant period between the pre- and post-training assessments in the aircraft, the results of the trial may be less clear due to the possibility of students forgetting. Kerlinger and Lee (2000) point out that from a pure experimental design perspective, the results of trials conducted using such a model may be confounded as factors other than the independent variable (aircraft/simulator) may influence the outcome.

9.6.2.4 Uncontrolled Transfer Model

There are situations where the creation of a control group is simply not feasible. Caro (1970) cites the extreme example of constructing a no-simulator control group for astronauts required to carry out lunar landings. A more frequently encountered situation is where, as training on a new training device is perceived as likely to be better, it would be politically unacceptable to disadvantage one group of students by not allowing them access to it. Kerlinger and Lee (2000) dismiss this model as scientifically worthless as there is 'virtually no control on other possible influences on outcome' (p. 469).

9.6.2.5 Simulator-to-Simulator (Quasi-Transfer) Model

In the simulator-to-simulator model, or quasi-transfer model as it is sometimes known, students' performance is evaluated in a higher-fidelity simulator than the one being evaluated. The validity of this model is based on the assumption of equivalence between the aircraft and the higher-fidelity simulator itself. Such an assumption may be tenuous in certain circumstances, and erroneous data may be produced. Caro (1970) suggests that the one situation where this approach would be valid is where the higher-fidelity simulator is the target environment, such as would be the case for a part-task training device designed to reduce time spent in the simulator.

9.6.2.6 Backward Transfer Model

This is an entirely different approach. A skilled operator, who has already mastered the skills to be inculcated in the operational aircraft, is required to perform the same skills in the simulator. If he or she is able to perform these skills without practice in the simulator, then 'backward transfer' has occurred and transfer from the simulator to the aircraft will therefore be positive. The potential flaws with this approach concern the nature of the skill set of the experienced operator. An experienced, highly skilled pilot may be able to perform adequately in a reduced cueing environment that is totally unsuitable for training. Hence, if there is such an inadequacy in the cueing environment, then it may not be revealed in this type of evaluation. In addition, it is also possible that the reason that an operator is able to perform to an acceptable standard in the simulator is that they have more generalised skills enabling them to overcome any limitations of the simulator. However, if the skilled operator *is not* able to perform well in the simulator, then there is likely to be something awry.

9.6.2.7 Simulator Performance Improvement Model

This final model is based on conducting pre- and post-training evaluations of performance in the simulator. Improvement in performance is attributed to the experience gained in the simulator. Since no evaluation is conducted in the aircraft, it can be argued that a trial of this kind can provide only relatively weak evidence as to the likelihood of transfer and therefore the effectiveness of the simulator. What it can do is act as a negative filter in the same way as some of the other models described: if no performance improvement occurs, it is unlikely that there will be positive transfer to the aircraft. In-simulator performance improvement is a necessary condition for effectiveness but does not provide conclusive evidence that transfer will take place. This model suffers all the same methodological weaknesses as the self-control model.

9.7 Measurement of Training Transfer and Cost-Effectiveness

When evaluating simulators against the criteria of training transfer and cost-effectiveness, AGARD (1980, p. 40) described three types of measurement that need to be considered:

- The amount of learning achieved in the simulator
- The amount of learning that transfers to the aircraft
- The extent of savings made as a result of using the simulator

Implicit within each of these measures is the requirement for assessment of student performance. However, since performance in a complex task such as the operation of a commercial aircraft is a multidimensional concept, any transfer-of-training measures based on a particular aspect of performance represent just one dimension of the transfer effect of the system being investigated. Morley (1995) developed a set of criteria for evaluating the suitability of potential training effectiveness measures. These criteria are summarised in Table 9.2.

TABLE 9.2

Criteria for Evaluating Potential Training Effectiveness Measures

Criteria	Definition
Sensitivity	Is the measure able to adequately differentiate between levels of the construct of interest?
Validity	(Content and face): Does the measure provide an indication of the construct which it is designed to measure (content validity) and does the measure appear to the participant to measure that which it is designed to measure (face validity)?
Reliability	Does the measure provide consistent results over time and between raters?
Diagnosticity	Does the measure provide any indication which aspects of the task or situation are contributing to the trainee score?
Utility	Can measures be applied in the actual training or operational context without compromising effectiveness or safety?
Scope	Do the measures selected provide an indication of how well the training is achieving each of its training goals?
Operational relevance	Does the measure provide an indication of operational readiness (the ultimate aim of training in a military context)?
Resources required	What sort of resources are required to use the measure? Are these resources readily available or are they prohibitively expensive to obtain?
Training effectiveness factor examined	Will the measures provide an indication of which individual and situational factors are affecting training effectiveness?

Source: Morley F. J. J., *Training effectiveness measures matrix and factors affecting training effectiveness*, DERA Bedford, Bedford, UK, 1995.

The type of measure has a significant effect on the validity and interpretation of the content of a transfer-of-training trial. Furthermore, training may either be conducted for a fixed time (or fixed number of trial repetitions) or until a trainee performance criterion standard is achieved (Liu et al. 2009a). Training to a criterion standard provides information about training duration, which may be used for cost-effectiveness calculations, but potentially requires more simulator time. In determining the degree of transfer, AGARD (1980) described two distinct approaches. In the first-shot approach, student pilot performance is assessed on initial exposure to the aircraft. The alternative is the trials-to-criterion approach, whereby the time taken in the aircraft for the student to reach criterion performance is measured. This second approach has the advantage of yielding further cost-effectiveness data but has a significant resource requirement given aircraft operating costs.

Once suitable performance data have been collected, there are several alternative methods for quantifying overall transfer and cost-effectiveness. The nature of the data and the research aim determine the choice of method. A range of common metrics follow.

9.7.1 Percentage Transfer

Proctor and Dutta (1995) identified percentage transfer as one of the most common indices; however, it can only be applied where relative performance measures between control and transfer groups have been made, such as during a first-shot trial. It is defined as follows:

$$\text{Percentage transfer} = \frac{P_{\text{transfer}} - P_{\text{control}}}{P_{\text{control}}} \times 100,$$

where
P_{control} = performance of the control group
P_{transfer} = performance of the transfer group

This equation simply expresses the difference in performance as a percentage of the performance of the control group. As P is the selected performance measure applied to the control and transfer (experimental) groups, if the transfer group performs better than the control group, they would be expected to get higher scores for P, resulting in a positive value for transfer. However, as stated earlier, P is a unidimensional measure reflecting only one aspect of pilot performance.

9.7.2 Percentage Savings

If a trials-to-criterion approach is adopted, Wickens and Hollands (2000) suggest that values for time taken to reach the criterion performance can be substituted into the original percentage transfer equation to give percentage savings in time, as follows:

$$\text{Percentage time saving} = \frac{T_{\text{control}} - T_{\text{transfer}}}{T_{\text{control}}} \times 100,$$

where
T_{control} = time for the control group to reach criterion
T_{transfer} = time for the transfer group to reach criterion

If the effect of using the device is to reduce the time that the transfer group take to reach criterion standard compared to the control, the formula will produce a positive value for the percentage time saving.

9.7.3 Transfer Effectiveness and Training Cost Ratios

Povenmire and Roscoe (1973) proposed an alternative calculation (the transfer effectiveness ratio [TER]) taking into account the amount of time spent in the simulator by the transfer group, hence giving an indication of relative efficiency in terms of the time spent in training by the two groups:

$$\text{TER} = \frac{\text{Time saving in reaching criterion performance in the aircraft}}{\text{Time taken by the transfer group in the simulator}},$$

which is calculated from the following:

$$\text{TER} = \frac{T_{\text{control}} - T_{\text{transfer}}}{T_{\text{transfer}} \text{ in simulator}},$$

where

T_{control} = time for the control group to reach criterion
T_{transfer} = time for the transfer group to reach criterion
T_{transfer} in simulator = time spent by the transfer group in the simulator

Where the time saved by the transfer group in the air is equal to the time spent in the simulator, the TER value will be +1.0. If the time spent in the simulator is less than the time saved in the air, then the TER value will be greater than +1.0. Conversely, if the time spent in the simulator is greater than the time saved in the air, the value will be less than +1.0.

Roscoe and Williges (1980) also proposed the training cost ratio (TCR) for comparing the simulator and aircraft operating costs:

$$\text{TCR} = \frac{\text{Training cost in the simulator}}{\text{Training cost in the aircraft}}.$$

When the value of TCR is above +1.0, this indicates that it is cheaper to conduct the training in the aircraft than in the simulator.

9.7.4 Incremental Transfer Effectiveness Ratio

Another question of interest is what gain in transfer is achieved for a given time spent in the simulator? This can be quantified using the incremental transfer effectiveness ratio (ITER) also proposed by Povenmire and Roscoe (1973). This measure is concerned solely with training on a given training device rather than in making a comparison with a control group that has not been trained on that simulator.

$$ITER = \frac{Y_{X-\Delta X} - Y_X}{\Delta X},$$

where

$Y_{X-\Delta X}$ = Time, trials or errors required for a control group to reach criterion standard after $X - \Delta X$ units of time on the device being investigated

Y_X = time, trials or errors required for an experimental group to reach criterion standard after $X - \Delta X$ units of time on the device being investigated

ΔX = incremental unit of time, trials or errors

If the same task is trained with increasing hours of simulator time, it is possible to determine the effect of each additional hour in the simulator on the time to reach the criterion performance. This can assist in the judgement of the optimum amount of time to be spent in the simulator.

9.7.5 Cost-Effectiveness Ratios

The cost-effectiveness ratio (CER) defined by Roscoe and Williges (1980) combines the cost and training effects of using a simulator:

$$CER = \frac{\text{Training effectiveness ratio (TER)}}{\text{Training cost ratio (TCR)}}.$$

Effectively, this becomes the following:

$$CER = \frac{\text{Cost saved in the air}}{\text{Cost of using the simulator}}.$$

Hence, once again, if the CER value exceeds +1.0, it is more cost-effective to use the simulator for that aspect of training than to use the real aircraft.

9.8 Some Factors Affecting Simulator Training Effectiveness

AGARD (1980) reported that the design of the simulator (particularly its fidelity), along with student motivation, had the most significant impacts on the effectiveness of delivering training. Moraal et al. (1989) identified the individual trainee, the trainer, training equipment, training programme and training environment as critical factors in determining training success. Salas et al. (1998) highlighted the importance of the design of the training being matched to the nature of the task to be trained. Figure 9.3 provides a synthesis of these perspectives.

FIGURE 9.3
Factors affecting simulator training effectiveness.

There are interactions with the task to be trained, the student character-istics and training syllabus. Alessi (1988) and Reigluth and Schwartz (1989) identified the nature of the task to be trained on the simulator as an impor-tant factor in its effectiveness in delivering training. They suggested that different features were required for different training tasks; for example training in psychomotor skills such as for aircraft handling placed greater requirements on the handling qualities of the simulator, whereas for cog-nitive tasks such as decision-making, more emphasis was required on the training scenario and the agents involved.

These authors also emphasised that the characteristics of the simulator, particularly its fidelity, and the manner in which it was utilised (the training programme characteristics or instructional overlay) were of fundamental importance. Furthermore, they argued that these two factors were inextri-cably linked. The instructional overlay encompassed both the design of the scenarios used for the training exercises and the nature of the feedback pro-vided to students. Alessi and Trollip (2001) described trainee feedback as being either natural or artificial. Natural feedback was the normal response of the system to the student's actions, whereas artificial feedback was in some way contrived, such as additional information fed back to the student through the simulation or inputs from the instructor.

The view that high-fidelity simulation is required for training trans-fer is predicated upon a behaviourist view of learning, with the student being regarded principally as a stimulus response mechanism (Gagne 1954). However, the cognitive perspective challenged the necessity for exact similarity between training and operational environments. He suggested that one of the features of simulators is that they may differ from the operational environment by omitting or distorting some elements. There is experimental evidence to support this proposition. The work of Lintern and Garrison (1992) demonstrated that low scene detail was more effective for training crosswind landing techniques than high scene detail.

As noted earlier, Alessi (1988) contended that the learning achieved through simulation was also dependent upon the experience level of the trainee. As a result, he proposed a more complex relationship between learning and simulator fidelity. He suggested that a lower level of simulation was required for a novice, thereby avoiding the possibility of overloading the student with excess information during the early stages of skill acquisition; as the student became more experienced, a higher level of simulation was more appropriate. The lines on the graph in Figure 9.4 relate learning to fidelity, illustrating that the relationship between learning in the simulator and training transfer is complex. As an example, point A on the graph may yield less learning but more transfer than point B.

Roscoe (1971) proposed a different model relating time spent in a simulator to the amount of transfer achieved. The model predicts a diminishing return for greater hours spent in the simulator, with overall percentage transfer eventually levelling off. Calculations of incremental TER can be used to determine the point beyond which simulator training is no longer cost-effective. This model was validated against trials conducted on Cherokee light aircraft with an experimental group trained on a GAT Link-1 trainer.

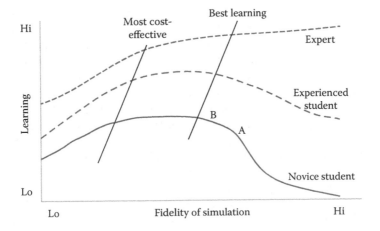

FIGURE 9.4
Fidelity and learning model. (From Alessi, S. M., *Journal of Computer-Based Instruction*, 15, 40–47, 1988.)

9.9 Conclusion: Problems of Transfer of Training in Commercial Aviation

The validation by Roscoe (1971) of his incremental transfer model using a Cherokee light aircraft trainer illustrates an interesting problem. Although this chapter has described the use of flight simulators for training commercial pilots and has described methods and equations for evaluating transfer of training, there have been few descriptions of studies employing any form of transfer of training. In fact, transfer-of-training studies undertaken in the current generation of high-fidelity simulators to prepare pilots to operate the current generation of sophisticated jet airliners are pretty well non-existent. The majority of studies have been undertaken either in a general aviation context or by the military, and even then, true full-scale transfer-of-training studies are rare. For example, in the military domain, Hays et al. (1992) reviewed 26 studies (19 fixed-wing jet aircraft and seven helicopters) which had sufficient information for statistical analysis. Rantanen and Talleur (2005) reviewed 19 studies from general aviation and the military between 1949 and 2005. The small number was as a result of the criteria employed to assess transfer of training, and only three studies employed a complete design utilising a control and experimental group.

Earlier, it was described how Johnston et al. (1997) demonstrated that CRM skills could be taught successfully using a desktop flight simulator. However, irrespective of the simulation technology underpinning the delivery of such programmes and even though they have been thought to provide a great many safety and economic benefits to airlines, this has actually been difficult to establish empirically (Edkins 2002).

Perhaps, the most contentious issue in the flight simulation fidelity debate has been on the use of motion platforms. Simulator motion seems to improve the acceptability of a flight simulator for its users (Reid and Nahon 1988). It enhances its motivational fidelity; however, it is debatable if motion actually improves trainee performance. Steurs et al. (2004) observed that motion cueing enhanced pilot ratings of the realism of a flight simulator but did not contribute to enhanced performance. Longridge et al. (2001) undertook a series of trials using a quasi-training transfer-of-training paradigm in a simulator using experienced crew members. Scenarios involving an engine failure on take-off (followed by either a rejected take-off or continuing) were flown with the motion either switched on and with it off. The results showed that the presence or absence of motion did not affect the performance or subjective perceptions in terms of workload, comfort or the acceptability of the simulator. Note also that only one specific aspect of the simulation of flight was being evaluated: the effect of motion. As described earlier, the operation of a commercial aircraft is a multidimensional concept, but any transfer-of-training measures are based on just one particular dimension of performance.

Pilot training is of paramount importance in maintaining aviation safety, and the provision of training courses and facilities represents a huge investment for airlines. At the time of writing, depending upon their exact specification, an approved FFS for training and licensing purposes will cost between $6 and $12 million (and can cost up to $20 million). In a large airline, these facilities will be supplemented by cockpit procedure trainers, other part-task trainers and computer-based training facilities for aircraft systems and procedures. There are really no issues concerning the training effectiveness of FFSs. It is almost inconceivable that airlines would either recommence pilot training in the aircraft itself or even commission a controlled study into evaluating training transfer using an FFS. What would happen if they were demonstrated to be ineffective or provide negative transfer? This is an unlikely scenario but poses an interesting question to the field.

References

Adorian, W. N., Staynes, M. B. E., and Bolton, M. 1979. The evolution of the flight simulator. In *Proceedings of the CAA Conference '50 Years of Flight Simulation'*, 1–23, Piccadilly Hotel, Piccadilly, London, 23–25 April 1979.

AGARD 1980. *Fidelity of Simulation for Pilot Training*. AGARD-AR-159. Neuilly-sur-Seine, France: NATO/OTAN.

Albett, R. 1989. The future for flight simulation in the Royal Aeronautical Society. In *1989 Spring Convention – Flight Simulation: Assessing the Benefits and Economics*. Royal Aeronautical Society, London.

Alessi, S. M. 1988. Fidelity in the design of instructional simulations. *Journal of Computer-Based Instruction*: 15, 40–47.

Alessi, S. M., and Trollip, S. R. 2001. *Multimedia for Learning* (3rd Edn). Boston, MA: Allyn and Bacon.

Allen, J. A. 1986. Maintenance training simulator fidelity and individual differences in transfer of training. *Human Factors*: 28, 497–509.

Andrews, D. H., Carroll, L. A., and Bell, H. H. 1995. Selective fidelity in training devices. Farnborough, UK: Military Simulation and Training, 3/95.

Bell, H. H., and Waag, W. L. 1998. Evaluating the effectiveness of flight simulators for training combat skills: A review. *International Journal of Aviation Psychology*: 8, 223–242.

Branson, R. K., Rayner, G. T., Cox, L., Furman, J. P., King, F. J., and Hannum, W. 1975. *Interservice Procedures for Instructional Systems Development: Executive Summary and Model*. Tallahassee, FL: Center for Educational Technology, Florida State University.

Brooks, P., and Arthur, J. 1997. Computer-based motorcycle training: The concept of motivational fidelity. In *Engineering Psychology and Cognitive Ergonomics (Volume One Transportation Systems)*, ed. D. Harris, 403–410. Aldershot, UK: Ashgate.

CAA (Civil Aviation Authority). 2003. *Joint aviation requirements flight crew licensing notes for the qualification and approval of flight navigation procedures trainers (FNPTs) and basic instrument training devices (BITDs). Standards Document 18 (Version 2).* London: CAA.

Caro, P. W. 1970. *Equipment-device commonality analysis and transfer of training* (Technical Report 70-7). Alexandria, VA: Human Resources Research Organisation, June 1970.

Caro, P. W. 1977. *Some factors influencing Air Force simulator effectiveness* (Report No. Hum-RRO-TR-77-2). Alexandra, VA: Human Resources Research Organization.

Code of Federal Regulations, Title 14, Part 60. Aeronautics and space: Flight simulation training device initial and continuing qualification and use. Washington, DC: National Archives and Records Administration. Available at http://www.gpo.gov/nara/cfr.

Code of Federal Regulations, Title 14, Part 61. Aeronautics and space: Certification: Pilots, flight instructors, and ground instructors. Washington, DC: National Archives and Records Administration. Available at http://www.gpo.gov/nara/cfr.

Edkins, G. D. 2002. A review of the benefits of aviation human factors training. *Human Factors and Aerospace Safety*: 2, 201–216.

FAA (Federal Aviation Administration). 1991a. *Advanced Qualification Program* (Advisory Circular AC 120-54). Washington, DC: US Department of Transportation.

FAA. 1991b. *Airplane Simulator Qualification* (Advisory Circular AC 120-40B). Washington, DC: US Department of Transportation.

Farmer, E., van Rooij, J., Riemersma, J., Jorna, P. G. A. M., and Moraal, J. 1999. *Handbook of Simulator Based Training*. Aldershot, UK: Ashgate.

Gagne, R. M. 1954. Training devices and simulators: Some research issues. *American Psychologist*: 9, 95–107.

Gross, D. C. 1999. Report from the Fidelity Implementation Study Group (paper 99S-SIW-167). In *Proceedings of Spring 1999 Simulation Interoperability Workshop*. Orlando, FL: Simulation Interoperability Standards Organization.

Haward, D. M. 1910. The Sander's Teacher. *Flight*: 2, 1006–1007.

Hays, R. T., Jacobs, J. W., Prince, C., and Salas, E. 1992. Flight simulator training effectiveness: A meta-analysis. *Military Psychology*: 4, 63–74.

International Civil Aviation Organization. 2009. *ICAO Document 9625 'Manual of Criteria for the Qualification of Flight Simulation Training – 3rd Edition'*. Montreal: International Civil Aviation Organization.

JAA (Joint Aviation Authorities). 1999. *Joint Airworthiness Requirement – Flight Crew Licensing – Aeroplanes* (JAR-FCL 1). Hoofdorp, Netherlands: JAA.

JAA. 2008. *JAR-FSTD A: Aeroplane Flight Simulation Training Devices*. Hoofdorp, Netherlands: JAA.

Johnston, N., McDonald, N., and Fuller, R. 1997. *Aviation Psychology in Practice*. Aldershot, UK: Avebury Aviation.

Kaiser, M. K., and Schroeder, J. A. 2003. Flights of fancy: The art and science of flight simulation. In *Principles and Practice of Aviation Psychology*, ed. P. S. Tsang and M. A. Vidulich, 435–471. Mahwah, NJ: Lawrence Erlbaum Associates.

Kerlinger, F. N., and Lee, H. B. 2000. *Foundations of Behavioural Research* (4th Edn). London: Wadsworth.

Lintern, G., and Garrison, W. V. 1992. Transfer effects of scene content and crosswind in landing instruction. *International Journal of Aviation Psychology*: 2, 244–255.

Liu, D., Blickensderfer, E. L, Macchiarella, N. D., and Vincenzi, D. A. 2009a. Transfer of training. In *Human Factors in Simulation and Training*, ed. D. A. Vincenzi, J. A. Wise, M. Mouloua and P. A. Hancock, 49–60. Boca Raton, FL: CRC Press.

Liu, D., Macchiarella, N. D., and Vincenzi, D. A. 2009b. Simulation fidelity. In *Human Factors in Simulation and Training*, ed. D. A. Vincenzi, J. A. Wise, M. Mouloua and P. A. Hancock, 61–73. Boca Raton, FL: CRC Press.

Longridge, T., Bürki-Cohen, J., Go, T. H., and Kendra, A. J. 2001. Simulator fidelity considerations for training and evaluation of today's airline pilots. In *Proceedings of the 11th International Symposium on Aviation Psychology*, ed. R. S. Jensen. Columbus, OH: Ohio State University.

Moraal, J., Stirling, B. S., and Butler, W. G. 1989. Effectiveness of training. In *The Value and Cost Effectiveness of Training*. NATO AC/243 (panel 7/RSG 15) D/4 Dec 1989.

Morley, F. J. J. 1995. *Training effectiveness measures matrix and factors affecting training effectiveness*. DERA Bedford Internal Report. Bedford, UK: DERA Bedford.

Moroney, W. F., and Moroney, B. W. 1999. Flight simulation. In *Handbook of Aviation Human Factors*, ed. D. J. Garland, J. A. Wise and V. D. Hopkin, 355–388. Mahwah, NJ: Lawrence Erlbaum.

Osgood, C. E. 1949. The similarity paradox in human learning and resolution. *Psychological Review*: 56, 132–143.

Pfeiffer, M. G., Horey, J. D., and Burrimas, S. K. 1991. Transfer of simulated instrument training to instrument and contact flight. *International Journal of Aviation Psychology*: 1, 219–229.

Povenmire, H. K., and Roscoe, S. N. 1973. Incremental transfer effectiveness of a ground-based general aviation trainer. *Human Factors*: 15, 534–542.

Proctor, R. W., and Dutta, A. 1995. *Skill Acquisition and Human Performance*. London: Sage.

Rantanen, E. M., and Talleur, D. A. 2005. Incremental transfer and cost effectiveness of ground-based flight trainers in university aviation programs. In *Proceedings of the Human Factors and Ergonomics Society 49 Annual Meeting*, 764–768. Santa Monica, CA: Human Factors and Ergonomics Society.

Reid, L. D., and Nahon, M. A. 1988. Response of airline pilots to variations in flight simulator motion algorithms. *Journal of Aircraft*: 25, 639–646.

Reigluth, C. M., and Schwartz, E. 1989. An instructional theory for the design of computer-based simulations. *Journal of Computer-Based Instruction*: 16, 1–10.

Rolfe, J. M. 1985. *Fidelity: An evaluation of the concept and the implications for the design, procurement, operation and evaluation of flight simulators*. Unpublished Defence Fellowship Thesis. Cambridge, UK: Wolfson College.

Rolfe, J. M., and Staples, K. J. 1986. *Flight Simulation*. Cambridge, UK: Cambridge University Press.

Roscoe, S. N. 1971. Incremental transfer effectiveness. *Human Factors*: 13, 561–567.

Roscoe, S. N., and Williges, B. H. 1980. Measurement of transfer of training. In *Aviation Psychology*, ed. S. N. Roscoe, 182–193. Ames, IA: Iowa State University Press.

Salas, E., Bowers, C. A., and Rodenizer, L. 1998. It is not how much you have but how you use it: Towards a rational use of simulation to support aviation training. *International Journal of Aviation Psychology*: 8, 179–208.

Schiewe, A. 1995. CRM training and transfer: The 'behavioral business card' as an example for the transition of plans into actual behavior. In *Aviation Psychology: Training and Selection*, ed. N. Johnston, R. Fuller and N. McDonald, 38–44. Aldershot, UK: Avebury Aviation.

Seamster, T. L., Redding, R. E., and Kaempf, G. L. 1998. *Applied Task Analysis in Aviation*. Aldershot, UK: Ashgate.

Steurs, M., Mulder, M., and van Paassen, M. M. 2004. A cybernetic approach to assess flight simulator fidelity. In *Proceedings of the AIAA Modelling and Simulation Technologies Conference and Exhibition* (Providence RI, 16–19 August 2004). Paper AIAA-2004-5442. Reston, VA: American Institute for Aeronautics and Astronautics.

Thorndike, E. L. 1903. *Educational Psychology*. New York: Lemche and Buechner.

Wickens, C. D., and Hollands, J. G. 2000. *Engineering Psychology and Human Performance (3rd Edition)*. Upper Saddle River, NJ: Prentice Hall.

10

Simulators for Aircraft Emergency Evacuations

Rebecca L. Grant, Dale Richards and Alex W. Stedmon

CONTENTS

10.1 Introduction

At 1527 hours on 15 January 2009, just after the take-off from LaGuardia Airport, New York, an Airbus A320-214 was involved in a multiple bird strike incident. The flight crew managed to successfully ditch the aircraft on the Hudson River. Fortunately, the event resulted in no fatalities and only a small number of occupants with serious injuries. The subsequent investigation concluded that the cabin crew had 'managed an effective and timely evacuation' (NTSB 2010, p. 106); however, issues in the evacuation were cited that meant that the occupant outcome could have been more serious. With water entering the rear of the aircraft, the cabin crew located in this area needed to redirect passengers away from their nearest exits towards the central and forward exits. Crowding in the aisle developed as passengers queued to use the smaller overwing exits, and therefore, passengers at the rear of the cabin were initially unable to move forward along the aisle (NTSB 2010). Cabin crew situated at the forward exits became aware of crowding, and passengers 'bottlenecked' around the overwing exits, so the crew then commanded some passengers to move along the aisle to the forward exits (NTSB 2010, p. 41). Most forward and mid-cabin passengers left the aircraft through their nearest exit; however, due to unusable exits at the rear of the cabin, some of the passengers who were located in the central and rear areas of the cabin evacuated via the forward exits (NTSB 2010).

Almost a quarter of a century before, at 0612 hours on 22 August 1985, another incident occurred when a Boeing 737-236 experienced an aborted take-off following an uncontained engine failure at Manchester Airport, United Kingdom. A part of the engine then struck and ruptured the fuel tank, and the leaking fuel resulted in an engine fire. The fire and toxic smoke engulfed the passenger compartment, resulting in 55 fatalities on board the aircraft. Due to the location of the fire, only three out of the six exits were available for use: the two forward exits and one overwing exit in the centre of the cabin. Delays in exiting the aircraft were specifically cited in relation to the high number of people who did not survive, with the passengers experiencing specific problems in accessing the forward and central overwing exits (Air Accidents Investigation Branch 1988).

Although accidents such as these tend to be very high profile, flying is relatively safe compared to other forms of transportation. Nevertheless, whilst the flight and cabin crew are trained in various emergency situations, both the earlier examples illustrate how an understanding of human behaviour in emergency evacuations is fundamental to the safety of passengers in such an event. The potential confusion and overall stress of such a situation can be compounded by other factors (e.g. fire, smoke, darkness, noise). Thus, the ability to leave the aircraft as quickly and as safely as possible – crucial to occupant survivability – may be hindered by the sudden unfamiliar circumstances and characteristics with which passengers are confronted. Quite literally, finding and getting through the exit door can be a matter of life and death once an evacuation order has been given.

From a passenger safety perspective, air accident investigators typically classify aviation accidents in relation to occupant survivability according to three main descriptors (European Transport Safety Council 1996):

- *Fatal/non-survivable accident* – where none of the passenger or crew occupants survive
- *Non-fatal/survivable accident* – where all of the passenger and crew occupants survive
- *Technically/partially survivable accident* – where some (at least one) passenger or crew member survives

It is in the latter two scenarios that the egress of crew and passengers is paramount within the field of emergency evacuation. A National Transportation Safety Board (NTSB) review of aircraft evacuations within the United States between September 1997 and June 1999 found that a total of 46 evacuations occurred, involving 2651 passengers and 195 crew (NTSB 2000). The majority of occupants (92%) were recorded as evacuating without injury. Of the remaining occupants, 6% experienced minor injuries and 2% sustained serious injuries (including death in 11 of the 62 cases). In some instances, the nature of the fatalities resulting from an aircraft emergency is directly associated with the evacuation procedure. Examples include the 1985 Manchester accident described earlier (Air Accidents Investigation Branch 1988) and an accident at 1920 hours on 2 June 1983 at Cincinnati, in the United States, where 23 of the passengers were unable to evacuate before being overcome by the effects of a hidden cabin fire (NTSB 1986).

The nature of passenger evacuation from aircraft has been investigated over the years in an attempt to best understand what design changes can be made to the aircraft, what formal procedures are needed to support passenger evacuation and what information that passengers require in order to increase the survivability of such events. Following a review of partially survivable accidents, Snow et al. (1970) identified a number of key factors that influenced the potential for passenger survival during an emergency evacuation: configuration and layout of the cabin, procedural guidelines carried out by the crew, environmental condition of the cabin (e.g. smoke, light, noise) and bio-behavioural indices (e.g. age, sex, fitness). This important work has provided a structured framework of significant factors in evacuations and has influenced much of the simulator research that has followed into passenger behaviour in emergency evacuations. Muir (2004) expanded these factors to include other aspects of aircraft design including crashworthiness, fire protection and evacuation aids (e.g. exit signs and emergency lighting).

Due to relatively few instances of aircraft accidents or incidents that result in passenger evacuations, though, gathering data is difficult and often piecemeal. Cabin simulators therefore remain one of the foremost and most practical approaches to develop and test the evacuation capability of an aircraft cabin. A range of simulator approaches can be taken to test the ability of passengers to evacuate an aircraft cabin, including full-scale aircraft evacuations, experimental testing of specific components using cabin simulators and evacuation modelling via computer simulation programmes. All of these methods provide different insights and understanding of passenger behaviour but share the common aim of attempting to predict likely outcomes so that strategies can be developed and tested in order to promote desirable behaviours whilst minimising undesirable behaviours. However, through all these approaches, an important research issue remains: how can such situations be investigated in a realistic manner when they do not occur very often in the real world and where it is extremely difficult to mimic the characteristics of these situations in laboratory trials? This chapter presents some of the research that has investigated cabin and passenger safety factors that influence evacuation performance.

10.2 Cabin Simulators and Evacuation Research Trials

Cabin evacuation research has been undertaken in a range of simulators, including decommissioned aircraft and cabin interior sections constructed for cabin crew training or evacuation research. The use of cabin simulators allows researchers to manipulate and control environmental conditions in order to enhance the realism of the evacuation scenario under test. This includes factors such as the cabin seating configurations, type and location of operational aircraft exits, width of aisles, reconfigurable internal features (e.g. galley areas), inflatable escape slides, levels of cabin lighting, introduction of emergency lighting systems, sound effects and audio recordings of engine noise and evacuation commands and the use of non-toxic theatrical smoke to simulate conditions of limited visibility. In addition, whilst simple performance metrics are generally used for data capture (e.g. evacuation times), simulators provide opportunities for the use of static video cameras to capture passenger behaviour during evacuations both inside the cabin and as they negotiate the exits and move away from the aircraft. Simulator studies also allow researchers to explore the participants' experiences during the evacuation.

Figures 10.1 through 10.4 show some of the cabin interior aspects inside a typical narrow-bodied (single-aisle) aircraft. Cabin simulators used for

FIGURE 10.1
View along the aisle of a typical narrow-bodied (single-aisle) aircraft.

FIGURE 10.2
Typical seat assembly (three seats) on a narrow-bodied aircraft.

FIGURE 10.3
Exit signs and overheard lockers on a typical narrow-bodied aircraft.

FIGURE 10.4
Example of a type I aircraft exit from inside the cabin. This type of exit is located on narrow-bodied aircraft. The specific design characteristics of the exit are dependent on the aircraft manufacturer.

human factors evacuation research are designed to simulate these environments and replicate key aircraft cabin attributes. Research can then be undertaken into passenger movement from the seats, movement along aisles and through cabin bulkheads, evacuation through exits and navigation aids designed to assist passengers during the evacuation process. Many of the evacuation trials highlighted in this chapter involve simulators designed to replicate key cabin features of narrow-bodied aircraft.

It has been observed that in some situations, passengers have evacuated the aircraft as a coordinated group in a fairly orderly manner (Muir et al. 1989). These occurrences tend to happen where the incident was not immediately life threatening and where passengers worked together collaboratively in an attempt to evacuate as quickly as possible (e.g. the A320 ditching on the Hudson River discussed earlier). However, in other accidents, the orderly process can break down and crowding or blockages have occurred at specific points throughout the cabin (as per the Boeing 737 accident at Manchester Airport). This often happened where the situation was perceived as life threatening (e.g. when smoke and fire were present). In these cases, passengers tend to be concerned with their own escape and that of their close companions and family members rather than the aircraft population as a whole. As a result, passengers tend to compete with each other to make their way to the exits (Muir et al. 1989). This behavioural response is not unique in aviation but can also be seen across other domains (see Stedmon et al. 2015 for a comparison between aviation and rail).

One of the key methodological challenges for researchers is how to simulate and replicate emergency situations in a way that promotes naturalistic behaviours. In any simulation, limitations to believability and realism will exist as participants will ultimately know that they are not experiencing things in the same way as they might in the real world and that the underlying dangers are not real or life threatening. Any research involving human participants in aircraft evacuation simulators therefore requires a balance between the necessary realism for the research to have relevance in the real world and safety and ethical factors in order to safeguard participant's well-being from both a physical perspective and a psychological perspective. These issues must take precedence in order that simulated aircraft evacuation trials do not put participants at an unacceptable level of risk.

In what is now regarded as a classic experiment, Muir et al. (1989) developed a novel approach in which participants were motivated to behave in a similar way as if they were experiencing a life-threatening evacuation. This is a particularly important issue within simulator studies of this kind. Across a range of domains (e.g. driver/rider behaviour, safety-critical operations and aviation evacuations), it is necessary to simulate situations that produce realistic behaviours that generalise to real-world situations. In this particular instance, investigating competitive behaviour was important in order to simulate the conditions where individual survival objectives may conflict with those of the overall group (Muir et al. 1996).

During these simulator evacuation trials, the participants were required to exit following a 'collaborative evacuation paradigm' (e.g. the same as those used in evacuation certification demonstrations). They were instructed to evacuate the aircraft cabin simulator as quickly as possible, and everyone was paid a standard attendance fee for taking part. However, in order to simulate more life-threatening situations, a 'competitive evacuation paradigm' was designed to simulate less altruistic individual behaviours seen in some real evacuations. In order to achieve this, the participants were again encouraged to exit as quickly as possible but also informed that in addition to the standard attendance payment, they would receive a bonus payment if they were within the first half of the participants to evacuate the cabin (Muir et al. 1989). This methodology has been used in a range of trials, with behaviours produced being similar to some of those witnessed in actual evacuations.

10.3 Computer Simulation for Aircraft Evacuation

In parallel to the physical trials within the aircraft evacuation simulation domain, computer modelling techniques have been developed to help understand aspects of passenger behaviour and movement during an emergency. One such simulation model that has been extensively discussed in the academic literature is airEXODUS, developed by Galea (2006) and Galea et al. (2005) from the Fire Safety Engineering Group at the University of Greenwich.

airEXODUS has been developed using insights gained from crew and passenger behaviour based on the data gathered from actual accident reports, survivor reports (including crew interviews and passenger questionnaires), regulatory certification evacuation demonstrations and simulated evacuation research trials to provide the model with a higher degree of realism (Galea et al. 2005). The airEXODUS model takes into account many of the possible interactions that the occupants will have with each other and the wider context of the emergency and environmental factors within which the situation is modelled. The model utilises a probabilistic approach to generating multiple responses and outcomes to reflect the uncertainty of human behaviour (Galea et al. 2005).

The use of behaviour modelling approaches such as this may also provide assistance in testing different aircraft design layouts at the design stage, before they are built. This approach may then be used to evaluate designs against regulatory compliance standards such as 90-second certification requirements (see Section 10.6), the evacuation dynamics, passenger exit usage, crew training, development of crew procedures and resolution of operational issues. Modelling techniques may also be used for post-incident analysis in order to build a better understanding of events during aircraft accident investigation (Galea 2006).

Although simulated computer models can provide powerful means of visualising, describing and predicting human behaviour in evacuations, it is important to note that they are limited by the data derived from behavioural research and the underlying assumptions that form the basis of the algorithms that they apply pertaining to human behaviour.

Galea et al. (2005) conducted a number of validation exercises using computational modelling on both narrow- and wide-bodied aircraft and found that airEXODUS was capable of predicting the total evacuation time for a certification demonstration to within an average of 5.4% or 3.8 seconds of the actual time taken during the real regulatory demonstration. It was also found that the model was able to reliably predict the likely evolution of the evacuation time across a given timeline. This suggests that such modelling can be useful and may assist in our understanding of the factors associated with different types of aircraft with different types of exit and with different levels of cabin crew assertiveness (Galea et al. 2005). Galea (2006) also suggested that by using appropriately developed and validated computer simulation models, the ability to process a large number of modelled trials not only provides the generation of data that can be applied to regulatory requirements but also removes the potential of injury or stress to the participants as well as many of the logistical challenges of conventional simulator evacuation trials. Clearly, there are benefits in the use of computer simulation models for assessing passenger evacuation. However, it is important to note that modelling cannot simply replace all physical simulation trials. The very nature of the data required to run the model when new designs, procedures or conditions are explored is derived from humans under simulated controlled evacuation events.

10.4 Leading Passengers to Floor-Level Exits

Whether the testing of new equipment or procedures is examined within live evacuation trials or computer modelling, it is important to focus on some of the key elements that are critical in facilitating safe egress from the aircraft. To this end, it is vital that we consider the different types of exit doors that exist on commercial aircraft. The different types of exits vary in size and location along the fuselage. The largest exit is a floor-level type A exit that must be at least 1.07 m (42 in.) in width and 1.83 m (72 in.) in height, whilst the smallest is a type IV exit that is usually located above the floor level in the centre of the cabin and must be at least 0.48 m (19 in.) in width and 0.66 m (26 in.) in height, with a step down from the exit not exceeding 0.91 m (36 in.) (CS-25.807; EASA 2015). In between these extremes, there are type I and II floor-level exits (smaller than the type A), the type III, which is a non-floor-level exit larger than the type IV, plus the ventral and tail cone exits on some aircraft types.

The published research from cabin simulators has addressed evacuation issues at floor and non-floor-level exits, with much on the floor-level exit simulations through type I exits. Typically in the event of an evacuation, type I exits (examples are shown in Figures 10.4 and 10.5) are opened by a member of cabin crew and a slide deploys to assist the occupants from the aircraft to the ground.

The aircraft category and the maximum number of occupants will determine the number and exit types on the aircraft. Typical exit configurations for a narrow-bodied (single-aisle) aircraft are shown in Figure 10.6, in this instance depicting a Boeing 737-700. The aircraft has four floor-level type I exits and two overwing type III exits (discussed in Section 10.5). Some other narrow-bodied aircraft have pairs of overwing exits near the centre of the cabin.

Exits 1 and 5 in Figure 10.6 are type I passenger exits, and exits 2 and 6 are type I service doors (which are narrower in width and lower in height compared to the other type I exits but would be used in an evacuation). Exits 3 and 4 are type III overwing exits.

As a result of the Manchester accident, Muir et al. (1989) investigated changes to egress rates in a series of simulated evacuations from a disused aircraft. In this series of trials, the simulator was a decommissioned Hawker

FIGURE 10.5
Example of a floor-level exit: a type I exit with boarding steps attached.

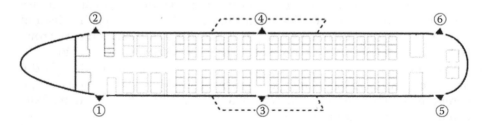

FIGURE 10.6
Example of cabin seating layout and emergency exit arrangement. (Courtesy of Aviation Safety Network, http://www.aviation-safety.net.)

Siddeley Trident 3B aircraft. The Trident was a narrow-bodied cabin, with an internal cabin width of 3.44 m (135 in.) and an internal cabin height of 2.02 m (80 in.) (The Trident Preservation Society 2010). The internal configuration replicated that of a traditional narrow-bodied layout, with a seat assembly containing three seats located in either side of the main aisle. The cabin interior included many of the original aircraft fixtures. For the research trials, the inflatable evacuation slides were substituted for ramps to assist passengers in travelling from the exit aperture to the ground (Muir et al. 1989).

These simulations tested variations in the cabin width between the bulkheads leading to the floor-level forward exits. It was reported that in the original accident, passengers displayed competitive behaviours that led to blockages whilst attempting to evacuate the burning aircraft. This behaviour was simulated using the competitive and collaborative evacuation experimental paradigms, and Muir et al. (1989) observed that as the passageway distance between the bulkheads was increased in size, the rate at which passengers exited the aircraft was increased as blockages were minimised. It was observed during the simulation that the crew sometimes experienced problems opening the exit doors and occasionally the cabin crew standing by the exits were actually pushed out and down the evacuation slides as the passengers rushed to exit the aircraft. When galley unit bulkheads were removed, the cabin crew had no protection from the evacuating passengers, and so, a recommendation was made that a minimum bulkhead width of 0.76 m (30 in.) should be maintained to ease the evacuation flow. Such simulated evacuations demonstrate that cabin designers need to consider the location of internal fixtures as these features affect cabin crew safety and passenger movement through an aircraft when evacuating in an emergency.

Following on from this research, Muir et al. (1990) and Muir et al. (1992) investigated the effect of bulkhead width on evacuation rate when non-toxic smoke was introduced to enhance the realism of the emergency situation by obscuring the vision of the participants trying to exit the cabin. These simulated studies were conducted in the same Trident 3B cabin as described by Muir et al. (1989). Overall, it was concluded that the presence of smoke reduced the rate at which the participants could evacuate and further supports the need for simulators to enhance the realism of situations in order that valid data be collected.

In further research, Muir and Cobbett (1995) investigated the role of cabin crew in supporting efficient evacuations. In these trials, the simulations were undertaken from a Boeing 737 cabin training simulator designed to replicate a cabin section of a typical narrow-bodied aircraft. As an example of the internal parameters of a cabin in this category, a Boeing 737-200 (the same as the one involved in the Manchester accident previously described) had an internal cabin height of 2.11 m (83 in.) and an internal cabin width of 3.54 m (139 in.). The aircraft had a single aisle of 0.51 m (20 in.) in width, with a standard economy configuration layout of a seat assembly with three seats in either side of the aisle (Boeing Commercial Airplanes 2013). For the simulated evacuations completed by Muir and Cobbett (1995), ten rows of six seats were installed in the simulator and the cabin simulator was fitted with four operational exits. At one end of the simulator was a type I exit; in the centre, one type III (overwing) exit; and at the other end of the simulator, a second type I exit and a service type I exit.

In the simulations, the participants were directed by cabin crew who were either assertive or non-assertive and, as a control condition, the participants were also required to evacuate without any cabin crew present. The results indicated that the presence and behaviour of the cabin crew had a significant effect on evacuation rates, with assertive crew producing the fastest evacuation rates compared to non-assertive crew. Interestingly, there was no significant difference between having non-assertive cabin crew and no crew at all (Muir and Cobbett 1995). This would suggest that the cabin crew play an important role in the instruction and guidance of passengers during evacuation events but that the crew also need to display confident evacuation behaviours (e.g. directing passengers to exits, following procedures).

10.5 Leading Passengers to the Non-Floor-Level Exits

Along with floor-level exits, non-floor-level exits can pose evacuation challenges due to the size of the exit and the adjacent space around the exit. Non-floor-level exits also bring additional operation issues due to their design. Galea (2003) observed that during evacuations, a high number of passengers used the central overwing exits as these were their nearest available exit. These are usually defined as type III exits and are a rectangular exit above floor level (similar to a type IV but larger) and located approximately in the centre of the fuselage either above or below the wings depending on the type of aircraft. The minimum dimensions of a type III must be 0.51 m (20 in.) in width and 0.91 m (36 in.) in height. The step-up to the exit inside the aircraft must be no greater than 0.51 m (20 in.) and the step down outside the exit must not be greater than 0.69 m (27 in.) (CS-25.807; EASA 2015). Type III exits are installed on aircraft with a maximum seating capacity of up to 299 seats (CS-25.807; EASA 2015). The published research from cabin simulators including exits similar to those in Figures 10.7 and 10.8 has addressed evacuation issues at type III (overwing exits), primarily access to the exit and passenger operation of the exit.

Work on evacuation through the type III exit was also conducted by Muir et al. (1989) in the Trident 3B simulator as part of the same research programme as the evacuation trials through floor-level exits. Muir et al. (1989) investigated the space that was available in the access row to the type III exit and, using several different seating configurations, concluded that a vertical distance (i.e. the minimum vertical projection between two seat rows at the

FIGURE 10.7
Example of a type III exit viewed from the outside of the aircraft.

FIGURE 10.8
Example of a type III overwing exit from inside the aircraft.

exit passageway according to Quigley et al. 2001) of 0.46 m (18 in.) was optimal for promoting efficient evacuation rates whilst at the same time avoiding crowding and bottlenecks. Further work supported this finding with the observation that a vertical projection distance between the exit seat rows of 0.33–0.46 m (13–18 in.) was optimal (Muir et al. 1990). However, in competitive behaviour trials simulating a life-threatening evacuation scenario, a wider vertical projection between the exit seat rows was associated with faster evacuation rates (Muir et al. 1992). Again, these simulations through the overwing exit were conducted in the Trident 3B cabin simulator.

As well as the UK-based work in this area, the Civil Aero-Medical Institute of the US FAA also undertook an extensive programme of experimental simulations to investigate passenger access to the overwing exits during an evacuation (Rasmussen and Chittum 1989; McLean et al. 1995; McLean et al. 2002). In one study (Rasmussen and Chittum 1989), a cabin simulator with 14 rows of triple seats installed in either side of the main aisle was used. A type III exit from a disused Boeing 720 aircraft was mounted in the simulator, which was modelled to represent a Boeing 727 in relation to overhead height restrictions in the seat rows adjacent to the exits. The simulator was designed so that the pathway width to the type III exit could be modified during the experimental work. Whilst the cabin simulator was modelled on a different aircraft type to the UK simulations, both simulators were modelling key attributes of narrow-bodied aircraft with overwing exits. Rasmussen and Chittum's (1989) study showed that egress was quickest when there was a 0.51 m (20 in.) vertical projection or when the outboard seat was removed and there were two 0.15 m (6 in.) passageways leading to the exit, compared to just a single passageway (Rasmussen and Chittum 1989). McLean et al. (1995), using a cabin simulator designed to replicate key features of narrow-bodied single-aisle aircraft with overwing exits, found that participants were slower to evacuate in simulated configurations where the vertical projection between the seats at the overwing exit row was narrow (e.g. 0.15–0.25 m or 6–10 in.) as well as when there was greater seat encroachment on the exit. McLean et al. (2002) went on to examine access to the type III exit in an extensive series of simulations from a typical narrow-bodied cabin, with three seats in either side of the main aisle and one operational type III exit on the starboard side of the cabin (leading out onto a sloped wing on the exterior of the simulator). Exit passageway configuration was one of the variables tested, but unexpectedly, given the previous research findings, it was found to have only a minimal impact on evacuation times, with anthropometric and bio-behavioural factors (e.g. participant waist size, gender and age) having a greater influence on evacuation rate. These findings emphasise the important influence of certain anthropometric characteristics of passengers on evacuation through a type III exit; however, these are not variables that can be controlled by the industry.

Having made their way to an exit, the assumption is that passengers will evacuate out of an open door, but one aspect of cabin design that has undergone a great deal of testing in experimental simulators is the manual operation of the passenger-operated type III exits (Fennell and Muir 1993; Cobbett et al. 1997; Wilson et al. 2004). Much of this research arose from observations of actual evacuations where doors either were not opened properly or took longer than expected to open before passengers could exit. The NTSB conducted a systematic review of aircraft evacuations and found physical difficulties operating this type of exit (e.g. manoeuvring the exit hatch and disposing of it) and a lack of understanding about the operation of the exit hatch mechanism (NTSB 2000). Simulator studies have therefore contributed to our understanding of the factors that may help or hinder passengers who are called upon to operate these exits in the event of an evacuation.

Traditional type III exits differ in their operation to floor-level exits in that once they are released, the hatch is not actually attached to the cabin fuselage by a hinge mechanism and will fall into the cabin (Wilson and Muir 2010). When opening a type III exit, the passenger first needs to check that outside of the aircraft is safe for the exit to be used (e.g. there is no overwing fire visible). The passenger is then required to pull on the release handle of the hatch, releasing the hatch from the aperture (see Figure 10.9). The passenger

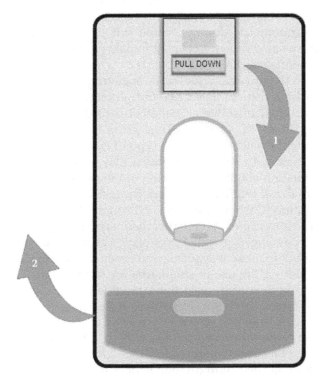

FIGURE 10.9
Typical operation of a type III exit.

operating the exit needs to physically manoeuvre and rotate the hatch that can weigh up to 29 kg (NTSB 2000). The hatch needs to be placed or disposed of in a suitable location where egress through the exit is not impeded. Within the United Kingdom, the advice on many safety cards is to dispose of the hatch outside the aircraft onto the wing; however, in other countries (including the United States), the advice is to dispose of the hatch either onto the aircraft wing or onto the cabin seats adjacent to the exit in line with their regulatory guidance.

A number of studies have been conducted into the potential improvements to the operation of type III exits (Fennell and Muir 1993; Cobbett et al. 2001; Wilson et al. 2007). Trials undertaken by Fennell and Muir (1993) used a Boeing 737 cabin simulator to examine the operation of the port side type III exit (this is the same simulator as described by Muir and Cobbett [1995]). However, given that the focus of the simulations was on the passenger-operated exit, in this series of exit operation trials, only five rows of aircraft triple seat assemblies were installed in either side of a single aisle in the centre of the cabin adjacent to the type III exit. The authors found that when the weight of the hatch was reduced to 12.5 kg and the distance between the seats in the exit row increased to 0.33 m (13 in.), exit operation time was significantly decreased. The change in weight of the hatch resulted in a benefit even when the seat space or passageway adjacent to the exit was impeded using an anthropometric dummy in the exit operator's seat. The lighter-weight hatch was also shown to benefit female passengers when they operated the exit (Fennell and Muir 1993).

In addition to the physical issues with operating traditional type III exits, past accidents have demonstrated that untrained passengers are not always able to operate the exit effectively. Cobbett et al. (2001) and Wilson et al. (2007) examined the nature of safety briefings given to passengers before being asked to operate the hatches located at type III exits. Given the non-intuitive nature of the task, both studies concluded that providing passengers with detailed information about the operation of the exit assisted them with the task. The more detail that was provided, the faster the exit was made available for evacuation (Cobbett et al. 2001; Wilson et al. 2007). This would suggest that in the absence of cabin crew, the instruction given to passengers located next to passenger-operated exits plays a critical role in the time taken to operate the exit to make it available and evacuate the aircraft.

When considering the use of simulators for such research, it is of interest to note that both studies used the same Boeing 737 cabin simulator as previously described; however, there were differences in the internal configurations between the two studies. Cobbett et al. (2001) used the standard narrow-bodied configuration with three seats in either side of the main aisle, whereas Wilson et al. (2007) simulated this condition but also completed some trials in a modified configuration simulating key internal features of a smaller transport aircraft (typically fewer than 44 seats). This modification was achieved with the installation of a false wall narrowing the cabin by approximately two passenger seats and reducing the cabin height by lowering the overhead lockers (the height from the base of the overhead lockers was 1.64 m [64.5 in.] in the standard configuration and reduced to 1.38 m [54 in.] in the smaller transport aircraft configuration). With the reduced cabin width, there were four seats across the cabin, two in either side of the central aisle. The ability to modify the internal aspects of any cabin simulator to meet industry needs and different research questions is of importance during simulator design and construction.

As a result of the issues surrounding type III exits in actual evacuations and exit operation findings from simulator trials, the European Aviation Safety Agency (EASA) now requires newly certificated aircraft types with overwing exits and seating configurations of at least 40 passengers to employ an overwing exit that is automatically disposed of once it has been opened (EASA 2008; EASA 2015). This ruling means that the physical and cognitive demands of operating traditional type III exits are minimised. The benefits in terms of exit operation time of one automatically disposed hatch design have been demonstrated in simulated exit operation experimental trials from the Boeing 737 cabin simulator previously outlined, in both the standard narrow-bodied internal configuration with three seats in either side of the aisle (Cobbett et al. 1997) and the smaller transport internal configuration with two seats in either side of the aisle and lowered cabin overhead space (Wilson et al. 2004).

10.6 Evacuation Certification Requirements

The importance of aircraft evacuation capabilities is recognised in airworthiness certification procedures. In terms of evacuation capability, manufacturers must ensure that in the case of a non-fatal or survivable landing, all occupants can exit the aircraft rapidly and without external assistance. The regulatory authority (i.e. EASA within Europe) issues strict regulations within Certification Specification Part 25.803 that require aircraft with a seating capacity of more than 44 passengers to demonstrate an emergency evacuation as part of the certification process. The evacuation certification demonstration is required to show that all passengers and crew can evacuate from the aircraft within 90 seconds, with cabin conditions designed to simulate an emergency and using only half the available exits (EASA 2015). This regulation sets a benchmark for new aircraft designs and lays down a detailed specification of the evacuation conditions.

These regulatory evacuation demonstrations are essentially simulations of an emergency scenario to establish that the aircraft can meet the certification requirements. However, by the time that a certification demonstration is required, there has been a significant investment in the design and development of the aircraft interior (and associated emergency equipment and procedures) to be used by passengers and crew.

As different aircraft vary in the characteristics of their exits (e.g. placement, size, operation), the demonstration exercise must include the use of at least one exit of each type installed. Modern aircraft design tends to place emergency exits in pairs, mirrored on the port and starboard sides of the fuselage (as in Figure 10.6). Galea (2006) has highlighted that this leads to the exits along one side of the fuselage being used to evacuate occupants for certification purposes. In reality, however, and as seen in the two examples at the start of this chapter, accident analysis has demonstrated that this is not the combination of available exits found in the majority of accidents (Togher et al. 2009).

The certification demonstration must also involve a representative number of occupants (passengers and crew) that reflect the aircraft operating at maximum capacity. Occupants are not allowed to practice any elements of the demonstration, nor have any aspects described to them in advance (above and beyond the standard safety briefing), such as which exits should be used during the evacuation exercise. The cabin crew who participate in demonstrations must be actual operating airline crew. The people acting as passengers may not take part if they are involved in the operational aspects of the aviation industry or have been involved in a similar demonstration within the previous six months. The regulations stipulate that those acting as passengers should be healthy and at least 40% of them are female. Furthermore, at least 35% of participants should be over the age of 50 years, and of this sample, at least 15% should be female. Due to ethical and safety concerns, infants are not involved in the demonstration; however, to simulate their presence, three passengers are required to evacuate with life-sized dolls representing infants of two years old or younger.

Whilst these regulations govern the certification demonstration, simulator studies are not bound by the same requirements and aim to have as a wider range of passengers participating if possible. However, given the physical and potentially psychological nature of evacuation simulation trials, not all typical passengers can be recruited for some trials, hence the benefits of combining actual trials with computer modelling.

The certification exercise must also be conducted with standard internal emergency lighting (e.g. illuminated exit signs, emergency cabin lighting and floor proximity emergency markings). Cabin luggage, blankets and other passenger items are distributed in aisles and exit access route locations to simulate minor obstructions that may occur as a result of items being displaced during an accident or rough landing.

In some circumstances, rather than the full-scale evacuation demonstration, the regulator may allow a manufacturer to use a 'combination of analysis and testing' if this 'will provide data equivalent to that which would be obtained by actual demonstration' (EASA 2015, p. 1-D-26). Due to the considerable cost in terms of time and economic factors associated with conducting an emergency evacuation demonstration, the use of sophisticated computer models using human behaviour data from previous simulator trials can considerably de-risk the design, development and certification process.

10.7 Issues of Ecological Validity in Simulators for Emergency Evacuations

Although we have outlined some of the main approaches to using simulators in passenger evacuation research, as with any form of simulation, it is important to consider underlying issues of ecological validity for investigating human behaviour in this way. It is crucial to achieve a balance between the realism of the experimental design that participants experience and ultimately ensure their safety. Issues of both physical and psychological well-being have to be considered, to ensure that the risks to those involved in the simulation are as low as possible without compromising the purpose of the research. When designing simulator studies, it is acknowledged that researchers have to make their best judgement in taking all experimental variables and test parameters into consideration. However, it would not be ethical, safe or feasible to expose participants to any experience that is too closely aligned to an actual accident (e.g. exposure to toxic gases, smoke, heat or significant physical or psychological distress).

Cabin simulators and the associated evacuation studies are made as representative as possible using realistic cabins, the presentation of pre-flight safety briefing demonstrations and safety cards and trained and uniformed cabin crew members displaying assertive behaviours when issuing commands and directions to encourage a timely and efficient evacuation. As is commonplace with experimental research, participants are not briefed in advance about some aspects of the specific evacuation scenario. This is done to minimise any influence that this may have on their subsequent behaviour. This information may include specific task details, procedures on evacuation and even which exits will be available (as would be the case during a realistic unplanned evacuation).

It is also important to acknowledge that there will be differences in the motivation of passengers taking part in a simulated evacuation trial compared with those unfortunate enough to be involved in a real evacuation situation. Participants adopting the role of passengers in an evacuation simulation know that they are not on a real flight and may not behave in the same way as they would in a situation where there is, or they perceive there to be, a real threat to their life or that of their travelling companions (i.e. family or friends). In order to enhance the ecological validity of the simulated evacuations in terms of the motivation of the participants to evacuate as quickly as possible, the competitive evacuation paradigm previously outlined (which involves providing additional bonus financial payments to those who evacuate in the first 50% of participants) may be used.

The competitive methodology can elicit different passenger behaviours from those typically found in collaborative trials (NTSB 2000). The NTSB provided valuable data from survivors of 46 evacuations, using self-completion questionnaires covering their behaviour during the evacuation and the behaviours of others at the time. Of the 331 passengers who answered, 75% agreed or strongly agreed that the other passengers were cooperative, with 13% disagreeing or strongly disagreeing and 12% remaining neutral on the issue. In the context of this discussion, some support is provided for the appropriateness of the cooperative behaviours observed during evacuation trials.

10.8 Conclusion

The aircraft upon which we travel every day are closely regulated to very high safety standards, and any modifications or changes in the design or procedure must reach a level that satisfies specific regulations. We have seen that aircraft emergency evacuations pose several challenges that we must examine and strive to address in order to make the evacuation as effective and efficient as possible. Simulation has been used within the area, with extensive research undertaken involving experimental trials from physical aircraft cabin simulators and computer modelling environments.

As has been discussed, cabin simulators can be used in experimental research to allow as much realism as ethically and practically possible, whilst at the same time giving the researchers control over the testing environment. Cabin simulators have been used to investigate the impact of a number of variables on passenger evacuation, including the internal cabin layout, passageway width of exits, the influence of exit type, location and size and the operation of exits and other safety-related equipment.

As we strive to ensure that passengers can safely exit an aircraft in the unlikely event of an emergency, it is important to recognise the limitations that all simulation methods possess as we cannot fully replicate the situation which we are trying to mitigate against. If simulating an evacuation, it is important to achieve a situation that best represents valid behaviours that reflect (as closely as possible) the actual event. However, it would be next to impossible to achieve an ethical and robust experimental environment that truly reflects the human behaviours associated with a life-threatening emergency situation, although much can be done to develop rigorous experimental designs and realistic test conditions. There are benefits in using different methodologies to address these factors and in some cases using combinations of different methods to assist in validating different approaches.

Whilst the use of simulation in the field of aircraft emergency evacuation has achieved a great deal, incidents and accidents continue to raise many questions and areas that need to be investigated from a human factors perspective, which can be done so with the aid of various forms of simulation.

References

Air Accidents Investigation Branch. 1988. *Report on the Accident to Boeing 737-236 series 1, G-BGJL at Manchester International Airport on 22 August 1985.* Aircraft accident report 8/88. London: Her Majesty's Stationary Office.

Boeing Commercial Airplanes. 2013. *737 Airplane Characteristics for Airport Planning.* Seattle, WA: Boeing Commercial Airplanes.

Cobbett, A.M., Jones, R.I., and Muir, H.C. 1997. *The Design and Evaluation of an Improvement to the Type III Exit Operating Mechanism.* CAA paper 97006. London: Civil Aviation Authority.

Cobbett, A.M., Liston, P., and Muir, H. 2001. *An Investigation into Methods of Briefing Passengers at Type III Exits.* CAA paper 2001/6. London: Civil Aviation Authority.

EASA (European Aviation Safety Agency). 2008. *Notice of Proposed Amendment No 2008-04 Type III Emergency Exit Access.* Cologne, Germany: EASA.

EASA. 2015. *Certification Specification 25 for Large Aeroplanes, Airworthiness Code.* Amendment 17, 15 July 2015. Cologne, Germany: EASA.

European Transport Safety Council. 1996. *Increasing the Survival Rate in Aircraft Accidents.* Brussels: European Transport Safety Council.

Fennell, P.J., and Muir, H.C. 1993. *The Influence of Hatch Weight and Seating Configuration on the Operation of a Type III Hatch.* CAA paper 93015. London: Civil Aviation Authority.

Galea, E. 2003. Passenger behaviour in emergency situations. In *Passenger Behaviour,* ed R. Bor, 128–182. Aldershot, UK: Ashgate.

Galea, E.R., Blake, S.J., and Lawrence, P.J. 2005. *Report on the Testing and Systematic Evaluation of the airEXODUS Aircraft Evacuation Model.* CAA paper 2004/05. London: Civil Aviation Authority.

Galea, E.R. 2006. Proposed methodology for the use of computer simulation to enhance aircraft evacuation certification. *Journal of Aircraft.* 43:5:1405–1413.

McLean, G.A., George, M.H., Chittum, C.B., and Funkhouser, G.E. 1995. *Aircraft Evacuations through Type III Exits I: Effects of Seat Placement at the Exit.* DOT/FAA/AM-95/22. Washington, DC: Office of Aerospace Medicine.

McLean, G.A., Corbett, C.L., Larcher, K.G. et al. 2002. *Access-to-Egress I: Interactive Effects of Factors That Control the Emergency Evacuation of Naive Passengers through the Transport Airplane Type III Overwing Exit.* DOT/FAA/AM-02/13. Washington, DC: Office of Aerospace Medicine.

Muir, H., Marrison, C., and Evans, A. 1989. *Aircraft Evacuations: The Effect of Passenger Motivation and Cabin Configuration Adjacent to the Exit.* CAA paper 89019. London: Civil Aviation Authority.

Muir, H., Marrison, C., and Evans, A. 1990. *Aircraft Evacuations: Preliminary Investigation of the Effect of Non-toxic Smoke and Cabin Configuration Adjacent to the Exit.* CAA paper 90013. London: Civil Aviation Authority.

Muir, H.C., Bottomley, D., and Hall, J. 1992. *Aircraft Evacuations: Competitive Evacuations in Conditions of Non-toxic Smoke*. CAA paper 92005. London: Civil Aviation Authority.

Muir, H., and Cobbett, A. 1995. *The Influence of Cabin Crew during Emergency Evacuations at Floor Level Exits*. CAA paper 95006, Part A. London: Civil Aviation Authority.

Muir, H.C., Bottomley, D.M., and Marrison, C. 1996. Effects of motivation and cabin configuration on emergency aircraft evacuation behaviour and rates of egress. *International Journal of Aviation Psychology*. 6:1:57–77.

Muir, H. 2004. In times of crisis how do passengers react? Paper presented at the 15th Westminster Lecture on Transport Safety. Parliamentary Advisory Council for Transport Safety, London.

NTSB (National Transportation Safety Board). 1986. *Aircraft Accident Report Air Canada Flight 797 McDonnell Douglas DC-9-32, C-FTLL Greater Cincinnati International Airport Covington, Kentucky June 2, 1983*. Washington, DC: NTSB.

NTSB. 2000. *Safety Study: Emergency Evacuation of Commercial Airplanes*. NTSB/SS-00/01. Washington, DC: NTSB.

NTSB. 2010. *Loss of Thrust in Both Engines after Encountering a Flock of Birds and Subsequent Ditching on the Hudson River, US Airways Flight 1549, Airbus A320-214, N106US, Weehawken, New Jersey, 15 January 2009*. Washington, DC: NTSB.

Quigley, C., Southall, D., Freer, M., Moody, A., and Porter, M. 2001. *Anthropometric Study to Update Minimum Aircraft Seating Standards*. Report prepared for the Joint Aviation Authorities. Loughborough, UK: ICE Ergonomics.

Rasmussen, P.G., and Chittum, C.B. 1989. *The Influence of Adjacent Seating Configurations on Egress through a Type III Emergency Exit*. DOT/FAA/AM-89/14. Washington, DC: Office of Aviation Medicine.

Snow, C.C., Carroll, J.J., and Allgood, M.A. 1970. *Survival in Emergency Escape from Passenger Aircraft*. AM70-16. Washington, DC: Office of Aviation Medicine.

Stedmon, A., Lawson, G., Lewis, L., Richards, D., and Grant, R. 2015. Human behaviour in emergency situations: Comparisons between aviation and rail domains. *Security Journal*. 9 November 2015. doi: 10.1057/sj.2015.34.

The Trident Preservation Society. 2010. Technical description. http://www.hs121.org .htm (accessed 17 August 2016).

Togher, M., Galea, E.R., and Lawrence, P.J. 2009. An investigation of passenger exit selection decisions in aircraft evacuation situations. Paper presented at The Fourth International Symposium on Human Behaviour in Fire. Cambridge, UK, 13–15 July 2009.

Wilson, R.L., Thomas, L.J., and Muir, H.C. 2004. Recent Transport Canada cabin safety research at Cranfield University. Paper presented at The Fourth Triennial International Aircraft Fire and Cabin Safety Research Conference. Lisbon, Portugal, 15–18 November 2004.

Wilson, R.L., Caird-Daley, A.K., and Muir, H.C. 2007. The effect of operators' briefing, cabin configuration and operating handle mechanism on type III exit operation. Paper presented at The Fifth Triennial International Aircraft Fire and Cabin Safety Research Conference. Atlantic City, NJ, 29 October–1 November 2007.

Wilson, R.L., and Muir, H.C. 2010. The effect of overwing hatch placement on evacuation from smaller transport aircraft. *Ergonomics*. 53:2:286–293.

Section V

Maritime

11

Maritime Research

Margareta Lützhöft, Paul Brown, Richard Dunham
and Wessel M. A. van Leeuwen

CONTENTS

11.1 Introduction

What constitutes a simulation and a simulator? In this context, a simulation is a combination of scenarios, human participants and models. A simulator provides the underpinning of the simulation, most often in the form of hardware and software. There are also a number of different types of simulator, usually divided in order of their fidelity, complexity and context coverage. We have full mission, multitask, limited task and special task bridge (or other control room) simulators and tabletop, software and games (e.g. Bergström et al., 2009). All have their uses, and the choice should be carefully considered in view of your goal and intended result, rather than utilising technology for its own sake.

However, a simulation must be validated to ensure the credibility or quality of results. We need to be confident that the elements of the simulation are behaving within an accurate degree of comparison to the real world (McIlroy et al., 1981), and to do so, we must validate such elements only within the context of that for which they are designed and for the particular research output that we desire (Balci & Sargent, 1981; Hora, 2015; Rehman & Pedersen, 2012; Sargent, 2013). In this chapter, we discuss the process of attaining quality results from research in simulation and provide tips for long simulator runs and prospective simulator buyers. We do not claim to offer a complete review of ship simulator research studies but provide a selection of representative areas of research studies.

11.2 History

The late 1960s saw the pioneering development of maritime simulation. SSPA Sweden (a ship research and development facility), the Netherlands Organisation for Applied Scientific Research and the Netherlands Ship Model Basin were all actively operating simulators with full wheel house panoramic capabilities (Eda, Guest & Puglisi, 1996). Whilst the focus was on the development of specialised training simulations for the maritime community using human-operated models for ship manoeuvring and radar simulations, research in marine simulations was inevitable, even if they did not know it at the time.

The earliest simulation developments include the Computer Aided Operations Research Facility (CAORF), operated by Marine Safety International at Kings Point, New York, which utilised various techniques with the aim of developing as realistic a simulated environment as technologies at the time would allow. CAORF used objective data derived from model testing in towing tank facilities to examine ship handling characteristics in various depths, with these data directly incorporated into the first of the computer-generated image models, a 763-feet tanker. To further investigate the validity of the model, subjective input was sought from the opinion of a well-respected New York harbour pilot. A series of trials concluded that the model was excessively manoeuvrable in shallow water and the model was amended accordingly. Shallow water effects (squat) in the CAORF facility are now well represented (Eda, Guest & Puglisi, 1996), as they are in other simulator facilities around the maritime world, and research using maritime simulators has been evolving ever since, still using a combination of objective and subjective analyses.

The US Maritime Administration and the US Coast Guard conducted extensive operational research into human performance during the 1970s and early 1980s by using a computer-based ship-bridge simulator constructed for that purpose (National Research Council Committee on Ship-Bridge Simulation Training, 1996). The Maritime Research Institute Netherlands (http://www.marin.nl) was founded in 1929 and expanded into navigational training in 1970. The International Marine Simulator Forum was established in 1978 and is an organisation for professionals in education, training, research and development in simulation.

Ship simulators can be regarded as a development of wave basins and towing tanks; following a growing need to more directly control or manoeuvre the ship, more end-user focus was needed. An early publication about ship manoeuvring simulators (Ivergard, 1975) describes how various types of simulators can be used for training and for studying hydrodynamics. Brief mention is made about studying human/machine experiments and bridge design, but the emphasis in the paper is still almost exclusively on simulators as training and learning tools. It is usual to differentiate between three areas of use for simulators: for technical development, for education and training and for experiments and research on the HMI (Ivergard, 1975). We can also distinguish the use of simulators for investigations of accidents. Furthermore, since Ivergard wrote his paper, we have expanded the areas of research from interfaces to a large range of subjects.

Much effort has been expended on making the visual view as realistic as possible, following feedback from seafarers on which aspects are useful for navigation and manoeuvring. Examples include the effect of currents on buoys and other navigational marks and the field of view from the bridge. A subsequent development has been on adding or networking simulators representing other workplaces and/or tasks on the ship. Examples include engine room and engine control room simulators; radio and communication; cargo handling; and various types of cranes and winches, dynamic positioning, tug and Vessel Traffic Services simulators (Lützhöft et al., 2010b; von Lukas, 2006). The focus at present is on interconnecting simulators to make a 'system simulation' or realistic workplace simulation environment including all operators and resources. At least two such projects have studied marine engineers, looking at benefits of training engineers on maintenance (Dimitrios, 2012) and the Project Horizon, described later in this chapter.

However, most examples are about bridge operations. These study the effects of new technology on the bridge, such as decision support or Electronic Chart Display and Information System (ECDIS), or the performance of crew under varying conditions of workload or fatigue. The next step is interconnecting networks of simulators in different locations. The Mona Lisa 2.0 project (http://monalisaproject.eu/) has interconnected simulator centres around Europe and globally to test e-navigation solutions.

11.3 Why Simulators?

As indicated in Chapter 12, simulators in the maritime environment have, to a large degree, been used for training, and the research aspects have come later. The alternative to a simulator, if one wants data on actual on-board processes, is to perform field studies on ships. They may be performed by researchers spending time on board collecting data through observations and interviews, by leaving or sending measuring instruments on board (ranging from logbooks to surveys to recording devices) or various combinations of this. In the strictest sense of a field study, one would expect a researcher to be on board for some measure of time. The take-home message from most field studies is that more time should be spent on board to cover more of the context; too many interacting variables make it hard to find that correlations, ships, crews and routes are all so different that drawing conclusions from any one ship or trip is almost impossible. On the other hand, rich data are collected; contextual data make it easier to interpret.

11.3.1 Benefits to Simulator Studies

There are a number of benefits to using simulators as opposed to real-world studies:

- Risk-free: Studies involving risk of accidents of any kind can safely be performed. Realistic hazards in high-risk operations are still present, which are important for fidelity, but the associated risk is absent.
- Ethics: Studies involving performance decrements are possible in simulators, whereas it would be nearly impossible to conduct these in real-life settings. Examples include alcohol, drug or medicine, sleep impairment, fatigue and motion sickness studies. Very stressful situations can also be explored, which would be both risky and perhaps unethical in real-world settings.
- Repeatability: A test can be repeated with or without variation almost endlessly. One should, of course, take into account participant dropout and/or learning effects.

- Controllability: A number of variables can be controlled and adapted to needs of the scenario: size of vessel, place of simulation/scenario, participants, weather, sea state, visibility and other environmental aspects.

- Data assimilation: Technical data, such as video or sound, are more simply and potentially more thoroughly recorded, with enhanced control of the placement of recording devices, which may not be possible on a real ship.

- Efficiency: Control of timeline (real time versus time segment); for example travelling 200 nautical miles at 10 knots need not take 20 hours nor is there a loss of time with a ship coming up to speed.

- Cost: Cost can be significant in the use of actual vessels for research purposes. There is a cost attached to the use of simulators, but compared to travelling to a real ship, the benefits of control, less risk and repeatability will outweigh this. For some studies, one may wish to validate in a ship environment, but all the groundwork can be done before, in simulators.

- Availability: Access to a specific vessel and personnel can be difficult in the shipping culture of maximising commercial gains. It may be that the research is dependent on availability of the vessel, rather than achieving the best research outcome at the appropriate time. However, availability may also be an issue for simulators. Simulators are mainly used for education purposes and likely to be well booked, and commercial projects may take priority, leaving the researcher with less favourable time slots.

11.3.2 Drawbacks to Simulator Studies

- Cost of validity: Developing maritime simulated models to accurately represent their real counterpart and specific requirements or answer particular questions can be a costly and difficult process to achieve. This can require significant resources, experimentation, data acquisition and input. Multipurpose maritime simulators can be limited in their ability to meet specific requirements such as bridge layout, size, hardware and software. Producing a valid area model with complete photorealistic visual features is an example.

- Fidelity: Important aspects of reality may not be accurately perceived by participants and/or a tendency to treat it like a very large and expensive computer game. However, it is often seen that after a short while, participants – especially those with more seafaring experience – will embrace the situation as if it were a real bridge watch. Often, expressions of relief or comments about sweaty palms are heard after a challenging scenario.

- Simulator sickness: None of the studies reviewed in this chapter have reported it, but there is anecdotal evidence. We may also assume that experienced seafarers seldom get motion sick, which might be a self-selecting trait. Some instances of simulator sickness are experienced, most likely due to the mixed signals – your eyes tell you that the ship is rolling and your vestibular system tells you that it is not. If the participant steps out briefly or looks at the floor, the feeling of sickness usually goes away. With a moving-platform simulator, it is a different story, of course. In a recent validation study by one of the authors of this chapter, a very experienced senior officer left the simulated bridge after suffering some ill effects and disorientation during sea trials even though the conditions being simulated were commonly encountered in real life with no effect on the officer in question.

11.4 Research Studies and Applications

In this section, we will focus on research as it is traditionally conceived – studies published in scientific journals. In the context of maritime research, there are elements of research in many studies in port development and waterway studies for the purpose of safe manoeuvring or accident investigation, but here, we will look mainly at human performance studies.

11.4.1 Fatigue

Project Horizon was a European Union–funded study of seafarer's fatigue and its consequences on performance using both bridge and engine control room simulators. The project involved a total of 90 experienced seafarers: 70 bridge officers and 20 engineers. Participants stayed for a period of a full week, living close to the simulators and operating under either a four-hour on/eight-hour off or a six-hour on/six-hour off watchkeeping regime. The simulated one-week journey took place in the North Sea and English Channel under average traffic density. For comparison, all watch standers navigated the same stretch of water during a watch, as the trip was 're-started' for each new officer (i.e. twice on the six-to-six watch and three times on the four-to-eight watch), and then, their position was moved forward to where they realistically would have been at the start of their next watch.

The aims of the project were to investigate how sleep, sleepiness, fatigue and neurobehavioral performance are affected in the different watch teams under the two different watchkeeping regimes. During several watches in the journey, participants' electroencephalography was recorded in order to obtain objective indications of sleepiness as well as to identify actual sleep.

Furthermore, reaction time was assessed at the start and end of every watch, using a psychomotor vigilance task (Lamond, Dawson & Roach, 2005). In addition to general questionnaires related to sleep and performance, sleepiness was rated every hour during the entire week by means of the Karolinska Sleepiness Scale (Åkerstedt & Gillberg, 1990).

Several clear results were obtained. Sleepiness and fatigue were shown to peak during night and early morning watches. Moreover, it was clearly observed that sleepiness and fatigue increased during the course of the watch. Overall, sleepiness and fatigue levels were higher in the six-hour on/six-hour off simulations than in the four-hour on/eight-hour off simulations. A final, highly interesting, observation was that a substantial amount of seafarers – in both bridge and engine room – actually fell asleep whilst on duty (van Leeuwen et al., 2013). This control over the conditions could not have been done in an on-board study, as all participants would have been on different parts of the journey and in different parts of their on-board period and subject to varying weather and traffic (Lützhöft et al., 2010a).

11.4.2 Evaluation of Technology

One study (Gould et al., 2009a, b) aimed to find the possible effects on objective and subjective performances due to different navigational aids (ECDIS versus paper charts) in high-speed archipelago navigation. Twenty naval cadets working in pairs completed a 50-nautical-mile run in a bridge simulator. The performance measures used were expert evaluation, cross track error, National Aeronautics and Space Administration Task Load Index ([NASA-TLX]; Hart & Staveland, 1988), heart rate variability and skin conductance. Differences were found in course keeping, where ECDIS groups were more efficient. However, the number of speed violations was higher in the ECDIS group. Other than that, no significant differences were found in the performance measured – that is, the more objective measures did not capture any effects.

Evaluations of new technology on ships have been mainly limited to studies of the effects of advanced navigation systems on bridge operations. Some have been general, studying design of displays (Sauer et al., 2002) and the number of aids available (Nilsson, Gärling & Lützhöft, 2009), whereas others were more specifically focused on, for example, ECDIS (Donderi et al., 2004). The studies all show that care needs to be taken when introducing new technology to a workplace but perhaps also that we need to get better at designing our experiments. Where Sauer et al. (2002) used non-seafarers in their test, Donderi et al. (2004) and Nilsson, Gärling & Lützhöft (2008, 2009) recruited seafarers. The Donderi et al. (2004) and Nilsson, Gärling & Lützhöft (2008, 2009) studies both used 39 participants, making them relatively large simulator studies; nevertheless, neither study showed any large effects. In Sauer et al. (2002) and Donderi et al. (2004) experiments, a few effects were noticeable in cross track error, a measure judged by many domain experts to say very little about human performance. The Donderi et al. (2004) study used six seafarers and at least reached some conclusions on how to combine information on the (then) new chart system ECDIS. In the Sauer et al. (2002) study, secondary tasks were included, whilst in the other two, realistic conditions were present.

11.4.3 Alcohol and Stimulant Studies

Two research groups have published results on simulator studies – one on the effects of low-dose alcohol exposure on 38 cadets (both deck and engine; Howland et al., 2000, 2001) and the other on the effects of alcohol on navigational ability of eight deck cadets (Kim et al., 2007). In the Howland et al. (2000, 2001) studies, the main results were that low levels of alcohol (between 0.04% and 0.05% blood alcohol concentration) had significant effects on ship operation and, interestingly enough, that cadets were unaware of the effects on their performance.

Kim et al. (2007) measured electrocardiogram, NASA-TLX and performance in the simulator. Some differences in heart rate between no dose and low dose (0.05/0.08%) were found; significant differences in mental workload were present in 0.0–0.08 and 0.05–0.08 comparisons. Higher levels of alcohol led to significantly longer task time and lower speed in the simulator performance measurements. In a steering test, course keeping was affected at higher levels of alcohol. The use of rudders was reduced, implying that the care and control of the ship were perceived as less important.

A further study (Marsden & Leach, 2000) studied the effects on seafarers of alcohol and caffeine, respectively, and in combination. Twelve experienced seafarers were tested on chart tasks, visual search tasks and a navigational problem-solving task but performed using charts and booklets – so it does not really qualify as a simulator study, but rather a simulation, in this review. Nevertheless, alcohol (75 mL) produced impairment in the performance of visual search and navigational problem-solving, whereas caffeine improved the performance of the visual search.

These are examples of studies that could not be performed in an operational setting on board and include one study that uses the interesting methodological combination of measuring effects and the awareness of effects.

11.4.4 Other Human Factors Research

Research on training in simulators is relatively new, and many initiatives are directed at researching simulator pedagogy. Hontvedt and Arnseth (2013) showed how maritime students (cadets) interacting with a maritime pilot in a simulation take on professional roles and make joint sense of the situation. This indicates that the realism of a simulation is constructed not only by hardware and software, but also by the participants and actors in a simulation (Lützhöft et al., 2010b).

Two studies on training are of particular interest. Chauvin, Clostermann and Hoc (2008, 2009) conducted two simulator studies on situation awareness of cadets. In the Chauvin, Clostermann and Hoc (2008) study, they found few significant results relating to situation awareness, but they saw that navigation rules and the interpretation of them were central. Over half of the participants broke the collision avoidance regulations, leading Chauvin, Clostermann and Hoc (2008, 2009) to devise a training programme to enhance their decision-making and hasten their skill acquisition in the simulator. In the Chauvin, Clostermann and Hoc (2009) paper, we are presented with the somewhat surprising result that the group receiving simulator and decision-making training was no better than the groups with more sea experience; in fact, the latter made safer decisions. So simulator training as a replacement for sea time looks like it still has some way to go.

A more recent study (Godwin et al., 2013) compared inexperienced boat drivers with experienced – in this case, smaller craft were used (e.g. powerboats). Eighteen of each group were asked to drive through varying sea states (wave heights) whilst boat speed and eye movements were measured. In general, experts maintained a higher speed, and in contrast to road driving, experts had longer fixation durations than novices. This is explained as being because the most dynamic aspect of a traffic scene ashore is the surroundings, whereas at sea, it is often the driving surface.

Hockey et al. (2003) conducted a study on 12 non-seafarers in a ship simulator to test the cognitive demands of collision avoidance. The study aimed to complement similar studies in other transport domains and contained two tasks: the primary to avoid collisions and the secondary to monitor a separate display and keep a number within limits. The main finding was that secondary tasks can be used to assess cognitive demands in a maritime simulator. This study probably does not tell us much about navigation in a realistic context.

Conversely, a study by Orlandi, Brooks and Bowles (2015) is an example of a very well-planned study with high reliability and validity. They compared plans for ship manoeuvres in port with the actual performance of those manoeuvres in a simulator. Ten marine pilots planned and subsequently performed easy and difficult manoeuvres, with variables such as cross track distance and estimated use of engine, thrusters and tugs, respectively. The results showed that the methodology is valid and that, for example, pilots showed a high level of adherence to plan, with variances mainly in forecasting ship position during the swing phase. Comprehensive attention to the applicability of results is evident, and outcomes for both measurements of ship handling simulations and for the performance of marine pilotage are included in the results.

11.5 Applications: Engineering and Development

Port and fairway development studies are often advertised by maritime institutes as an area of research and use bridge simulators as the tool for such research. These simulators are primarily designed as instructional tools and generally meet the standard on maritime simulator systems from Det Norske Veritas Germanischer Lloyd (DNV-GL, 2014) designed to allow a flag state to approve the simulator installation as meeting the International Maritime Organization (IMO) training requirements. The research methodology is generally to carry out a number of entries and/or exits from a new port to determine such things as required channel depth and width, navigation mark type and position, environmental constraints, ship size limits, towage requirements and trial manoeuvres.

In 2011, The International Association of Marine Aids to Navigation and Lighthouse Authorities published guideline 1058, entitled *Use of Simulation as a Tool for Waterway Design and Aids to Navigation Planning* (IALA, 2011). This states that 'the purpose of simulation in AtoN (Aids to Navigation) planning and waterway design is to test, demonstrate and document various scenarios for deployment of various AtoN and waterway design under different conditions with the aim of identifying optimal operational safety and efficiency'. This implies that the test methodology will consist of approaches and departures to the port or fairway varying the environmental conditions (wind direction and strength, current direction and strength, height of tide, light conditions etc.) to determine optimal conditions. The results will be dependent on the accuracy of the models on which the simulation is based. In general, the accuracy of the models is gauged on how an experienced mariner feels that the model reflects the behaviour of a real vessel. This does not mean that the results obtained are unreliable due to the subjectivity of the model validation, since they will assist in decision-making and will help indicate the limitations of the trial.

Simulators are sometimes used as part of an investigation into the causes of an accident. Let us take as an example a collision event, in which detail of the positions, courses and speeds of the two vessels involved have been obtained from their voyage data recorders. The situation is re-created in the simulator and the exercise is run. For a collision to occur, the person in charge on both vessels must have taken actions contrary to the International Regulations for Preventing Collisions at Sea 1972 (Colregs) (International Maritime Organization, 1972). By running the exercise a number of times, it is possible to review the latest time at which either the course or speed of the vessel could have been adjusted, and the collision avoided, or to find the action necessary to avoid the collision, say, 10 minutes before the collision. Similarly, in the case of grounding, it may show that the earlier action, either changing course or speed, could have avoided the grounding or that given the particular series of events (emergency or environmental), the grounding could not have been avoided. This, again, is an intuitive result and may not need the use of a simulator to come to this conclusion. However, the simulator might be of use in showing that the navigation marks in the vicinity could be amended to allow the mariner earlier warning that the situation is becoming dangerous.

11.5.1 Two-Step Validation: Science Meets Engineering

Two worlds are meeting here: engineering and science. Both worlds are necessary for the making of good research results. The engineering speaks about verification, validation and fidelity of the models – mostly of the hardware and software. In science, similar concepts are used, for similar purposes, but the overlap is not complete. Science uses reliability, validity and sometimes objectivity. We have to apply a two-step process to make good science. Our instrument, the simulator, must be calibrated – thus the verification and validation of models and the need for high-fidelity models especially, but not exclusively, for those elements with which participants interact. Then, we can design the experiment, following the quality control of science.

11.5.2 True Simulator

The true *simulator* appears to mean many different things to many different professionals in simulation research. Here, we discuss definitions from publications and methodologies related to the development and application of the true *simulation* for research. These include accuracy, credibility, fidelity, validity and reliability.

Accuracy is about exactness (for example, compare predicted trajectory with actual) (National Research Council, 1996). If simulators are to be used for research, then we need to be able to rely on the information and data that are extracted from them and their integrated models. The credibility of the simulation and associated results is paramount, and the research output must be trustworthy.

Fidelity has been broadly defined as the degree of similarity between the simulated and the operational situations (Thomas, 2003). This broad definition has been further divided into a number of separating characteristics that can each be an individual focus of attention during research. Firstly is the notion of physical fidelity, which includes visual, kinaesthetic and spatial aspects. Secondly, functional fidelity refers to the degree of accuracy in system operation (Hays & Singer, 1989; Thomas, 2003). Thirdly, psychological fidelity refers to the degree of perceived realism, for which many extend this definition to fidelity as a whole. Finally, task fidelity has been defined as the degree to which a simulation is able to re-create the actual parameters of the operational mission (Macfarlane, 1997; Thomas, 2003) and is often dissolved within physical and functional fidelity.

Fidelity is a key focus point for effective implementation of maritime simulators for research and as such requires careful and specific consideration. A high-fidelity research environment may be specifically important to achieve the desired outputs of research results, but to achieve such an environment may come at considerable cost. This cost may be able to be considered or factored in during the research design and funding acquisition.

Reliability is addressed through model verification and validation, which is an important aspect of the model development process. Model verification has been defined as 'ensuring that the computer program of the computerized model and its implementation are correct', whilst model validation has been defined as the 'substantiation that a model within its domain of applicability possesses a satisfactory range of accuracy consistent with the intended application of the model' (Sargent, 2013). The elements of a simulation that require attention are the accuracy and fidelity of the following objective judgements (National Research Council, 1996):

- Image portrayal, including the content, quality, field and depth of view and movement of the visual scene
- The predicted ship trajectories based on hydrodynamic and aerodynamic modelling
- Own-ship (ship model) characteristics
- The operational scenarios used for evaluating or assessing

Simulation fidelity and credibility are intrinsically important considerations of any simulation system that is to be used, or is designed for research purposes, and it makes sense that this consideration is best made during the experimental design and achieved with a robust verification and validation process during this phase. It is commonly recognised by the modelling and simulation community that simulation fidelity is an essential vehicle in properly assessing the validity and reliability of simulation results and is one of the main cost drivers of any model or simulation development (Roza, 2005). There is a tendency to assume that the higher the fidelity of a simulator, the better it is for training purposes (Dahlström et al., 2009). The same argument might be applied to research purposes, taking into account the goal of the research. Dahlström et al. (2009) argued that a midlevel simulation (as opposed to a high-fidelity one) led to a focus on generic skills rather than procedural skills.

A recent research project at the Australian Maritime College investigated the importance of validity in regard to model integration in maritime simulators. The simulation used incorporated a number of individual models such as the ship model (hydrostatics, hydrodynamics and visual representation), the area model (charted area and visual representation), the climatic model (meteorology, currents and tides as examples) and the simulator layout (controls, electronic navigation aids). This study aimed to investigate the importance of validating individual models towards the complete validity of the simulation as a whole. It is suggested that validation of marine simulators or simulations might follow the approach of the commercial air carrier industry, with both objective and subjective validations (National Research Council, 1996).

For training purposes, fidelity and credibility are addressed in standards imposed by authorities and classification societies, such as those posed by the IMO, within the convention for Standards of Training, Certification and Watchkeeping and those detailed by the DNV-GL Classification Society. Standards are yet to be implemented for research simulators. In order to be able to perform research and collect valid and reliable data, we may have to bring in equipment (for example data collection or storage) because even if the simulator is good for training, that does not necessarily make it good for research. This additional equipment is not included in any validation process, which again may not matter, because to the best of our knowledge, no accreditation process exists for research simulators.

11.5.3 Validation Methodology

The validation methods chosen inevitably shape the approach taken to collecting research data. Since the earliest documented developments in maritime simulated models, various objective and subjective methods have been used. Many of the methods discussed in this section have been defined by Sargent (2013). Before we go there, let us consider what we wish to validate. Is it the simulation itself? Is it the results? Is it the applicability of those results, for publication, real-world application or both?

11.5.4 Objective Validation

Objective methodological approaches used in the development of maritime-relevant models draw on historical data, derived directly from the actual vessel for which the model is being developed. The inclusion of Global Positioning System (GPS) and more recently differential GPS (more accurate GPS) to modern navigation systems has provided more simple means of direct measurement and access to vast amounts of ship performance data that can be directly acquired for mathematical analysis and incorporation into models. For area model development, there are often much accurate historical data to be found. This includes weather and climatic data, tidal information and current movements. Whether these types of data, once included into the area model, have the correct interactive effect on other models requires further validation techniques to be used.

11.5.4.1 Extreme Condition Test

An extreme condition test can be used for any extreme and unlikely combination of levels of factors in the system. This method has been used in ongoing research at Australian Maritime College into the performance of simulated ice breaking. If the simulated icebreaker is able to uninhibitedly progress through impossible ice conditions, then there is certainty that the model(s) demonstrates significant flaws.

11.5.4.2 Event Validity

The events of occurrences of the simulation model are compared to those of the real system to determine whether they are similar. For example compare the number of helm movements required to navigate through set environmental conditions or the density of traffic in a specific waterway.

11.5.4.3 Data Relationship Correctness

Data relationship correctness requires data to have the proper values regarding relationships that occur within a type of data and between and amongst different types of data. It may be that a large amount of data is available from the actual ship, for example, but whether all of those data can be used in simulated model design or whether it is possible to incorporate all of the data as it exists in the real ship without it affecting other model data is an issue. Modelling professionals often need to decide which data to use and, often in the field of hydrodynamics, trade-off input of various data to best meet comparative outputs from the actual ship.

11.5.5 Subjective Validation

Subjective validation of the simulator and simulation is suggested to assess the behaviours in the simulator as compared to the real world. This may work for training purposes, but many research projects aim either to test new equipment for which no behaviour is well known or to assess the effect of changed work circumstances, whether induced by external circumstances such as drugs or internal factors such as stress or fatigue. Therefore, looking for similar behaviour will not suffice.

One can use experts of varying background: one or more instructors will be present, one or more experienced seafarers, a researcher experienced in experimental design and one or more researchers to collect data. Depending on the goal of the project, it may be valuable to include secondary stakeholders such as union representatives to ensure face validity of the results.

11.5.5.1 Face Validity

Individuals who are knowledgeable about the system are asked whether the model and/or its behaviour is reasonable. One way of measuring this is to apply usability or quality in use measures. Usability is defined as efficiency, effectiveness and satisfaction (ISO 9241-210), whilst the definition of quality in use is efficiency, effectiveness, satisfaction, freedom from risk and context coverage (ISO 25010). In the current context, freedom from risk has been interpreted to mean safety and context coverage as a variation of realism. A subjective metric called the quality in use scoring scale (QIUSS; developed by Brian Sherwood Jones under a Creative Commons licence) measures efficiency, effectiveness, satisfaction and safety (http://www.processforusability .co.uk/). Two of the present authors have built on the original QIUSS to also include realism and applied it to ship model validation studies. The face validity of, for example, interfaces, bridge or control room design has to be judged in accordance with the purpose of the study. If a generic study is being designed, then the interface need not look exactly as the real ship, but if a particular ship is being modelled, then the interface should be mimicked as closely as possible.

11.5.5.2 Internal Validity

Whilst it is expected in maritime simulations that repetitive performance of set values will be consistent in line with mathematically modelled algorithms, stochastic performance may occur as a result of operator input. Maritime ship models are designed to offer consistent interactions, but often, variability can be integrated into the simulation by the operator, such as variable tolerances in sea and weather conditions and changing tides.

11.5.5.3 Parameter Variability–Sensitivity Analysis

This technique consists of changing the values of the input and internal parameters of a model to determine the effect upon the behaviour or output of the model. The same relationships should occur in the model as in the real system. This technique can be used qualitatively – directions only of outputs – and quantitatively, in both directions and (precise) magnitudes of outputs. Those parameters that are sensitive and may cause significant changes in the behaviour or output of the model should be made sufficiently accurate prior to using the model.

In the case of making a 'good enough' simulator, we could also apply the well-known Turing test: individuals who are knowledgeable about the operations of the system being modelled are asked whether they can discriminate between real system and model outputs.

11.5.6 Link between Engineering and Science

In order to ascertain valid and reliable research in simulators, one piece is missing from the puzzle. We should include the participants, who in a research situation may vary from students without seafaring experience through cadets to experienced seafarers. More generally speaking, does the performance in the simulator match the performance in the real world?

In this particular setting, we must include the simulator operator into the experimental system. In many experiments, all is planned in advance, and when the simulation run starts, no changes are made. In a maritime simulation, changes may have to be introduced into the simulation as it is being run due to unexpected actions or events. This is an experiment with several dynamic variables where one of the most dynamic is the seafarer/participant. At sea, there are many ways to solve an upcoming problem. Some solutions are better than others, but most follow the rules and lead the navigator out of trouble. The experience of the participant, of course, has an effect on the validity and credibility of the results. It is quite common to use students, sometimes even non-seafaring, for simulator studies. In an attempt to counter the non-seafaring experience, introductions to the maritime simulator and navigation rules are made. Using students may be acceptable in traditional experiments, but using this approach in maritime simulator research is not acceptable for any validity. It is akin to giving a person without a driver's licence 30 minutes of driver training, putting them in a driving simulator, letting them drive and publishing the results as valid.

11.5.7 True Simulation

The concepts applied to a simulation should be the scientific ones: validity, reliability, objectivity and credibility. A number of maritime simulator research projects have the final aim of gathering data and results that are practically applicable to the maritime domain and its operators. Furthermore, to achieve uptake of and trust in results on potentially sensitive issues such as fatigue or shore-based assistance, experience shows that an appropriate level of realism needs to be applied. What is *appropriate* may differ from issue to issue, but the Project Horizon employed the use of a maritime stakeholder group for the planning and performance of the fatigue studies. The group was extensively involved in discussions of trade-offs necessary to get valid and reliable as well as applicable and trustworthy results.

A large number of studies focus on the measurement of (human) performance in the simulator. Very often, this is about navigational performance, couched in the terminology of less human error. The aim may be to show better performance when introducing new systems (Kircher et al., 2011) or decreased performance in studies of fatigue (Kircher & Lützhöft, 2011; van Leeuwen et al., 2013) or alcohol (Howland et al., 2001; Kim et al., 2007). Regardless of the objective, the phenomenon of performance is difficult to measure in a clear and unambiguous way. Some navigational acts may in the short term have an apparent positive effect, but the longer-term effect may be a more complex situation to handle. Hence, it is very difficult to quantify performance from a navigational perspective, where right and wrong are hard to identify and quantify in isolation from the context and task. At times, one may have to bend or break the rules to be safe, which is hard to judge if realism (or ecological validity) is more important than experimental control.

A common issue in comparative studies is how to design scenarios so as to get a measurable effect. Low measure sensitivity and/or the expertise of professional seafarers makes it very hard to pick up differences between assumed low- and high-workload scenarios. To get a measurable difference, the scenario with high workload has to be ramped up to almost unrealistic circumstance (high speed, a lot of traffic, communication and events to handle). All these aspects make it more difficult to validate measures and/or results in on-board studies.

Project Horizon, for instance, aimed to identify navigational performance impairments after night-time overtime work but, to a large extent, failed to do so due to this difficulty of quantifying what is to be considered good or bad performance. Furthermore, variability exists between different assessors. Some may consider a given manoeuvre an act of good performance, whereas others may consider the same manoeuvre an act of poor performance. Even with training to achieve inter-rater reliability and agreements on margins and limits, actions are always contextual. Therefore, as long as no incidents or accidents occur, it is rather difficult to quantify navigational performance within the normal range. Excessive deficits in performance (e.g. near collisions), although useful for quantification purposes, may not occur frequently enough to be useful in distinguishing a good from a worse performer. What we may consider looking for are not absolute values but comparisons, as in 'this performance was better than the last one', which is what many studies aim to find out.

Some results indicate that using the collision avoidance regulations (International Maritime Organization, 1972) is a reasonable but not perfect basis for judgement of some aspects of performance (Kircher & Lützhöft, 2011). However, this does not address the whole spectrum of good seamanship. The Colregs are written in a way which necessitates judgement, as they contain very few 'hard numbers'. Judgement reflects experience to judge what a safe margin is, be it distance or time, and rules of thumb develop. On the other hand, with experience, one may assume that some experience erodes the margins and, some might say, margins of safety. A safe distance when two ships meet may be affected by the size, speed and manoeuvrability of the ships, other traffic, depth and width of navigable water, weather, visibility etc. – which then has a follow-on effect when judging whether an action was indeed good performance.

Another example is a performance rating scoring system (e.g. Howland et al., 2001), in which the authors clearly state that the severity of a given error is defined by context, thus being a major or minor error depending on the circumstances.

Kim et al. (2007) measured time to complete a circular track, speed and course keeping. The course keeping was analysed by looking at both the radius of the track and the use of rudders. The issue with course keeping is that sometimes, it is safer to leave the course to avoid an incident. Staying on course is also a non-event and does not really tell us much about performance.

A study on Royal Norwegian Navy cadets to measure the effect of navigation method (paper chart or ECDIS) on performance showed lower workload effects than expected (Gould et al., 2009b). The same is true for a study comparing the relevance of navigational aids for decision-making (Nilsson, Gärling & Lützhöft, 2008), which found few significant effects. In both cases, it was concluded that seafarers are able to handle a relatively high workload, and to be able to find differences, some extra task or stressor would have to be introduced. Then, the problem is that we step away significantly from the normal work situation of a seafarer and lose ecological as well as face validity.

11.6 Practical Considerations

Most simulators appear to be made more for training than for research. One issue is that in some simulators, it is difficult to get data out of the internal database to use for scientific analysis. When it can be extracted, it may be in a format that needs extra work to be practically usable. Furthermore, with increasing systemic simulations, many parts, players or simulators are used and it can be difficult to synchronise or even time stamp for later synchronisation the different parts (e.g. ship and tug simulators). The operator of the simulator is often a technician or a trainer, who may be untrained in scientific rigor. We have seen examples of trainers adapting scenarios or events during the runs in order to make it more challenging with loss of control as a result. Obviously, this is a task of the team leader to ensure that the research team is briefed thoroughly but also shows something that may be unique to training – the practice of increasing difficulty for the students to increase the learning effect.

If you are planning a large and extended simulator study, you are invited to take note of these lessons learned – mainly from Project Horizon.

11.6.1 Briefing

Briefing is of vital importance for a simulator-based research project to succeed. Preparations for a successful briefing should include all relevant aspects that are normally included in any other types of research project. However, due to the likely lack of experience of participating in research projects from the side of the seafarer, extra care should be taken. The same applies for simulator and other assisting personnel. Possible lack of experience should be tackled by thorough and careful briefing about what the upcoming project is about and what information is and is not to be given to the participants in the project. Many questions will undoubtedly come from the participants, and careful preparations are needed to answer those in a coordinated and mutually satisfactory way.

11.6.2 Working Schedules

Larger experiments may include unusual working times for the personnel involved, for instance, working around the clock, in shifts and during unfavourable times, such as at night. If a voyage in the simulator is to take place for a prolonged period, this clearly implies that personnel (researchers and simulator instructors) are to be present continuously, which needs careful planning and making sure that there are substitutes available, should something happen. Phone lists are a good idea. Food, refreshments and a place to rest are needed. Also, consider whether medical or psychological personnel should be available in some capacity.

11.6.3 Simulator Burnout

Larger experiments do put a burden not only on the personnel, but also on the simulator itself. For Project Horizon, for instance, the simulators had to be up and running for a full week, without any interruptions. Careful testing of whether the simulator is capable of doing so is therefore an important step in preparing for the experiment (for instance, with respect to the heat, the simulator is generating or the lifespan of certain parts such as projectors and/or bulbs). It might therefore be worthwhile considering the presence, either on call or not, of a representative from the manufacturer.

11.6.4 Environment

A simulator is always part of a larger environment, and it should be noted that the environment may affect the experiment going on in the simulator. Alarms may sound or drills take place, cleaners may come in, and students may rush around etc. These possible disturbances should be dealt with in advance and arranged so that they disturb the experiment to a minimum. Also, make sure that services which differ between night and day are managed; for instance, fans/heating/cooling and lights may be switched off at night, whereas security guards and alarms may be active at night.

11.6.5 Participants

Aside from normal due diligence as regard experiments, large simulator studies may need a few extra considerations. A phone and contact list (including next of kin) can be useful, and if possible, consider including substitutes. Some practical points regard the use of personal phones during the study and when to allow participants to return home, which is especially important after a fatigue or similarly taxing study. Should a participant be allowed to drive their own car home, for instance? Regarding long experiments, one might expect participant burnout. However, only one participant (out of 90) left the Horizon experiment prematurely, after about three days, most likely due to not being used to the six-hour on/six-hour off watch system.

11.7 Summary

Continuing on with the division between simulator and simulation, we provide a list of aspects to consider when planning to acquire a simulator. We then conclude by discussing simulation and how to increase the quality of output from work in simulators:

1. Who are the manufacturers and what are their policies?
2. What type and certification of system are needed?
3. What are the visual display system requirements and desired field of view?
4. Who are the other customers of this brand and what is the system used for?
5. What are the training/research needs and cost for internal technical and operating staff?
6. What software upgrade agreements are available and at what cost, timing and method?
7. What hardware upgrade agreements are available and what type of additions and flexibility?
8. Which parameters can the operator adjust?
9. What models are available and at what cost and what is the possibility of in-house programming?
10. What is recorded internally? Can it be accessed? Can other measures be easily added?
11. What is the output format? Can it be exported easily for data analysis?
12. What synchronisation capability is there and is time stamping of data available when using multiple sources or simulators?
13. Is it possible to connect to other simulators or equipment of other brands?
14. How complex do you need it to be?
15. What degree of fidelity is necessary?

A common criticism is that most simulators only represent one manufacturer (and thus, only a limited number of workplace(s) – ships) and will not provide generalised training or generalisable research. A few ways of addressing this are as follows:

- Making simulators more flexible may be easier in the future with software and displays.
- Making manufacturers more harmonised in at least certain aspects of their design.
- Creating an S-mode* – a standard basic design in which most values and information are designed to jointly agreed best practice.

Additionally, what makes a good simulation for research? The simulator, participants and 'real actors', scenarios and models? Is greater fidelity necessarily better? What methodologies of validation are used?

Generally, we need to put more thought into our experiments and examine our assumptions. It is also clear that the simulator allows us to use methods that cannot be used with ease in an on-board setting, such as secondary tasks. One thing is for certain: if we are to trust research results (that is, promote a degree of credibility to the simulation output), then sound validation process and assessment are essential. What must be remembered is that such assessment, like fidelity, comes at a price and must be considered and defined in the planning and costing of the project.

Whilst many simulators are approved for training by regulatory bodies, that does not necessarily mean that the same simulator is valid for the research that is proposed. To validate the simulation, then we need to look closely at the discussed elements which it is composed of and determine validation methods that can be applied to each. Ship models in particular can undertake a robust validation process incorporating both objective and subjective approaches. Presence, be it the participants' feeling of being on the real ship or in the real environment (Slater & Wilbur, 1997), requires subjective analysis. Simulator hardware controls can be objectively compared to their actual counterparts, whilst the layout in comparison to the real ship can be subjectively assessed. One area that poses further research work and consideration in maritime simulation is where the two validation methodologies are utilised in research and deliver differences in results. What is more important is that objective analyses between simulator elements are correct or subjectively that the participant has credibility in the output of the simulation for robust results?

Think about the two-step process – a simulator is a tool that must be calibrated before you start a simulation, and the use of experienced personnel

---

* S means standard and is an initiative initiated by the Nautical Institute and adopted by the IMO.

is part of the simulation. Looking forward, simulations will become more multifaceted to address issues and questions on systemic levels, considering socio-technical systems, holistic simulations and distributed teams, to mention a few. Interconnections, whether technical or human, will be more common which adds complexity to planning and performance of studies and a need to ensure valid and applicable outcomes.

References

Åkerstedt, T., & Gillberg, M. (1990). Subjective and objective sleepiness in the active individual. *International Journal of Neuroscience, 52*, 29–37.

Balci, O., & Sargent, R. G. (1981). A methodology for cost-risk analysis in the statistical validation of simulation models. *Communications of the ACM, 24*, 190–197.

Bergström, J., Dahlström, N., Van Winsen, R., Lützhöft, M., Dekker, S., & Nyce, J. (2009). Rule- and role-retreat: An empirical study of procedures and resilience. *Journal of Maritime Research, IV*, 75–90.

Chauvin, C., Clostermann, J. P., & Hoc, J. M. (2008). Situation awareness and the decision-making process in a dynamic situation: Avoiding collisions at sea. *Journal of Cognitive Engineering and Decision Making, 2*, 1–23.

Chauvin, C., Clostermann, J. P., & Hoc, J. M. (2009). Impact of training programs on decision-making and situation awareness of trainee watch officers. *Safety Science, 47*, 1222–1231.

Dahlström, N., Dekker, S., van Winsen, R., & Nyce, J. (2009). Fidelity and validity of simulator training. *Theoretical Issues in Ergonomics Science, 10*, 305–314.

Dimitrios, G. (2012). Engine control simulator is a tool for preventive maintenance. *Journal of Maritime Research, IX*, 39–44.

DNV-GL (Det Norske Veritas Germanischer Lloyd). (2014). Maritime simulator systems (Vol. DNVGL-ST-0033:2014-08). Retrieved from: https://rules.dnvgl.com /docs/pdf/DNVGL/ST/2014-08/ DNVGL-ST-0033.pdf.

Donderi, D. C., Mercer, R., Hong, B., & Skinner, D. (2004). Simulated navigation performance with Electronic Chart and Display Systems (ECDIS). *Journal of Navigation, 57*, 189–202.

Eda, H., Guest, F. E., & Puglisi, J. J. (1996). Twenty years of marine simulator (CAORF) operations: Lessons learned during these years. Paper presented at the *Proceedings of MARSIM*. MARSIM '96, Intl Conf on Marine Simulation and Manoeuvrability, 9–13 September 1996; Copenhagen, Denmark. Procs. Ed by M.S. Chislett. Publ by A.A. Balkema, Rotterdam, The Netherlands, 1996, ISBN 90 5410 831 2, p 3 [10 p, 14 ref, 9 fig].

Godwin, H. J., Hyde, S., Taunton, D., Calver, J., Blake, J. I. R., & Liversedge, S. P. (2013). The influence of expertise on maritime driving behaviour. *Applied Cognitive Psychology, 27*, 483–492.

Gould, K. S., Hirvonen, K., Koefoed, V. F., Røed, B. K., Sallinen, M., Holm, A., Bridger, R. S., Moen, B. E. (2009a). Effects of 60 hours of total sleep deprivation on two methods of high-speed ship navigation. *Ergonomics, 52*, 1469–1486.

Gould, K. S., Røed, B. K., Saus, E.-R., Koefoed, V. F., Bridger, R. S., & Moen, B. E. (2009b). Effects of navigation method on workload and performance in simulated high-speed ship navigation. *Applied Ergonomics*, 40, 103–114.

Hart, S. G., & Staveland, L. E. (1988). Development of NASA-TLX (Task Load Index): Results of empirical and theoretical research. *Human Mental Workload*, 1, 139–183.

Hays, R. T., & Singer, M. J. (1989). *Simulation Fidelity in Training System Design: Bridging the Gap between Reality and Training.* Springer Science & Business Media, Berlin, Germany.

Hockey, G. R. J., Healey, A., Crawshaw, M., Wastell, D. G., & Sauer, J. (2003). Cognitive demands of collision avoidance in simulated ship control. *Human Factors: Journal of the Human Factors and Ergonomics Society*, 45, 252–265.

Hontvedt, M., & Arnseth, H. C. (2013). On the bridge to learn: Analysing the social organization of nautical instruction in a ship simulator. *Computer-Supported Collaborative Learning*, 8, 89–112.

Hora, J. (2015). A review of performance criteria to validate simulation models. *Expert Systems*, 32, 578–595.

Howland, J., Rohsenow, D. J., Cote, J., Siegel, M., & Mangione, T. W. (2000). Effects of low-dose alcohol exposure on simulated merchant ship handling power plant operation by maritime cadets. *Addiction*, 95, 719–726.

Howland, J., Rohsenow, D. J., Cote, J., Gomez, B., Mangione, T. W., & Laramie, A. K. (2001). Effects of low-dose alcohol exposure on simulated merchant ship piloting by maritime cadets. *Accident Analysis and Prevention*, 33, 257–265.

IALA (International Association of Marine Aids to Navigation and Lighthouse Authorities). (2011). *Use of Simulation as a Tool for Waterway Design and Aids to Navigation Planning* (2nd ed., Vol. IALA Guideline No. 1058). Saint-Germain-en-Laye, France: IALA/AISM.

International Maritime Organization. (1972). *Convention on the International Regulations for Preventing Collisions at Sea (Colregs).* London: International Maritime Organization.

Ivergard, T. (1975). The use of ship manoeuvre simulators. *Journal of Navigation*, 3, 358–362.

Kim, H., Yang, C. S., Lee, B. W., Yang, Y. H., & Hong, S. (2007). Alcohol effects on navigational ability using ship handling simulator. *International Journal of Industrial Ergonomics*, 37, 733–743.

Kircher, A., & Lützhöft, M. (2011). Performance of seafarers during extended simulation runs. Paper presented at the RINA – *Human Factors in Ship Design and Operation*, 16–17 November 2011, London, UK.

Kircher, A., van Westrenen, F. C., Söderberg, H., & Lützhöft, M. (2011). Behaviour of deck officers with new assistance systems in the maritime domain. Paper presented at the HFES – *Human Factors and Ergonomics Society Europe Chapter Conference*, 19–21 October 2011, Leeds, UK.

Lamond N., Dawson D., & Roach, G. D. (2005). Fatigue assessment in the field: Validation of a hand-held electronic psychomotor vigilance task. *Aviation Space Environmental Medicine*, 76, 486–489.

Lützhöft, M., Dahlgren, A., Kircher, A., Thorslund, B., & Gillberg, M. (2010a). Fatigue at sea in Swedish shipping – A field study. *American Journal of Industrial Medicine*, 53, 733–740.

Lützhöft, M., Porathe, T., Jenvald, J., & Dahlman, J. (2010b). System simulations for safety. In O. Turan, J. Bos, J. Stark & J. L. Colwell (Eds.), *Human Performance at Sea* (pp. 3–10). Glasgow, UK: University of Strathclyde.

Macfarlane, R. (1997). Simulation as an instructional procedure. In G. J. F. Hunt (Ed.), *Designing Instruction for Human Factors Training in Aviation* (pp. 59–93). Aldershot, UK: Ashgate.

Marsden, G., & Leach, J. (2000). Effects of alcohol and caffeine on maritime navigational skills. *Ergonomics, 43*, 17–26.

McIlroy, H., Grossman, H., Eda, H., & Shizume, P. (1981). Validation procedures for ship motion and human perception. Paper presented at the Second International Conference on Marine Simulation Symposium (MARSIM 81), 1–5 June 1981, National Maritime Research Center, Kings Point, NY.

National Research Council. (1996). *Simulated Voyages: Using Simulation Technology to Train and License Mariners*. Washington, DC: National Academies Press.

National Research Council Committee on Ship-Bridge Simulation Training. (1996). *Simulated Voyages: Using Simulation Technology to Train and License Mariners.* Washington, DC: National Academies Press.

Nilsson, R., Lützhöft, M., & Gärling T. (2008). Fairway navigation—Observing safety-related performance in a bridge simulator. *TransNav, the International Journal on Marine Navigation and Safety of Sea Transportation, 2*, 17–21.

Nilsson, R., Gärling, T., & Lützhöft, M. (2009). An experimental simulation study of advanced decision support system for ship navigation. *Transportation Research Part F: Traffic Psychology and Behaviour, 12*, 188–197.

Orlandi, L., Brooks, B., & Bowles, M. (2015). A comparison of marine pilots' planning and manoeuvring skills: Uncovering mental models to assess shiphandling and explore expertise. *Journal of Navigation, 68*, 897–914.

Rehman, M., & Pedersen, S. A. (2012). Validation of simulation models. *Journal of Experimental & Theoretical Artificial Intelligence, 24*, 351–363.

Roza, Z. C. (2005). *Simulation Fidelity Theory and Practice*. Doctoral Thesis. DUP Science. Retrieved from http://repository.tudelft.nl/islandora/object/uuid:a5afd816-4b04 -459d-b54c-da93e3b8a0d7? collection=research.

Sargent, R. G. (2013). Verification and validation of simulation models. *Journal of Simulation, 7*, 12–24.

Sauer, J., Wastell, D. G., Hockey, R. J., Crawshaw, M. C., Ishak, M., & Downing, J. C. (2002). Effects of display design on performance in a simulated ship navigation environment. *Ergonomics, 45*, 329–349.

Slater, M., & Wilbur, S. (1997). A framework for immersive virtual environments (FIVE): Speculations on the role of presence in virtual environments. *Presence: Teleoperators and Virtual Environments, 6*, 603–616.

Thomas, M. J. (2003). Operational fidelity in simulation-based training: The use of data from threat and error management analysis in instructional systems design. In *Proceedings of SimTecT2003: Simulation Conference* (pp. 91–95). Adelaide, Australia: Simulation Industry Association of Australia.

van Leeuwen, W. M. A., Kircher, A., Dahlgren, A., Lützhöft, M., Barnett, M., Kecklund, G., & Åkerstedt, T. (2013). Sleep, sleepiness, and neurobehavioral performance while on watch in a simulated 4 hours on/8 hours off maritime watch system. *Chronobiology International, 30*, 1108–1115.

von Lukas, U. (2006). Virtual and augmented reality in the maritime industry. In Z. Weiss (Ed.), *Virtual Design and Automation: New Trends in Collaborative Product Design* (pp. 193–201). Poznan, Poland: Publishing House of Poznan University of Technology.

12

Maritime Bridge Crew Training

Kjell Ivar Øvergård, Linda Johnstone Sorensen, Magnus Hontvedt,
Paul Nikolai Smit and Salman Nazir

CONTENTS

12.1 Background and History of Maritime Training

Seafarers belong to a profession with deep historical roots, with distinctive strategies for navigation and teamwork. Within the Western world, seafarers have traditionally been educated through participatory learning on vessels and ships as part of their vocation. During the eighteenth century, in the British Navy, sailors went to sea as boys, usually aged 12, serving on deck where they performed a range of duties concerned with the operation and upkeep of deck department areas and equipment. They could be rated as seafarers from when they were about 16 years old, having mastered all the skills associated with their rank. To rise to the rank of an officer, the sailor had to pass an examination of all aspects of seamanship. The training of seafarers was very much on the job, and long durations of service were required to prove one's worth (Lavery, 1998). During the twentieth century, however, formal education became increasingly important as seafarers faced new demands for formal courses and certificates. Within the field of shipping, professional conduct follows strict procedures and a hierarchical division of labour, as well as cultural norms developed throughout history.

The nautical profession is recognised by a high degree of teamwork, as an individual alone cannot operate a large ship. Consequently, the joint ability of the crew and its associated teamwork is a key focus in training. Teamwork on board the bridge of a ship involves the distribution and coordination of work tasks involved in successful navigation, manoeuvring and supervision of technical systems. The bridge crew functions as a cognitive and computational system that exceeds the individual's cognitive abilities where the human operators partake in historically defined activity systems (Hutchins, 1995). The training of seafarers thereby involves increasing the proficiency for participating in sociotechnical systems and operating the tools that mediate professional action.

Today, nautical education facilities provide seafarers the competence needed to obtain the mandatory certificates before they start working at sea. To do this, simulators that mimic the maritime working environment or that mimic a particular subsystem commonly found aboard a vessel are routinely used.

Additionally, professional seafarers often undergo regular training past their initial qualification – for maintaining and increasing their expertise – and this training is also often given in simulator training centres that allow for repetition of tasks that would otherwise be too dangerous or costly to perform aboard a real vessel. Although the training of seafarers has relocated from practice on ships to classrooms, some of the profession's traditional models for learning through participation may still be evident – for instance, in the extensive use of simulators in vocational schooling and professional training of seafarers.

Simulators allow for the rendering of a real-life-like training ground that allows for participation in these work practices without endangering capital, lives or the environment. In the maritime profession, ship simulators are used within educational and follow-up training for professionals, which may have a range of objectives such as high-speed navigation for rescue vessels or transferring goods between large ships at sea. Various simulators are used for learning the key aspects of the maritime domain, from technical skills to bridge team management. The simulators range from plain desktop simulators to full-mission bridge simulators where a physical replication of the bridge of a ship is surrounded by large visual displays.

This chapter will provide a short description of the characteristics of work related to maritime navigation and manoeuvring. We will then give an overview of the development and use of simulators in maritime training, how different types of simulators relate to different kinds of training, and it will touch upon a large number of training practices and objectives. To provide an overview of the field, we will point out some common features and strategies in training, such as when to use full-mission simulators, which are considered to be especially beneficial for working on team-based activities and for training skills in realistic settings (Vincenzi et al., 2009).

12.2 Characteristics of Work in the Maritime Domain

12.2.1 Individual versus Team Skills in the Maritime Domain

The distinction between individual and team skills is important for maritime training. *Individual training* pertains to the individual seafarer's attainment of skills and competence, acquired through a range of learning methods, including practicing technical and non-technical skills. Whilst all teams consist of individuals, the focus on team training is more important in the maritime domain due to the fact that navigation and manoeuvring a vessel require collaboration both within and between teams of personnel (navigators, engineers, deck personnel and other personnel groups such as chefs and waiters). Because of the implicit requirement for collaboration in the maritime domain, this chapter will to a large degree focus on *team training*.

Team training refers to training where individual student seafarers take on the functional roles of the bridge, that is, the captain, chief officer, second officer, dynamic position operator and navigator. The training in this instance will focus not only on the display of necessary technical competence but also on the required ability to function as a team by, for instance, delivering the right information to the right person at the right time on the bridge to ensure efficient operation of crucial manoeuvring tasks (e.g. sailing a route out of a busy port). Similarly, there is a need to identify the differences between technical and non-technical skills when considering maritime simulator-based training.

12.2.2 Characteristics of Teamwork in the Maritime Domain

The maritime domain has some distinguishing features that are of relevance to simulator-based team training. The main features that we will discuss are lack of standardisation of interfaces and technology, complex and variable team compositions, dynamically changing work teams, cross-disciplinary teams, geographically distributed teams and the complexity of operations.

12.2.2.1 Lack of Standardisation of Interfaces and Technology

The training of seafarers becomes particularly challenging by the fact that there is no standardisation of technologies or HMIs in maritime control rooms. Bridges and engine control rooms have been (and still are) built up of hardware and software from multiple manufacturers, with the result that a lot of systems have been added to ship control rooms without considering the needs of seafarers, thus increasing workload and reducing usability of the total system (Lützhöft and Lundh, 2009). As a result, no two ships, not even sister ships, have identical layouts and interface designs. This trend creates challenges for simulator training because it is impossible to make a generic simulator environment that is isomorphic to a particular crew's actual working environment. However, the trend of Integrated Bridge Systems points towards a greater unification of information from sensors for display at centralised modules in control rooms (Lützhöft and Dekker, 2002). This means, usually, that one manufacturer of hardware and software supplies the complete control systems and interfaces for a vessel, although instances where different hardware and software manufacturers are integrated into one display are also possible. The maritime industry, however, is still far away from the standardisation of HMIs seen in commercial aviation.

12.2.2.1.1 Complex and Variable Work Team Composition

Crew composition in the maritime domain possesses higher variability than that of most other transport sectors. Maritime crew composition can range from a single navigator on the bridge in the case of a small coastal voyage passenger ferry to a large and complex work organisation with a number of different work teams. Common work teams are the *bridge* that can have a large number of people at work (sometimes as many as 10 people), the *engine room* (with up to 5 people at work) and the *deck* (with up to 6 or even more people). In addition, there might be other specialised teams such as remotely operated vehicle (ROV) operators and operation supervisors who usually work from different locations on the vessel.

12.2.2.2 Changing Constituents of Work Teams

The number and constituents of a bridge team might also be altered by the addition of a pilot depending on the areas in which the ship is navigating; in specific coastal areas, vessels are obligated to have a maritime pilot attending. The pilot will access the vessel before the vessel enters a designated zone demarcating the free waterways and where piloting is compulsory and will stay with the vessel until it has cleared the designated waterways. The inclusion of the pilot as a new team member may lead to new safety challenges for the bridge team (see e.g. Nuutinen and Norros, 2009).

12.2.2.3 Cross-Disciplinary Teams

Maritime crews can also include personnel without a maritime background who are not trained in accordance with Standards for Training, Certification and Watchkeeping (STCW), but who still form a vital part of the team in the performance of certain maritime operations (e.g. ROV operator, navigators, field experts, dynamic positioning [DP] operators, crane operators, offshore installation managers and bridge operation supervisors).

12.2.2.4 Geographical Distribution of Teams

In addition to the number of different people on the bridge of a ship, it is also common that teams and team members can be located in different areas on the ship (e.g. the aft and fore bridge, engine room, ROV control room, aft deck of the ship and so on). Also, multiple vessels often cooperate and coordinate their activities in maritime operations necessitating remote or integrated operations, which place high demands on communication and information sharing during the performance of complex tasks (e.g. drilling operations, anchor handling). Finally, the bridge crew must communicate with people on land such as the harbour administration, vessel traffic supervisor (VTS), piloting services and the cargo or ship owner.

12.2.2.5 Complexity of Operations

The number of interacting vessels, the distributed nature of maritime operations, high levels of automation, highly variable operational environments and multiple people/teams involved make maritime operations highly complex (Burns and Hajdukiewicz, 2004).

12.3 Development and Use of Simulators in Maritime Training

12.3.1 IMO Regulations on Training and Simulator Use

The minimum standards for seafarers' competencies are defined by the International Maritime Organisation's (IMO) *International Convention on Standards of Training, Certification and Watchkeeping for Seafarers* (IMO, 2011). (IMO is the United Nations' specialised agency responsible for the safety and security of shipping and the prevention of marine pollution by ships.) Education in accordance with STCW is required only when issuing certificates for operating ships exceeding 500 gross tonnes. The STCW is an internationally agreed convention that regulates four aspects of training (see IMO, 2011):

- The necessary competencies that each seafarer must show
- The knowledge, understanding and proficiency required of each seafarer
- The methods that the seafarers can use to demonstrate competence
- The methods for evaluating/assessing the seafarer's competence

Even though the STCW establishes internationally common guidelines for the education of seafarers, there remain some challenges as STCW contains general descriptions of the performance output of training but does not stipulate specific behaviours to be displayed. Nor does STCW describe any validated methods for assessment of the required skills and behaviours associated with competent seafarers. This shortcoming of STCW allows for a subjective interpretation of the regulations and can thus lead to different training content and certification practices between nations (Barnett et al., 2002). Indeed, there is currently no tradition for evaluating the competence of seafarers after certification. Requirements involve *proof of participation* (e.g. hours, days or months aboard a vessel; see IMO, 2011) rather than *proof of competence* (e.g. actual skills, knowledge and attitudes). Similarly, despite the extensive use of simulators in training, there are few empirical studies of their outcome on actual job performance (Nazir et al., 2013).

12.3.2 Specific Maritime Skill Training

Technical skills in the maritime domain involve the operation of technical equipment as well as the observation of standard operating procedures (Flin et al., 2008). Technical skills are often seen as the 'hard' skills that to some extent define a profession (e.g. blacksmiths heat and shape iron into tools and nurses tend to wounds). For seafarers, technical skills would also involve the ability to plot and plan a journey, to know and to follow the regulations for preventing collisions at sea (IMO, 1972) and to pinpoint the position of the vessel at any given moment in paper charts or in the Electronic Chart Display and Information System (ECDIS).

Professional expertise thus involves a range of technical skills and motor abilities. Technical skill training is subject to different training designs within maritime education, and it is a common practice to train technical skills separately – often on part-task trainers. Such technical skills are needed to manoeuvre a vessel or operate technical systems and are often characterised in opposition to the non-technical skills described earlier. Handling technological tools and quick responses to a changing environment are in many cases crucial for effective navigation, and skill learning in ship simulators is commonly achieved through repeated exposure to a certain task. Desktop simulators or part-task trainers are often used to provide opportunities for individual training with a high number of repetitions. The IMO has made model training courses for a number of these subsystems such as radar (IMO, 1999), Global Maritime Distress and Safety System (GMDSS; IMO, 1997) and ECDIS (IMO, 2012). These model courses are made to enable organisations to give training in accordance with the requirements in the STCW convention (IMO, 2011). Training on these subsystems is often given using part-task simulators (see e.g. Karlsson, 2011; Øvergård and Smit, 2014).

12.3.3 Team Training in Maritime Simulators

In contrast to technical skills, *non-technical skills* are those which are of importance for maintaining a safe and efficient voyage. These skills involve situation awareness, decision-making, communication, teamwork, leadership, stress management and coping with uncertainty (Flin et al., 2008). A growing proportion of maritime accidents have been shown to involve failures in some form of non-technical skills rather than technical failures (Hetherington et al., 2006). The focus on the training of non-technical skills has increased from the field's beginning in the early 1980s when the aviation world started to develop crew resource management (CRM) (Helmreich et al., 1999). Other professions, such as medicine, have since followed suit (Fletcher et al., 2003). Until 2010, the international maritime industry focused only on the training of technical skills. Non-technical skills such as resource management, leadership and decision-making were made mandatory only in 2010 when the STCW was amended. The STCW contains the international competence requirements for seafarers and defines the necessary minimum skills and knowledge that seafarers must have (IMO, 2011). In essence, and particularly in the maritime domain, these non-technical skills relate to teamwork.

The first simulator-based maritime team training with a panoramic visual representation of the environment was established at Warsash Maritime Academy in Southampton, United Kingdom, in the early 1970s (Barnett et al., 2003; Hayward and Lowe, 2010). Early simulator-based team training focused on route planning and pilot–master cooperation (Barnett et al., 2003), which in turn developed into the Bridge Team Management courses that are still given today. In 2010, the STCW (IMO, 2011) was updated to include elements such as Bridge Resource Management (BRM), Leadership and training with ECDIS. These additional requirements introduced the understanding that there is a difference between pure technical skills that pertain to individual skill sets (Flin et al., 2008; IMO, 2011) and the training of skills for cooperation and coordination of work (the so-called non-technical skills).

The focus on non-technical skills in the maritime industry is evident in the BRM training that was developed as an initiative between the maritime industry and the former Scandinavian Airlines Flight Academy (now Oxford Aviation Academy) in the early 1990s. This course was based upon the assumption that findings and training techniques in the CRM courses of the aviation industry were applicable to the maritime industry (Hayward and Lowe, 2010).

More recent developments in maritime CRM training have been in the Maritime Resource Management (MRM) training course of the Swedish Club Academy. This training course focuses on CRM training in a distinct maritime environment (Hayward and Lowe, 2010) and goes beyond the focus of BRM on the roles specific to the bridge, to include all roles involved in ensuring safety at sea, such as the engineers, superintendents, occasional on-board pilots and so on. An MRM course by Swedish Club covers three to five days and involves courses on attitudes and management skill, situation awareness, cultural awareness, communication and briefings, authority and assertiveness, challenge and response, short-term strategy, workload, automation awareness, state of the ship, human involvement in error, management styles, judgement and decision-making, leadership in emergencies and crisis and crowd management. The MRM training course involves workshops with computer-based training, classroom lectures and the use of films (Hayward and Lowe, 2010).

The overall idea with non-technical skill training in bridge simulators is to prepare participants for the challenges involved in working as a team. Simulator training also gives participants an opportunity to realise and practice these skills in an environment similar to a real-life bridge. Scenarios can range from everyday routine work to unexpected critical situations. Learning to use profession-specific artefacts is a key element of novices being socialised into communities of practice (Lave and Wenger, 1991). The training of seafarers involves socialising novices into a specific communicative work system and operating the artefacts that mediate professional action. Rather than being mere physical locations, such workplaces have been delineated as actively constituted fields of perception and interaction that are continuously maintained by the participants and which involve a range of coordinating actions (Heath and Luff, 1996). From this perspective, workplaces may facilitate efforts that exceed the individual's accumulated capacity, as their performance is linked to participation in larger work systems. Accordingly, not only do they executing professional techniques, but they also become participants in social structures where the techniques become inseparable from the social doings (Lave and Wenger, 1991; Hutchins, 1995).

Several studies have identified factors that are commonly related to ship accidents, such as fatigue, lack of situational awareness, lack of teamwork and poor decision-making (see e.g. Hetherington et al., 2006). However, even if the role of human factors in accidents is known, it is not obvious how training in these factors can be effectively accomplished.

12.3.4 Research on Simulator-Based Bridge Team Training

Ship simulators have been used as test beds for research in the human factors of navigation and manoeuvring, but to our knowledge, few empirical studies on bridge simulator training exist. Some examples include research on the cognitive demands associated with collision avoidance (Robert et al., 2003), assessing electronic chart and ECDIS compared to paper (map) chart navigation (Donderi et al., 2004) or the effects of display design and navigation system complexity on performance in a simulated ship navigation environment (Sauer et al., 2002; Nilsson et al., 2009).

Øvergård et al. (2005) evaluated objective differences between real- and simulator-based military high-speed craft navigation with the Combat Boat 90 (CB90). The CB90 is a military high-speed craft capable of speeds in excess of 35 knots. Swedish recruits formed two-person teams and navigated a route in the archipelago outside Gothenburg. Two days later, they navigated the same route in a CB90 full-mission simulator. Three different route segments of differing complexity were analysed. Results indicated that the complexity involved in the navigational task affected the choice of speed and trajectory variation. Speed and trajectory variability was lower in real navigation than in simulator-based navigation in complex segments. For simple segments, the opposite was true; higher speeds and larger trajectory variability were observed in the real navigation as compared to the simulator-based navigation. Data from questionnaires indicated that a lack of experienced danger in simulator-based navigation could have led to the observed speed trajectory pattern (Øvergård et al., 2005). A negative effect of this lack of emotional engagement or sense of danger is that it can make simulator training less than representative of actual behaviours. The extent of this possible effect is currently not known.

In the context of maritime vocational education, Hontvedt and Arnseth (2013) showed how a group of nautical students, together with a professional maritime pilot, enacted professional roles and collaboratively solved tasks as part of a larger scenario of a cruise ship entering the Oslofjord in Norway. Interaction analysis was used for examining how a simulated context for learning to navigate was socially cocreated in a full-mission ship simulator and how opportunities for learning were entwined in this collaborative activity. The study demonstrated how the crew's role-playing became a resource for learning in a work-like setting. However, this study also showed that it is challenging for the participants to notice and adopt professional actions that were performed by the experienced pilot in the evolving scenario and to later confront these situated experiences in the debriefing.

Gould et al. (2009) examined seafarers' mental workload and performance in simulated high-speed ship navigation. Two navigation methods were compared, based on ECDIS and a conventional system using paper charts. Naval cadets navigated in high-fidelity simulators with varying levels of difficulty, and the researchers took on a triangulated measurement strategy, concurrently assessing performance, subjective workload and psychophysiological activation (such as heart rate variability and skin conductance). They found that ECDIS navigation improved course-keeping performance but reduced the total amount of communication on the bridge, thus changing the strategies for effective navigation. Also, conventional paper charts seemed to be associated with higher workload, although these differences were not statistically significant.

Other studies within the educational field discuss how to delineate objective criteria for assessment (Kobayashi, 2005). They address future needs for ship simulator training, such as the degree of transfer of learning between the simulator and the real work setting, the assessment of non-technical skills and behavioural markers for expertise and how organisational culture impacts on accident causation (Barnett et al., 2002). However, these studies relied on literature reviews and descriptions of educational programmes more than on empirical observations.

Despite the focus on BRM training, little research has been presented to show the influences of these initiatives. However, a few studies have reported that CRM training led to a reduction of incident and accident rates, as reported for Maersk (Byrdorf [1998], referenced in Hayward and Lowe, 2010). Similar findings have been found in a process industry, where teamwork also plays a major role in ensuring safe and efficient performance (Nazir et al., 2015). For example, Komulainen et al. (2012) found that the use of high-fidelity operator training simulators in the oil and gas sector led to an estimated saving of $15.3 million per plant per year due to a reduction in the number of unplanned shutdowns and saved time on commissioning and start-ups after major modifications.

12.4 Matching Types of Simulator with Training Needs

Simulators have been in use for maritime manoeuvring and navigational training for several decades (Barnett et al., 2003). The ability of the simulator to represent the visual and dynamic properties of the vessel and the surrounding environment has improved greatly (Nazir et al., 2014). The degree to which the simulated experiences match 'the real thing' is often conceptualised as its level of *simulator fidelity*. Fidelity is a key concept used to describe the resemblance of a simulator with the real work setting (Dahai et al., 2009;

Dahlstrom et al., 2009; see also Chapter 1). Fidelity is often conceptualised as *high* or *low*, and such general levels are frequently connected to learning efficiency (Alessi, 1988). Whilst high-fidelity simulators increase the ability to recreate realistic maritime operations, low-fidelity simulators can support the training of specific skills, such as particular navigational skills (Dahai et al., 2009; Dahlstrom et al., 2009).

The focus on the representation of the external environment has led to a quite one-sided approach to structure the physical environment in order to create life-like scenarios that are thought to facilitate learning and transfer of learning. Another, partially overlooked, element is that learning and simulator training also consist of structuring of the social environment, that is, training on patterns of communication and cooperation (Kozlowski and Deshon, 2004).

12.4.1 Types of Simulators in Use for Team Training

There are a large number of simulator types available for maritime team training (see Figures 12.1 and 12.2). The IMO's classification of simulators is as follows (see e.g. Committee on Ship-Bridge Simulation Training, 1996):

- *Full mission* (a simulator which can simulate full visual navigation bridge operations and that has capability for advanced manoeuvring and pilotage in restricted waterways)
- *Multitask* (can simulate full visual navigation but does not support advanced manoeuvring in restricted waterways)

FIGURE 12.1
Full-mission ship bridge simulator Horten at the University College of Southeast Norway. Picture by authors.

FIGURE 12.2
Full-mission ship bridge simulator SimSam at University College of Southeast Norway. The bridge simulator is shown with a 180-degree angle of vision. Picture by authors.

- *Limited task* (capable of simulating an environment for limited navigation and collision avoidance)
- *Special task* (a simulator that mimics the workings of a particular subsystem where this system is not located on the bridge)

We will next describe the major categories of simulators that can be used for team training; see Table 12.1 for an overview of different types of maritime simulators used for team training.

Some simulators of the limited-task or special-task type can also be used in parts of team training if they are integrated with other simulators; see Table 12.2 for an overview. Images of two computer-based special-task simulators for ECDIS and dynamic position training are shown in Figure 12.3.

Figure 12.3 shows a screenshot of a Kongsberg Maritime K-SIM (DP) simulator (left) and a Kongsberg Maritime ECDIS simulator (right).

12.5 Role of Simulators in Formal Competence Programmes

Navigation is primarily a practical profession – meaning that practical experience is paramount to the ability for safe and efficient navigation (Committee on Ship-Bridge Simulation Training, 1996). As discussed earlier, the on-board training that was given prior to the formalisation of maritime

TABLE 12.1

Overview of Different Types of Simulators Suitable for Team Training

Simulator Type	Training Purpose	Contents of Representation	Suitability for Team Training	Fidelity	No.
Bridge simulator[a]	Navigation, manoeuvring, route planning and preparation, team training	Bridge of the ship, often with exterior view between 60 and 360 degrees	High	High	1–6
Engine room simulator[a]	Engine supervision	HMI in control rooms	Medium, high if connected to bridge simulators	High	1–4
Offshore simulator aft bridge[b]	Anchor handling, manoeuvring, team training (can work as a full-mission simulator)	Bridge of ship and HMI, view of back deck and sea plus external objects such as other vessels and rigs	High	High	2–4
Seismic aft deck operator simulator[b]	Maintenance and repair work on seismic cables and related equipment, training on retracting and extending seismic cables, remote control of the winches and equipment failure scenarios	Visualisation of back deck, streamers and sea behind vessel	Medium/high	High	3–6

Note: No. refers to the number of people that can be involved in each exercise. The higher number is a suggestion for maximum persons that should partake in any one simulator during training.

[a] Control room simulators. The most costly and most versatile simulators often contain several single-system simulators integrated with bridge or machine room interfaces.

[b] Special purpose simulators designed for particular work tasks.

TABLE 12.2

Overview of Simulators that can be Used for Team Training if Integrated or Combined with Other Simulators

Simulator Type	Training Purpose	Contents of Representation	Suitability for Team Training	Fidelity	No.
ECDIS	Route planning, chart plotting, use of radar	ECDIS HMI, otherwise electronic charts and related technologies	Low; high if integrated in bridge simulator	High	1–2
DP simulator	Vessel station keeping, manoeuvring, handling of critical situations	Representation of HMI related to DP	Medium; high if integrated in a full-mission bridge simulator	High	1–2
VTS simulator	Communication with ships, control and surveillance of coastal area/port	Shows VTS interface, ECDIS plot over coastal waters and automatic identification system information of ships	Low; but medium/high if taken as a part of multiteam cooperation scenario	Medium	1–2
GMDSS simulator	Radio communication	HMI radio interface	Suitable for communication training; otherwise low, if not integrated in full-mission simulator	Medium	1
Crane simulator	Crane operations	Crane interface and visualisation of external world and dynamics, cooperation with other personnel (e.g. deck crew)	Low; high if integrated with other simulators	Medium	1

Note: *No.* refers to the number of people that can be involved in each exercise. The higher number is a suggestion for maximum persons that should partake in any one simulator during training. For ECDIS and GMDSS training, the systems used are the same as those aboard vessels and it is only the training situations that are simulated.

FIGURE 12.3
Left, Kongsberg Maritime K-Sim® DP Advanced simulator; *right*, ECDIS console in the K-Sim Polaris Ship Bridge simulator. Pictures by authors.

education in the twentieth century (Lavery, 1998) must now be offered at a maritime educational institution. This means that the practical on-board training related to navigation must be given either by using a school ship or by using simulators. According to STCW, a total of 60% of the required competencies (as defined by STCW) allow for the use of simulators to demonstrate these competencies. However, it is not required for a seafarer to undergo simulator training at all as long as the demonstration of competencies is in accordance with STCW. Seafarers must show their competence to manoeuvre and handle a ship in all conditions through one of three possible methods: (1) through approved in-training experience (e.g. aboard a real ship), (2) through approved simulator training or (3) in an approved manned scale ship model (IMO, 2011, p. 116).

Still, there are many reasons for choosing a simulator as a training and assessment tool: in particular, the lack of risk during training, the cost-effectiveness (related to training aboard a real vessel) and the possibility of quick repetition and of recording scenarios to use in debriefing settings (Committee on Ship-Bridge Simulation Training, 1996). These are some of the reasons why the Norwegian bachelor education model involves a great deal of simulator use (20–100% of courses depending on the operational focus and the competencies) in the training and certification of deck officers and engine officers. This allows students to be acquainted with work practices on the bridge or in the machine room prior to their first experiences in a job at sea.

12.6 Case Studies on the Use of Simulators in Maritime Training

A realistic scenario gives opportunities for technical training, honing critical skills required in a range of operations. It is common that company procedures are used in the design of scenarios and preparation for the exercises. Each company can have their own procedures, and simulator training must sometimes be designed or altered to accommodate that fact. The procedures are often used as 'benchmarks' that form the baseline for the assessment of the observed behaviour during simulator training.

Roles are normally allocated in the pre-training briefing. The functional roles are similar to those found in operations on international vessels. During the exercises, crews on two vessels will swap bridges. This is done to allow both teams to understand the operational challenges as seen from the perspective of other vessels. This is expected to give a better basis for cooperation and improve each member's understanding of the challenges that each role and work team face during these operations. During the exercise, the participants observe each other and are observed by experienced training personnel. The main purposes for the observations are to evaluate how the BRM principles are executed and to consider technical performance. These observations give the foundation for the debriefing. The participants are encouraged to think aloud as they perform and to guide their teammates where appropriate.

The four case studies presented here provide examples of MRM training sessions at the University College Southeast Norway, Bakkenteigen Campus, Horten, Norway, and describe specific aspects of maritime simulator training. The first three case studies are taken from a 2013–2014 training course in replenishment at sea (RAS) with seismic and support vessels given. The fourth case study is from an advanced course in navigation for final-year bachelor students in navigation. Because of the complexities and the need for teamwork on each bridge as well as between the two vessels, the training sessions were designed to train the skills involved in MRM in scenarios that have a close resemblance to real-life work during navigation and manoeuvring operations.

12.6.1 Case Studies 1–3: MRM during RAS Operations

RAS is a process of transferring goods (fuel, food and so on) from one ship to another during movement. Usually, the goods are moved between the vessels by using a cable or a crane. Using a cable requires that the two ships hold a fixed distance between them for a prolonged time. The use of a crane requires that the vessel makes an alongside mooring at speed. RAS involving seismic vessels must occur either from the side or from the beam of the ship because a normal maritime seismic operation involves trailing 6–12 km long cables spread out behind the vessel with a total width of 2000 m. Sharp deviations in heading or speed are inadvisable as it can lead to the streamers being tangled up – effectively destroying the equipment and halting the production. Recovery of the streamers before replenishment is not an option as it is time consuming, and the company is not paid for this work. The only solution is that the seismic vessel must keep a fixed speed and heading for prolonged times – even during a RAS operation. The support vessel therefore is faced with significant challenges with regard to manoeuvring, even more so than the challenges faced by the seismic vessel. The safety-related challenges for the support vessel involve, amongst others, the possibility of collision with the seismic vessel or the streamers, losing position, breaking the cable and rupturing fuel hoses.

12.6.1.1 Vessels and Teams Involved

The training scenarios involve two ships – a seismic vessel and a support vessel – and the two bridge teams of each of these vessels. The training is designed as an integrated operation involving the two bridge teams, who must collaborate and coordinate their efforts to perform the work tasks safely and efficiently. In addition to the bridge crew of the vessel, engineers, some from the seismic department on board and some from the company management, participate in the training. Bridge crew and engineers usually constitute a total of 8–12 persons (3 to 5 persons on the bridge of the seismic vessel and around 3 persons on the bridge of the support vessel). The exact number of people may change from scenario to scenario depending on the task demands and crew composition.

12.6.1.2 Simulator Set-Up

Exercises are performed on two bridge simulators that operate in the same simulator scenario where both vessels are visible to each other. The support vessel has a small bridge with a 360-degree view. The support vessel is equipped with control interfaces consisting of rudder and thruster control, ECDIS and a communication module with very high-frequency (VHF) radio and internal communication. The seismic vessel has a full bridge (except for GMDSS equipment) with a 240-degree view. This set-up allows for the training of the two bridge teams in joint operations (see Figure 12.4).

12.6.1.3 Simulator Familiarisation

All courses start with a technical familiarisation with the simulators. Participants receive a theoretical introduction in the use of azimuths, propeller and rudders in towing operations and are encouraged to attempt manoeuvres and experiment in the simulator familiarisation. The intention of having such familiarisation is to enable students to focus on other elements than the strictly technical skills involved in sailing and controlling a vessel.

12.6.1.4 Individual–Group–Plenary Method

After familiarisation, the participants partake in a short session where they share their experiences and opinions on critical incidents and errors using

FIGURE 12.4
Two separate bridge teams in joint operation at the SimSam full-mission bridge simulator.
Pictures by authors.

the individual–group–plenary (IGP) reflection methodology (Gausdal, 2015).
The IGP is a method for individual and group reflection – thereby provok-
ing individual responsibility and a group consensus on best practice. This
allows the participants to externalise implicit knowledge on human involvement in teamwork, thereby establishing a focus on the non-technical aspects
of navigation and manoeuvring.

12.6.1.5 Preparations and Briefing

The preparation and briefing for each exercise are done by presentations,
videos, whiteboard, flipcharts, introductions, demonstrations and case study
review. This prepares the students for the application of their theoretical

knowledge in the simulator scenarios as well as aiding competence transfer between students. In the briefing, participants will be prepared for the use of simulators as an effective tool for training for a real operation. The participants are instructed to act as they would in a real situation, but with awareness of the limitations of the simulator.

12.6.1.6 Debriefing Common for Cases 1–3

The debriefings start with a recall of the participants' personal perceptions of the exercise. All observations, both of the participants and the observers, are discussed with particular attention being paid to the execution and role of the BRM principles. The recall and observations are the basis against which the training outcome is evaluated. The aim of the debriefing is to cement the externalised learning and to provide clarifications where needed for deepening of students' knowledge and skills. Video-based replay of the scenarios is used for clarification during debriefing but is primarily used to support advancement of the students' technical skill (see e.g. Øvergård [2008] for a similar method).

12.6.2 Case 1: Port Manoeuvring and Handing over the Ship to Pilot

The aims of the first exercise are to train communication protocol and procedures when the captain welcomes the pilot to the bridge and to give an understanding of the safety challenges involved in having a pilot aboard the ship. The exercise starts with a common brief of the scenario outline highlighting the learning goals with emphasis on the non-technical skill performance expectations, in line with the STCW (IMO, 2011). Technical performance includes procedure and checklist awareness, identifying possible hazards, 'what if' thinking and risk assessment and organisation of communication procedures and equipment. The focus on procedures means that this course involves to some extent technical skills; however, it is the cooperation between multiple persons and handling the changed crew composition that are the foci of this training session. The scenario takes about 60 minutes.

The scenario starts with the bridge team aboard the seismic vessel contacting VTS central to notify that they will accept a pilot. The pilot (played by one of the course participants) soon enters the bridge, and the bridge team then proceeds to inform them about the vessel characteristics and capabilities in accordance with the communication protocol. The pilot then proceeds to perform the standard duties and to communicate with the VTS central, port authorities and other vessels nearby. Having a course participant playing the pilot gives the participants an insight into the other work tasks and responsibilities involved in manoeuvring and navigation of the vessel. Target behaviours in this scenario involve following communication protocol and procedures. A subject matter expert in navigation and CRM assesses performance. Observed errors that could have potential negative consequences are followed by a stop in the training scenario, and the participants are simply asked to replay the last part before the error occurred. Multiple errors will lead to instructions on correct behaviour until they successfully finish the scenario.

12.6.3 Case 2: Approaching Another Vessel to Make an Alongside Mooring

In another simulator scenario, the work task involved is an alongside mooring at speed. The seismic vessel normally moves at a constant speed, whilst the support vessel needs to manoeuvre alongside the seismic vessel and then make contact before the ships are moored together whilst travelling at a speed of 4.5 knots. The BRM skills of communication, planning, risk assessment and risk management and resource management are the foci of this exercise that lasts between 1 and 1.5 hours.

The scenario starts with the seismic and the support vessel distanced approximately 0.2 nautical miles apart whilst moving at the same heading in open waters. The support vessel must then manoeuvre closer to the seismic vessel and make a soft contact – a task that requires technical skills with respect to manoeuvring. Before the vessels make contact, deck personnel on both vessels must also be coordinated with the bridge teams in order to allow for safe and effective mooring. During the simulator exercise, the deck personnel communicate with the bridge personnel via internal communication. The scenario ends when the fuel hose is hooked up and the bunkering has started.

Target behaviours during this exercise involve efficient and smooth communication in accordance with communication procedures, effective planning and use of available resources as well as an assessment of the risks involved in this operation. The generality of the STCW requirements (Barnett et al., 2003) in combination with the lack of specific procedures for this operation means that a subject matter expert in navigation and CRM does the assessment. Errors that can have negative consequences during the operation are followed by a brief stop of the simulation and a request to replay the scenario from the point before the error.

12.6.4 Case 3: In-Line Fuelling with Handling of a Critical Incident

The third scenario on the RAS course during seismic operations involves an in-line fuelling procedure. In an in-line fuelling procedure, the support vessel manoeuvres in front of the seismic vessel at a distance of about 0.2 nautical miles and a hose are used to transport fuel from the support vessel to the seismic vessel. This requires that the exact relative distance between the vessels be maintained whilst the ships move at the same speed. The fuel hose is transferred from the support vessel to the seismic vessel by a rope connected to a buoy. After the fuel hose is connected and the fuel has begun to be pumped over, the support vessel experiences an engine malfunction, which reduces the ability of the support vessel to maintain its forward speed. The consequences of the engine failure – which are not directly apparent to the bridge crews – are so serious that the in-line fuelling operation must be aborted at some point. A number of challenges face the participants in this scenario. First, the two vessels might collide. Second, the support vessel can be passed by the seismic vessel and can end up being tangled up in the streamer trailing behind the seismic vessel. Third, the cable and fuel hose must be disconnected at the right time in order to avoid spillage of heavy fuel oil. The targeted behaviours in this scenario are related to risk assessment, resource management and procedures during critical incidents. Since the consequences of the engine malfunction are not directly apparent to the bridge crew, timing and performance during the disconnect procedure are of interest.

12.6.4.1 Effectiveness of Simulator Training in Cases 1–3

The effectiveness of simulator training and CRM training in particular is a challenging topic to investigate, and findings are mixed across and within work domains (Salas et al., 2006). For example Byrdorf (1996; cited in Hayward and Lowe, 2010) indicated that BRM can reduce maritime causalities, whilst O'Connor (2011) found that BRM training had little impact on attitudes and safety.

Hence, a recent study investigated the effects of training in the cases described previously (Larsen, 2015). Findings indicated learning effects on both individual and organisational levels. On an individual level, the results indicated an improved use of team members' knowledge, improved communication amongst team members and making it easier for lower-ranking team members to speak up – leading to a more open work atmosphere. On the organisational level, the training courses led to new routines/procedures for manoeuvring and for in-line fuelling – and these procedures were then trained and tested during simulator training (Larsen, 2015), thus having a probable impact on real-world behaviour.

12.6.5 Case 4: CRM Training during Port Departure in Poor Weather Conditions

To meet STCW requirements (IMO, 2011), simulator courses at the University College of Southeast Norway are focussed towards collaborative problem-solving, which is a key aspect of CRM (Flin et al., 2008). Training is primarily for fixed teams with changing roles, thereby allowing people to become accustomed to each other and to facilitate the establishment of team cohesion as this would enable teams to be more effective (Bird, 1977). The exercises are designed to give the students high workloads with increasing elements of stress and time pressure. Sound BRM principles underpin all exercises in the sense that the students will not be able to manage the accomplishment of the tasks that they are given without effective planning, task allocation, teamwork and communication. One specific exercise is related to the planning and execution of a safe departure from the Gothenburg area. The exercise is demanding and requires adherence to the International Regulations for Preventing Collisions at Sea 1972 (Colregs) and use of radar/automatic radar plotting aid (ARPA) under the following factors: night passage, drifting ice floes, navigation marks out of position, poor visibility, heavy traffic and VTS reporting. The objectives are to train the following BRM skills: allocation assignment and prioritisation of resources, effective communication, assertiveness and leadership and obtaining and discussing situational awareness. The exercise has six major parts: (1) pre-planning/preparations, (2) briefing, (3) planning, (4) execution of plan, (5) evaluation and (6) debriefing.

12.6.5.1 Pre-Planning/Preparations

The exercises including its area and surrounding waters, ship type, scope and objectives are published one week in advance. Students are encouraged to start pre-planning and preparations for the exercise as soon as the scope of the exercise is published.

12.6.5.2 Briefing

The briefing lasts for approximately 20 minutes and focuses on the task at hand as well as the elements that the student needs to know prior to starting the exercise such as the characteristics of the harbour, sailing routes, particularities of terrain and VTS communication procedures.

12.6.5.3 Planning

Following the briefing, the students are given 20–30 minutes to prepare for the voyage. The preparations are done at the bridge as it involves use of the systems and equipment at the bridge. Students are required to plan the following:

- The *route* has to be planned by use of paper charts or ECDIS to identify and prepare the optimal route to manoeuvre the ship out of the harbour. STCW highlights route planning as a specific skill required by all certified deck officers (IMO, 2011).
- *Identifying dangers* including calculating distance to dangers en route to ensure that planned manoeuvres are possible to do with adequate safety margins including appropriate alarm settings.
- *Identifying safe speeds* for different parts of the planned route.
- *Planning resource management* involving roles and task sharing, the use of different technical aids/tools at different points in time during

the manoeuvre(s) and information distribution between crew members and during different parts of the operation.

- *Identifying alternate plans* in case of failures, blackouts or other incidents.
- *Create checklist* for identifying tasks and sequences of actions related to the manoeuvre.
- *Control of equipment* is done by testing the control systems for failures and choosing the correct system settings. This part also includes identification of erroneous system settings (conditions/states/modes) that have been introduced by the instructors prior to the exercise.

12.6.5.4 Execution

The execution and monitoring of the exercise take between 60 and 80 minutes. The trainees/students are expected to perform the following tasks:

- Plot the planned passage.
- Conduct the planned passage with due regard to regulations in the VTS.
- Use ARPA to obtain information of the movement of other vessels.
- Analyse and interpret traffic situations.
- Use parallel indexing to ensure a safe distance to the coastline.
- On radar, put barriers along the coast.
- Use fixed targets to ground stabilised true vectors.
- Alterations of speed and/or course should be in accordance with collision regulations (IMO, 1972).
- Focus on BRM and on keeping each other updated.

12.6.5.5 Evaluation

Evaluation is achieved by a subject matter expert who holds a certificate as 'deck officer class 1 master mariner' according to STCW regulations. The evaluation includes checking the following:

- Information from radar and ARPA is correctly interpreted and analysed, taking into account limitations of the equipment and prevailing circumstances and conditions.
- Action taken to avoid a close encounter or collision with other vessels is in accordance with collision regulations (IMO, 1972).

- Decisions to amend course and/or speed are both timely and in accordance with accepted navigation practice.
- Adjustments made to the course and speed of the ship maintain safety of navigation.
- Communication is clear, concise and acknowledged at all times in a seamanlike manner.
- Manoeuvring signals and fog signals are made at the appropriate time and are in accordance with collision regulations and local regulations.
- BRM principles are followed.

The lack of specific behavioural markers for maritime teamwork in the STCW regulations and in research literature in general means that the evaluations are based on the expert's judgement and experience and relate to the suitability of the plan, the execution of the plan as seen in the route taken and the relationship between performance and task workload.

12.6.5.6 Debriefing

Students are given 5–10 minutes for internal debriefing within the bridge teams as a self-assessment of their performance, followed by a 20-minute debriefing with all participants, including discussions and replay of the bridge team's performance during the exercise.

12.7 Future Challenges for Simulator-Based Maritime Crew Training

Within the domain of simulator training, simulations are considered to provide risk-free training of critical situations such as accidents. Simulators also provide opportunities for repetition and organising of activities that are not possible in actual work settings – such as the opportunity to 'freeze' scenarios for discussion or instruction as well as take the system to the boundaries, and beyond, of what it can take (Committee on Ship-Bridge Simulation Training, 1996), thereby providing students with vital critical incident management and stress coping skills.

Future work in simulator-based crew training should focus on solving the major challenges. For maritime simulator-based team training, the challenge is not so much in the integration between simulators, but more related to the assessment of simulator-based training and more specifically the assessment of non-technical skills related to CRM/MRM. This is not a uniquely maritime challenge; as has been noted by Flin and Martin (2001) and Nazir and Manca (2014), there is little research on the use of behavioural markers to evaluate CRM training and the associated reliability and validity in the use of such scoring systems. This is particularly so for the maritime applications of behavioural markers in CRM/MRM training as the research on this is simply non-existent.

One possible approach to the assessment of teamwork comes from literature on distributed situation awareness (DSA), where knowledge sharing and communication underpinning distributed situational relevance are highly correlated ($r = .923$) with team performance in experimental tasks (Sorensen and Stanton, 2013). Replications in a maritime high-speed craft navigation task have unfortunately shown lower associations ($r = .349$) between the relevance and correctness of communication and navigational performance (Øvergård et al., 2015). Hence, measurement of DSA can be one fruitful way of assessing teamwork in multiteam cooperation. However, there are limitations of the current DSA and situation awareness assessments. These limitations involve lack of robustness, non-dependence on time, invasiveness of real-world work tasks, evaluation of SA after the task is finished, lack of consistency and lack of consideration of HMI (Salmon et al., 2009).

As with the behavioural markers of CRM (Flin and Martin, 2001), the measurement of DSA in maritime training would require the development of a methodology to ensure reliability and validity of CRM/MRM and teamwork assessments. With the 2010 amendments to STCW focusing on non-technical skills (IMO, 2011), there is a great need for research on assessment methods for non-technical skills and teamwork. Lacking structured ways of assessing CRM/MRM and teamwork, organisers of simulator training often put considerable effort into debriefing, meta-reflection and scaffolding in forms of instructor, peer or technological support.

Future research and improvements on simulator training can profit from experiments where new types of instructional scaffolds are applied – such as video debriefing or actual video material from real-life work practices (Øvergård, 2008). Empirical studies focusing on the outcome of simulators in training on-the-job performance have also been called for (Nazir et al., 2013). In particular, research on simulators lacks studies that report longitudinal changes of competence and that measure development across different types of simulators – as well as actual participation aboard ships.

Acknowledgements

The authors want to thank Thomas Førlie and Martin Birkeland at the University College of Southeast Norway (Høgskolen i Sørøst-Norge [HSN]) for supplying us with detailed information about their simulator training practices, and special thanks go to Haakon Thorvaldsen for providing us with images of the simulators at HSN.

References

Alessi, S. M. (1988) Fidelity in the design of instructional simulations, *Journal of Computer-Based Instruction*, vol. 15, pp. 40–47.

Barnett, M., Gatfield, D., and Habberley, J. (2002) Shipboard crisis management: A case study, *Proceedings of the Human Factors in Ship Design and Operation Conference.* Available at http://ssudl.solent.ac.uk/432/1/RINA202002.pdf (last checked 27 January 2014).

Barnett, M., Gatfield, D., and Pekcan, C. (2003) A research agenda in maritime Crew Resource Management, *Proceedings of the International Conference on Team Resource Management in the 21st century.* Available at http://ssudl.solent.ac.uk/433/1/Embry -20Riddle-202003.pdf (last checked 27 January 2014).

Bird, A. M. (1977) Leadership and cohesion within successful and unsuccessful teams: Perceptions of coaches and players. In Landers, D. M. and Christina, R. W. (eds.), *Psychology of Motor Behavior and Sport*, vol. 2 (pp. 176–182). Campaign, IL: Human Kinetics Books.

Burns, C. M., and Hajdukiewicz, J. R. (2004) *Ecological Interface Design*. Boca Raton, FL: CRC Press.

Committee on Ship-Bridge Simulation Training (1996) *Simulated Voyages: Using Simulation Technology to Train and Licence Mariners*. Washington, DC: National Academies Press.

Dahai, L., Nikolas, M., and Dennis, V. (2009) Simulation fidelity. In Vincenzi, D. A. et al. (eds.), *Human Factors in Simulation and Training* (pp. 61–73). Boca Raton, FL: CRC Press.

Dahlstrom, N., Dekker, S., van Winsen, R., and Nyce, J. (2009) Fidelity and validity of simulator training, *Theoretical Issues in Ergonomics Science*, vol. 10, pp. 305–314.

Donderi, D. C., Mercer, R., Blair Hong, M., and Skinner, D. (2004) Simulated navigation performance with marine electronic chart and information display systems (ECDIS), *Journal of Navigation*, vol. 57, pp. 189–202.

Fletcher, G., Flin, R., McGeorge, P., Glavin, R., Maran, N., and Patey, R. (2003) Anaesthetists' non-technical skills (ANTS): Evaluation of behavioural marker system, *British Journal of Anaesthesia*, vol. 90, pp. 580–588.

Flin, R., and Martin, L. (2001) Behavioral markers for crew resource management: A review of current practice, *The International Journal of Aviation Psychology*, vol. 11, pp. 95–118.

Flin, R., O'Connor, R., and Crichton, M. (2008) *Safety at the Sharp End: A Guide to Non-technical Skills*. Farnham, UK: Ashgate.

Gausdal, A. H. (2015) Methods for developing innovative SME networks. *Journal of Knowledge Economy*, vol. 6, Iss. 4, pp. 978–1000.

Gould, K. S., Røed, B. K., Saus, E.-R., Koefoed, V. F., Bridger, R. S., and Moen, B. E. (2009) Effects of navigation method on workload and performance in simulated high-speed ship navigation, *Applied Ergonomics*, vol. 40, pp. 103–114.

Hayward, B. J., and Lowe, R. L. (2010) Migration of crew resource management training. In Kanki, B., Helmreich, R., and Anca, J. (eds.), *Crew Resource Management* (pp. 317–342). San Diego, CA: Academic Press.

Heath, C., and Luff, P. (1996) Convergent activities: Line control and passenger information on the London Underground. In Engeström, Y., and Middleton, D. (eds.), *Cognition and Communication at Work* (pp. 96–129). Cambridge, UK: Cambridge University Press.

Helmreich, L., Merrit, A. C., and Wilhelm, J. A. (1999) The evolution of crew resource management training in commercial aviation, *The International Journal of Aviation Psychology*, vol. 9, pp. 19–32.

Hetherington, C., Flin, R., and Mearns, K. (2006) Safety in shipping: The human element, *Journal of Safety Research*, vol. 37, pp. 401–411.

Hontvedt, M., and Arnseth, H. C. (2013) On the bridge to learn: Analysing the social organization of nautical instruction in a ship simulator, *International Journal of Computer-Supported Collaborative Learning*, vol. 8, pp. 89–112.

Hutchins, E. (1995) *Cognition in the Wild*. Cambridge, MA: MIT Press.

IMO (International Maritime Organization) (1972) *Convention on the International Regulations for Preventing Collisions as Sea*. London: IMO.

IMO (1997) *General Operators Certificate for the Global Maritime Distress and Safety System*. IMO Model Course 1.25. London: IMO.

IMO (1999) *Radar Navigation Operational Level*. IMO Model Course 1.07. London: IMO.

IMO (2011) *International Convention on Standards of Training, Certification and Watchkeeping for Seafarers, Including 2010 Manila Amendments*. London: IMO.

IMO (2012) *The Operational Use of Electronic Chart Display and Information Systems*. Model Course 1.27. London: IMO.

Karlsson, T. (2011) *The importance of structured briefings & debriefings for objective evaluation of ARPA simulator training*. Report No. NM 11/20, Chalmers University of Technology, Department of Marine and Shipping Technology, Gothenburg, Sweden. Available at http://publications.lib.chalmers.se/records/fulltext/162667.pdf (last checked 21 April 2016).

Kobayashi, H. (2005) Use of simulators in assessment, learning and teaching of mariners, *WMU Journal of Maritime Affairs*, vol. 4, pp. 57–75.

Komulainen, TM, Sannerud, R., Nordsteien, B., and Nordhus, H. (2012) Economic benefits of training simulators, *World Oil*, pp. R61–R65. Available at https://oda.hio.no/jspui/bitstream/10642/1544/1/939013.pdf (last checked 20 January 2014).

Kozlowski, S. W. J., and DeShon, R. P. (2004) A psychological fidelity approach to simulation-based training: Theory, research, and principles. In Schiflett, S. G., Elliott, L. R., Salas, E., and Coovert, M. D. (eds.), *Scaled Worlds: Development, Validation, and Applications* (pp. 75–99). Burlington, VT: Ashgate.

Larsen, L. B. (2015) *Tailored courses – A road to learning for shipping companies*. Unpublished master thesis at the Department of Maritime Technology and Innovation, Buskerud and Vestfold University College, Kongsberg, Norway.

Lave, J., and Wenger, E. (1991) *Situated learning: Legitimate peripheral participation*. Cambridge, UK: Cambridge University Press.

Lavery, B. (ed.) (1998) *Shipboard Life and Organisation 1731–1815*. Aldershot, UK: Ashgate.

Lützhöft, M., and Dekker, S. W. A. (2002) On our watch: Automation on the bridge, *Journal of Navigation*, vol. 55, pp. 83–96.

Lützhöft, M., and Lundh, M. (2009) Maritime application of control systems. In Ivergård, T., and Hunt, B. (eds.), *Handbook of Control Room Design and Ergonomics – A Perspective for the Future* (pp. 227–261). Boca Raton, FL: CRC Press.

Nazir, S., Colombo, S., and Manca, D. (2013) Testing and analyzing different training methods for industrial operators: An experimental approach, *Computer Aided Chemical Engineering*, vol. 32, pp. 667–672.

Nazir, S., and Manca, D. (2014) How a plant simulator can improve industrial safety, *Process Safety Progress*, vol. 34, pp. 237–243.

Nazir, S., Kluge, A., and Manca, D. (2014) Automation in process industry: Cure or curse? How can training improve operator's performance, *Computer Aided Chemical Engineering*, vol. 33, pp. 889–894.

Nazir, S., Sorensen, L. J. Øvergård, K. I., and Manca, D. (2015) Impact of training methods on distributed situation awareness of industrial operators, *Safety Science*, vol. 73, pp. 136–145.

Nilsson, R., Gärling, T., and Lützhöft, M. (2009) An experimental simulation study of advanced decision support system for ship navigation, *Transportation Research Part F: Traffic Psychology and Behaviour*, vol. 12, pp. 188–197.

Nuutinen, M., and Norros, L. (2009) Core task analysis in accident investigation: Analysis of maritime accidents in piloting situations, *Cognition, Technology and Work*, vol. 11, pp. 129–150.

O'Connor, P. (2011) Assessing the effectiveness of bridge resource management training, *The International Journal of Aviation Psychology*, vol. 21, pp. 357–374.

Øvergård, K. I., Bjørkli, C. A., Hoff, T., and Dahlman, J. (2005) Comparison of trajectory variation and speed for real and simulator-based high-speed navigation. In Fostervold, K. I. et al. (eds.), *Proceedings of the 37th Annual Conference of the Nordic Ergonomic Society, Ergonomics as a Tool in Future Development and Value Creation* (pp. 275–279). Oslo, Norway: Nordic Ergonomics Society.

Øvergård, K. I. (2008) A video-based phenomenological method for evaluation of driving experience in staged or simulated environments. In Hoff, T., and Bjørkli, C. A. (eds.), *Embodied Minds – Technical Environments* (pp. 259–277). Trondheim, Norway: Tapir Akademiske.

Øvergård, K. I., and Smit, P. N. (2014) Effects of sea experience and computer confidence on ECDIS training, *Journal of Maritime Research*, vol. 11, pp. 25–31.

Øvergård, K. I., Nielsen, A. R., Nazir, S., and Sorensen, L. J. (2015) Assessing navigational teamwork through the situational correctness and relevance of communication, *Procedia Manufacturing*, vol. 3, pp. 2589–2596.

Robert, G., Hockey, J., Healey, A., Crawshaw, M., Wastell, D. G., and Sauer, J. (2003) Cognitive demands of collision avoidance in simulated ship control, *Human Factors*, vol. 45, pp. 252–265.

Salas, E., Wilson, K. A., Burke, C. S., and Wightman, D. C. (2006) Does crew resource management training work? An update, and extension and some critical needs, *Human Factors*, vol. 48, pp. 392–412.

Salmon, P. M., Stanton, N. A., Walker, G. H., and Jenkins D. P. (2009) *Distributed Situation Awareness*. Farnham, UK: Ashgate.

Sauer, J., Wastell, D. G., Hockey, G. R. J., Crawshaw, C. M., Ishak, M., and Downing, J. C. (2002) Effects of display design on performance in a simulated ship navigation environment, *Ergonomics*, vol. 45, pp. 329–347.

Sorensen, L. J., and Stanton, N. A. (2013) Y is best: How distributed situational awareness is mediated by organisational structure and correlated with task success, *Safety Science*, vol. 56, pp. 72–79.

Vincenzi, D. A., Wise, J. A., Mouloua, M., and Hancock, P. A. (eds.) (2009) *Human Factors in Simulation and Training*. Boca Raton, FL: CRC Press.

Section VI

Conclusions

13

Evolution of Simulators in Transportation Human Factors

Michael G. Lenné and Mark S. Young

CONTENTS

13.1 Introduction

One of the aims of this book is to provide opportunities to learn from simulator-based applications across domains to see what lessons could be gleaned – and to see if there are aspects of simulator set-up and functionality, performance assessment and use generally – that can accelerate efficacy in one or more domains. The book draws upon the expertise and experience of many of the chapter authors and, therefore, highlights different aspects of simulation use rather than covering every research and training application.

It is clear from the contributions in this book that the application of simulation across transport domains varies considerably. Sources of this variance cut across the physical infrastructure involved, the system flexibility and cost, the use cases (namely, research versus training), the types of technical and non-technical skills under assessment, the assessment methods and the use of other actors/confederates, amongst others. This final chapter begins by considering applications in training and research drawing upon the material presented in this book as well as the authors' own experiences and prior work. Some guidelines for simulator specification are presented followed by some thoughts on how the field might evolve over coming years.

13.2 Applications in Training

The introductory discussion in Chapter 1 highlights that the roots of simulation in transport stem from the aviation sector. We should therefore not be surprised that the use of simulation to support training is the most established in aviation. The earliest flight simulators were developed in the interwar years as a cost-effective means of training military personnel when real aircraft were not available or were in short supply. Chapter 9 presents a rich history of how simulator use developed for commercial flight training in the context of a longstanding and strong regulatory framework to support its use. Aviation today relies heavily on the use of simulators to train pilots, more so than the other domains covered in this book. As noted in our previous work (Salmon et al. 2009), high-fidelity simulators are widely used for all aspects of aviation training, including advanced training, re-training, periodic assessment and tests of fitness for duty. A range of aviation tasks are trained via simulation including basic stick-and-rudder skills, combat manoeuvring skills, instrument training, decision-making and CRM (Salas et al. 1998). The universal acceptance of simulators for training within aviation is based on early research, experience, ability to simulate emergencies safely and in detail and – importantly – cost considerations. Unsurprisingly, then, most of the evidence supporting the use of simulators for training purposes has emerged from the aviation industry; in turn, the reputation of simulation in the aviation training domain largely rests on the traditional training research conducted more than 20 years ago (as shown in training and simulation compendiums such as Swezey and Andrews 2001). However, more recent applied research does provide further demonstration of the efficacy of simulator training in aviation (Taylor et al. 1999; Lee 2005); such research is discussed in detail in Chapter 9.

Although simulators are increasingly being applied for train operator training (see Chapter 6), evaluation of the effectiveness of such simulators has received only limited attention when compared to the aviation domain. Further, of the evaluation studies undertaken, the majority have relied on questionnaires, interviews and operational experience of the transport organisation, whilst formal transfer-of-training studies have yet to be conducted. Schmitz and Maag (2008) for example examined the role of training simulators in a number of European rail systems, in order to benchmark the status of driver training and to compare the current practice with the perceived needs of stakeholders. Whilst providing strong support for simulation, it was concluded that simulation needs to be used in concert with other training methods to produce highly qualified driver personnel. Chapter 6 describes some formal mechanisms for integrating simulator-based training into broader programmes (Chapter 6).

For automotive applications, simulator training also has many advantages to offer over training in an actual vehicle. Evaluations have shown that simulator-based driver training is effective, depending on the task and the simulator used (Kaptein et al. 1996). Whilst studies reporting the beneficial effects of simulator-based training of heavy-vehicle operators at the entry-level exist, those that draw on objectively derived data are few in number (Mitsopoulos-Rubens et al. 2013). Clearly, there is a need for well-designed studies which aim to provide a comprehensive and objective assessment of the quality of simulator-based training for heavy-vehicle operators and that provide adequate coverage and assessment of both technical and non-technical skill areas. An area gaining in momentum is the use of simulation to train existing heavy-vehicle operators in the principles of eco-driving. Encouragingly, studies showing real-world improvements in fuel efficiency and that draw on objectively derived data already exist. These studies constitute good, growing support for the continued development and implementation of simulation-based fuel efficiency training for existing operators of heavy vehicles (see Chapter 3). Further research efforts in this area can help to identify with greater certainty the precise mechanisms underlying the observed positive effects and to explore in depth any potential consequent effects on safety.

The application of simulation to novice driver training should also be noted (see Goode et al. 2013; Lenné et al. 2017). Currently, in the Netherlands alone, over 100 simulators are dedicated to novice driver training (Kappé and Emmerik 2005). Over the last decade or so, there has been considerable interest, particularly in Europe, regarding the use of simulation for general driver training. De Groot et al. (2007), for example, suggested that increasing numbers of driving schools are integrating simulators into their training curricula. Accordingly, various general driver training simulators now exist across several different countries. Disappointingly, there has been a lack of evaluation studies investigating the effectiveness of driving simulators for novice driver training; however, there is sufficient related research to suggest that well-designed instructional programmes involving simulators, where emphasis is given to specific training in risk perception and awareness, are likely to provide a real benefit. The success of computer-based packages with some simulator-like features such as those developed by Regan et al. (2000) and Fisher et al. (2006) provides a pointer to the potential for simulation in this area.

13.2.1 Strengths and Weaknesses

There are many advantages associated with the use of simulators for operator training within complex socio-technical systems. Although the initial outlay is often considerable in terms of time and financial resources, the use of simulation for training invariably saves money due to a reduction in costs associated with training and the reduction in operating costs of the actual system for training purposes (e.g. Hancock 2009). Thompson et al. (2009), for example, pointed out that military flight simulators operate at approximately

5–20% of the cost of operating the actual aircraft and that research has demonstrated that using simulators in the initial stages of training reduces the number of flight hours required in the actual aircraft to achieve proficiency. Tasks can also be repeatedly practiced in a realistic but safe context until an accepted level of competence is achieved.

Many of the strengths of simulators for training are noted in Chapter 1. The high degree of control over training scenarios is a key benefit; in contrast to normal operations, full training control is, in principle, possible in the simulator, as there is the ability to specify precisely the type and level of operator task elements and demand. Emergency and non-routine scenarios can also be practiced without exposure to the high-risk levels typically associated with such scenarios. Given enough practice, automatic behavioural response patterns can be learned, as is done in the aviation industry. The data recording and performance measurement capabilities of simulators are also a key strength as they permit the accurate measurement of trainee performance and the provision of detailed feedback.

The degree of flexibility offered by simulation is also attractive; scenarios can be structured and modified according to training requirements. Inadequate task performance, such as errors, can also be demonstrated, which allows operators to understand the associated consequences and to learn error recovery strategies. Thompson et al. (2009) also discussed the environmental ecology associated with simulators; since simulators reduce the operational hours of the actual system, the running costs of the system, maintenance costs and wear and tear are reduced. In addition, loss of revenue, due to the system being in operation for training purposes only, is also reduced.

On the downside, there are some disadvantages associated with the use of simulators for training. Building and developing simulated environments is expensive and time consuming; the initial cost associated with purchasing a new simulator can be high, and the resources invested in terms of time and money when installing, testing and accepting the simulator are also typically high. High-fidelity simulators also often have special installation and housing requirements, such as air-conditioning facilities. In addition, the maintenance and running costs associated with simulators, particularly high-fidelity ones, can also be substantial. Cost–benefit studies can help present the economic case for simulation here.

As reviewed by Salmon et al. (2009), a major challenge associated with the use of simulators noted in our work for training purposes is the fact that performance in the simulator may not reflect operator performance in the real world (Thompson et al. 2009), and therefore, proficiency in a simulator does not always equate to proficiency in the operational system. In addition to the issue of simulator fidelity, that being the extent to which performance in the simulator is representative of real-world performance, there are a range of factors that might lead to enhanced performance in the simulator. These include reduced stress levels, trainee motivation, review of policies and procedures immediately prior to training and the lack of adverse performance shaping factors such as fatigue, complacency and boredom (Thompson et al. 2009). As noted in Chapter 6, the problem of simulator sickness, caused by discrepancies between visual and vestibular cues, also presents a significant issue when using simulators for training purposes (Thompson et al. 2009). Simulator sickness may affect some operators' ability to proceed with training, and it is even suggested that trainees may devise strategies to compensate for, or avoid, simulator sickness that actually leads to performance decrements when using the operational system (Thompson et al. 2009). Despite these disadvantages, however, it is generally accepted that the advantages associated with the use of simulators for training outweigh the disadvantages.

13.2.2 Summary – Training

Overall, although evaluation evidence is limited, the literature is strongly supportive of the effectiveness of simulators for training operators across the transportation domains. Simulator-based training applications cover initial training, instrument training, normal and emergency procedure training, and upgrade and transition and refresher training (Thompson et al. 2009). In addition, simulators have also been used extensively for training specific cognitive skills, including decision-making, situation awareness (e.g. Strater and Bolstad 2009), hazard perception (e.g. Regan et al. 2000) and teamwork (Lenné et al. 2011b; Goode et al. 2013).

Whilst guidance exists on how best to match simulator capabilities and task type with training needs, there is still a gap in knowledge regarding how much simulation (i.e. what proportion of a broader training programme should be simulation based) is both necessary and sufficient to maximise the overall effectiveness of a training programme. Another key question is where best to insert simulator use into the training programme. In the absence of such knowledge, the issue is typically addressed on pragmatic grounds. This then leads to the question of where should the focus of the simulator-based training component lie for a simulator platform of given characteristics, the length of time that can reasonably be allocated to simulator-based training and the purpose of the training and the target trainee audience.

13.3 Applications in Research

Whilst the previous section covers the application of simulators in training applications, much of the underpinning work is, of course, drawn from research studies using simulation. The focus of this section is on the application of simulation to research applications beyond training programmes per se. There is, of course, a related body of research in road safety that uses simulation to address the differences in skill, typically referred to as hazard perception, between novice and experienced drivers and riders (see Chapter 4). Noteworthy here is that a focus on more cognitive than technical skills is likely to reflect not only the interests of the road safety research groups involved, but also the technical limitations of simulators, in particular motorcycle simulators, to provide the full range of cues needed for research that focuses on some technical skills of handling a motorcycle.

In contrast to the training-based applications that have underpinned the introduction and continued adoption of simulation in aviation and rail sectors, in the automotive sector, the primary application of simulators has been for research. In part, this reflects the large number of road safety-related research groups around the world, whilst in part, it reflects the advances in technology and, correspondingly, that simulators have become more affordable and are being used by an increasing number of researchers worldwide. In some areas of research, computer-based driving simulator tasks, often using systems with low functionality (as distinct from fidelity), are being used because the experimenters seek to use a complex task or an enhanced tracking-type task (Lenné et al. 2011a).

Extensive research has used simulators to better understand car driver behaviour in the face of different forms of impairment, performance-shaping factors and technologies (see Chapter 2). This type of research, whilst typically measuring aspects of individual task performance, is almost always strongly linked to an underlying theory. This in itself is a good example of the major advantage of simulation in that a high degree of control and depth of performance measurement affords strong opportunities for hypothesis testing. Driving simulators have been used to provide information on the behaviour of drivers in particular situations (e.g. see Fischer et al. 2011). In our research, for example we have examined underlying mechanisms related to priming and expectancy in how drivers interact with in-vehicle systems and other road users (Lenné et al. 2008; Beanland et al. 2014). The issues highlighted in this book focus more on the requirements for research simulators and practical guidance on how to design and conduct such studies (Chapters 2 and 4).

In the rail signalling domain, the small amount of research that has been conducted tends to have focused on the HMI, technology evaluations and infrastructure changes where key objectives include measuring situation awareness by using simulation (Chapter 7). Note here that the simulators used are not typically developed for research purposes, and this could certainly be one factor constraining research in this field. In the rail signalling context, it has been suggested that simulators should allow easier modification of timetables and the simulated infrastructure as well as supporting communications between different controllers. The linking of train driver simulators with signalling simulators is noted in Chapter 7 as a future pursuit to provide more authentic research and training opportunities. Similar issues are apparent in the maritime field, where the potential value of providing training to teams that are located in different areas of ships has been noted (see Chapter 12).

Finally, one interesting aspect highlighted in Chapter 8 is that in aviation, there is a very tight coupling of research outcomes into training, as simulator-based research has been used to explore pilot proficiency in aviation. A further notable aspect of aviation simulators is the importance of motion and the novel approaches that are used to recreate the appropriate physical forces.

13.4 Practicalities of Using Simulators

The historical use of simulators also influences the simulation infrastructure. In aviation, for example there has been an emphasis on full motion and full replicas of the real thing. This is likely driven by the predominant use of simulators in aviation training, and perhaps, research simulators have been somewhat forced to keep pace. Contrast this with road safety, where the predominant use is for research, and not training, which may have opened up the field to consider lower-cost options.

Issues of reliability, validity and fidelity are not new and are discussed at length in this book (see Chapter 1). An interesting and less documented issue is that of presence, as described by Burnett et al. (this volume) in Chapter 2. They propose that to be fully present in a simulation, an operator should not be aware of the display screens, lack of real motion, awareness of the experimenter or any recording equipment during the simulated experience.

It would seem reasonable to propose also that an enabler for presence from the behavioural viewpoint is that the operator is provided with sufficient freedom of action in the simulation to allow typical response tendencies to be shown (Lenné et al. 2011a). For understandable reasons relating to experimental control and data analysis, behaviour is often constrained in the scenario design process so that only specific stimuli are available to the operator and, it follows, so that specific responses can be observed. This is particularly the case in research studies where participants also have expectations – and these may compromise the objectives that we have as researchers. It is often assumed that subjective expectancies do not vary across events within an experiment and that expectancy does not interact with any of the independent variables under examination. But we know that expectancy has major influences on perception, decision-making and action selection; hence, these would seem important assumptions to test.

In Chapter 2, Burnett et al. argued that achieving presence is rarely the end goal of a simulation, yet it is widely accepted as necessary. Providing a greater sense of presence should evoke more realistic performance, which in turn should have a positive impact on validity. This would appear to be a concept that is worthy of further exploration in terms of its impact on validity – and this is ultimately determined by the specification of the simulator.

13.4.1 Simulator Requirement Specification: Considerations

In defining or developing a simulator, the first step would be a functional requirement specification. Within transport simulation, design solutions are usually achieved through the integration of underlying hardware (often taken from, or mimicking, an actual system) coupled with digital media and computer processing to support an interactive user experience (often coupled with motion platforms) (Stedmon et al. 2012). Considering the factors underpinning the development and use of simulators, it could be argued that investment is better spent on software improvements (such as the simulator engine and the visual database) rather than hardware (once a certain level of fidelity is achieved). Given the advances in software and PC technology in recent years, good-quality simulations can be acquired for reasonable sums of money.

Considerations that apply to both research and training simulators include whether the simulator should represent generic or specific scenarios, visual databases, dynamic (vehicle) models and hardware equipment (controls, displays, layouts etc.). Visual databases can be expensive to develop and maintain (i.e. keeping them up-to-date), but a geo-specific database can offer advantages in for example route and environment familiarity for operators. On the other hand, it limits the usage of the simulator unless additional databases are developed. The same is true for hardware (i.e. cab/cockpit equipment) – the decision being whether to adopt a specific vehicle or whether to adopt a generic layout. Again, it is a case of the right tool for the right job. Another key question that is covered in Chapter 1 is the relative merits of motion systems in light of their relative cost in the risks of simulator sickness.

In any case, it is useful to have the capability to future-proof the simulator by adding more scenarios and databases, preferably via a toolkit or library on the user interface without recourse to programming from the supplier. The simulator should also include the flexibility to create and edit visual objects, scenarios, vehicle models etc. Operation of the database editor should not require specific expertise in computer modelling, making it suitable for use by operators and infrastructure engineers who are familiar with PC operation.

There should also be the capability to add or integrate with peripheral systems, in particular the (hardware) instrument panels and displays of the vehicle, but this could also include novel technologies and interfaces such as automation. These could potentially be simulated on the graphical display or alternatively presented using original hardware or other external equipment that is then connected to the inputs and outputs of the simulator, triggering (or being triggered by) events in the simulated scenario and being synchronised with the data outputs of the simulator. As well as vehicle systems, such peripheral integration could include research or training manipulations, such as the inclusion of a secondary task.

13.4.2 Simulator Requirement Specification: Template

The following presents an outline of areas to consider in developing a functional specification for a simulator. This specification is deliberately generic and can be adapted for research or training purposes in any of the transportation modes. It is intended as a guide only.

- Hardware
  - PC based
  - Fully functional replica/generic controls and displays (although some may be simulated on touch screens)
  - Realistic cab/cockpit/control environment, including haptic feedback through controls
  - Capability to reconfigure the physical environment (controls, layout etc.)
  - Simulator control desk (with interfaces for developing, modifying and monitoring the scenario in use, including the live visual scene and replica instruments, plus closed-circuit television monitoring of participants)
  - Two-way communication system between control desk and simulator
  - Environmental control inside the simulator (heating/air conditioning, lighting)
- Visual system
  - Wide field of view, encompassing peripheral vision (i.e. up to 180 degrees horizontal and 45 degrees vertical)
  - Potential for rear-view/side-view images (e.g. in a driving simulator – which may either be a virtual mirror on the forward screen or an additional back projector to be viewed in the real mirror)
- Audio system
  - Three-dimensional sound reproduction
  - Low-frequency speaker for vibrations
- Motion system (where applicable)
  - Acceleration rates to reflect normal and extreme operations (within the bounds of safety)
  - Inclusion of an emergency stop control

- Set-up
  - Input and recording of date/time (automated), operator and participant ID (incremented automatically) and scenario descriptor
  - Set trial conditions (route, event script, environmental conditions, duration)
  - Save conditions
  - Play/pause/resume/stop simulation
- Database editor
  - Graphical user interface (point and click, drag and drop)
  - Library of environmental objects, including static objects (e.g. signs, infrastructure) and animated objects (e.g. vehicles, people), plus artificial intelligence in other vehicles to interact with during the simulation
  - Library of routes, including capability to customise such routes and create new routes from a library of route components
  - Selection of contexts/background scenery (urban, rural etc.)
  - Selection of and editable vehicle dynamic models (i.e. vehicle performance characteristics)
  - Capability to preview databases during the development process, including keyboard control of the simulator
- Scenario editor
  - Capability to introduce faults, hazards, emergencies or other events – both in real time (by the experimenter/trainer) and programmable into a script, triggered by time, distance, location or condition (e.g. vehicle speed)
  - Capability to simulate a variety of environmental conditions (lighting, weather, day/night conditions, vehicle lighting etc.)
  - Capability to add/integrate secondary tasks
  - Capability to include automation or other novel technologies
  - Definable responses to collision scenarios (e.g. simulated collision, simulation ends, no response)
- Data collection
  - Ability and storage space to record data from extended length trials (e.g. where investigating fatigue or vigilance)
  - Definable sampling rate
  - Range of variables both continuous (e.g. time, distance, speed, position, control inputs) and discrete (collisions, events, conditional criteria, secondary tasks)
  - Data file should also include the programmed script events

- Data file should be time-stamped for synchronisation with external devices (e.g. vehicle instruments, eye tracker, video recording)
  - Data viewer available in simulator interface, but data should also be exportable to spreadsheets or statistical packages
  - Replay facility including full visual reproduction and data log
- Accommodation
  - Simulator room (which may need structural strengthening, depending on the nature of the simulator itself)
  - Control workstation (ideally in a separate room)
  - Observation station with replay and debriefing facilities (e.g. for groups of trainees)
  - Briefing and debriefing space
  - Toilet, catering and medical facilities
- Peripherals
  - Eye tracker
  - Physiological measuring equipment
- Documentation
  - Operation and maintenance manuals
  - Training manuals
  - List of spare parts

For the sake of empirical control, a research simulator needs more flexibility in the manipulation of scenarios, visual environments and even the physical layout of the cab, as well as data recording, output and analysis for a range of performance and event variables. Many training simulators do not offer such flexibility for research. A key question is what aspects of the task should be represented and whether it should be specific or an abstraction.

The requirements for training should be defined through a training need analysis. This should comprise a task analysis and a definition of the training objectives, including the difficulty, importance and frequency of the tasks, as well as the baseline skills of trainees, the success criteria and any standards that performance is measured against. These factors should be compared against potential delivery methods (i.e. types of simulator), remembering that a simulator need not be a high-fidelity full simulator (Salmon et al. 2009).

13.5 Conclusion

The use of simulators for both research and training applications varies significantly across the transport domains covered in this book. The extent to which each domain has embraced each application is influenced by many factors. Several opportunities for the future are raised by the contributors of this book.

The important role for simulation from a training perspective is evident in that it provides one of the only means for operators to experience novel and infrequent tasks and events in environments where operations are becoming safer and tasks becoming more automated. A key question in any training application should centre on the success of the training programme however success may be defined. Chapters 3 and 9 present different models that can be used to assess training effectiveness, and it is clear that this remains a clear need for further investigation. Showing simply that training is cheaper using a simulator compared to an operational aircraft or ship is not overly insightful and is only a small part of the business case. Linking total costs invested in training to a positive and sustained transfer to the real world remains the type of data needed to strengthen the business case for using simulation in training.

Distributed simulation as a concept has been discussed for many years, but it is yet to really take-off in either research or training applications. Perhaps, the value proposition here is not yet clear. Having confederates in a simulation – for example having a flight instructor play the role of an air traffic controller during a flight training exercise – is not new. What would be the value added for a flight training exercise to have a controller at an air traffic control console playing that role versus the flight instructor from the back of the flight simulator? This is not to say that it would not be a benefit, but it should be quantifiable to a reasonable extent before building such a training platform. Chapter 7 notes the potential of distributed simulation to better support training for rail signallers, and again, building a business case for this capability would likely be required before investments are made.

From a research viewpoint, a further opportunity is to incorporate additional types of simulation that support more systemic evaluations of performance. For example Chapter 10 describes an example where computer simulation and modelling have been used to complement human-in-the-loop trials for testing the efficacy of different approaches to passenger evacuation from aircraft in emergency situations. The value in integrating different classes of simulation has also been noted in road safety applications (Young et al. 2014). Road safety simulation models aim to provide a platform for assessing and predicting the safety performance of drivers, vehicles and the transport system. This requires an accurate representation of the behaviour and character of each of these system components. As such, the type of simulator-based research described in this book has great potential to enhance the utility of simulation models more generally.

There are, of course, issues that cut across both training and research applications. One of these issues relates to simulator set-up and fidelity, notably with respect to the issue of motion. For example it is noted that motion is not critical for the vast majority of train driver tasks (Chapter 6), but it is more critical for aviation. The underlying theme here, which is noted in several chapters (Chapters 3, 6, 9 and 12), is that the analysis of training needs should dictate the requirement for not only motion but also other fundamental requirements. For aviation, where motion is more important for training, it is clear that significant work has been undertaken to examine different methods of recreating different motion profiles (Chapter 8).

A second issue that cuts across both training and research applications is performance measurement. Whilst measuring individual performance is quite mature, the analysis of non-technical skills in teams remains an area where further development is needed. Whilst a number of behavioural markers are used in CRM training as applied in aviation and maritime environments, better behavioural markers of crew performance are needed to provide greater training value (Chapter 12). The current focus here is for aviation and maritime teams who are colocated. Other theoretical approaches such as distributed situation awareness and associated measurement approaches may need to be considered for geographically separated teams such as a driver and a guard on a train. Performance measurement is one example where ongoing research could feed directly into training applications.

In closing, at the conclusion of the first chapter, we stated our hope that readers would find value and inspiration in learning from the experiences of those in other modes and applications outside of their own domain. We hope to have gone some way towards achieving this and will watch with great interest to see how the use of simulation continues to evolve across these varied transportation domains and applications.

References

Beanland, V., Lenné, M. G., and Underwood, G. 2014. Safety in numbers: Target prevalence affects detection of vehicles during simulated driving. *Attention, Perception and Psychophysics, 76*: 805–813.

Burnett, G., Irune, A., and Mowforth, A. 2007. Driving simulator sickness and validity: How important is it to use real car cabins? *Advances in Transportation Studies,* 33–42.

De Groot, S., De Winter, J. C. F., Mulder, M., and Wieringa, P. A. 2007. Didactics in simulator-based driver training: Current state of affairs and future potential. *Proceedings of the Driving Simulation Conference – North America,* Iowa City, IA, 12–14 September 2007.

Fischer, D. L., Rizzo, M., Caird, J. K., and Lee, J. D. 2011. *Handbook of Driving Simulation for Engineering, Medicine and Psychology.* New York: CRC Press.

Fisher, D. L., Pollatsek, A. P., and Pradhan, A. 2006. Can novice drivers be trained for information that will reduce their likelihood of a crash? *Injury Prevention, 12:* 25–29.

Goode, N., Salmon, P. M., and Lenné, M. G. 2013. Simulator-based driver and vehicle crew training: Applications, efficacy and future directions. *Applied Ergonomics*, 44: 435–444.

Hancock, P. A. 2009. The future of simulation. In D. Vincenzi, J. Wise, M. Mouloua and P. A. Hancock (Eds.). *Human Factors in Simulation and Training* (169–186). Boca Raton, FL: CRC Press.

Kappé, B., and Emmerik, M. L. 2005. The use of driving simulators for initial driver training and testing. Report No. 75151.01. The Hague, Netherlands: TNO Defence, Security and Safety.

Kaptein, N. A., Theeuwes, J., and Van Der Horst, R. 1996. Driver simulator validity: Some considerations. *Transportation Research Record*, 1550: 30–36.

Lee, A. T. 2005. *Flight Simulation*. Aldershot, UK: Ashgate.

Lenné, M. G., Mulvihill, C. M., Triggs, T. J., Regan, M. A., and Corben, B. F. 2008. Detection of emergency vehicles: Driver responses to advanced warning in a driving simulator. *Human Factors*, 50: 135–144.

Lenné, M. G., Groeger, J. G., and Triggs, T. J. 2011a. Contemporary use of simulation in traffic psychology research: Bringing home the Bacon? *Transportation Research Part F: Traffic Psychology and Behaviour*, 14: 431–434.

Lenné, M. G., Liu, C. C., Salmon, P. M., Holden, M., and Moss, S. 2011b. Minimising the risks and distractions for young drivers and their passengers: An evaluation of a novel driver-passenger training program. *Transportation Research Part F: Traffic Psychology and Behaviour*, 14: 447–455.

Lenné, M. G., Mitsopoulos-Rubens, E., and Mulvihill, C. M. 2017. Simulation-based training for novice car drivers and motorcycle riders: Critical knowledge gaps and opportunities. In D. L. Fischer, J. K. Caird, W. J. Horrey and L. T. Trick (Eds.). *The Handbook of Teen and Novice Drivers* (319–335). Boca Raton, FL: CRC Press.

Mitsopoulos-Rubens, E., Lenné, M. G., and Salmon, P. M. 2013. Effectiveness of simulator-based training for heavy vehicle operators: What do we know and what do we still need to know? *Proceedings of the 2013 Australasian Road Safety Research, Policing and Education Conference*, Brisbane, 28–30 August 2013.

Regan, M. A., Triggs, T. J., and Godley, S. T. 2000. Evaluation of a novice driver CD-ROM based training program: A simulator study. *Proceedings of the IEA 2000/HFES 2000 Congress* (pp. 2-334–2-337), San Diego, CA.

Salas, E., Bowers, C. A., and Prince, C. 1998. On simulation and training in aviation. *International Journal of Aviation Psychology*, 8: 195–196.

Salmon, P. M., Mitsopoulos-Rubens, E., Lenné, M. G., Triggs, T. J., and Wallace, P. 2009. Novel training technologies and future vehicle operator training: Review and recommendations. Report prepared for the Defence Science and Technology Organisation (DSTO). Clayton, Australia: Monash University Accident Research Centre.

Schmitz, M., and Maag, C. 2008. Benchmarking report on computer-based railway training in Europe. 2TRAIN Project report. Wuerzburg, Germany: University of Wuerzburg, Center for Trafffic Sciences, July.

Stedmon, A. W., Brickell, E., Hancox, M., Noble, J., and Rice, D. 2012. MotorcycleSim: A user-centred approach in developing a simulator for motorcycle ergonomics and rider human factors research. *Advances in Transportation Studies*, A27: 31–48.

Strater, L. D., and Bolstad, C. A. 2009. Simulation-based situation awareness training. In D. Vincenzi, J. Wise, M. Mouloua and P. A. Hancock (Eds.). *Human Factors in Simulation and Training* (129–148). Boca Raton, FL: CRC Press.

Swezey, R. W., and Andrews, D. H. 2001. *Readings in Training and Simulation: A 30-Year Perspective.* Santa Monica, CA: Human Factors and Ergonomic Society.

Taylor, H. L., Lintern, G., Hulins, C. L., Talleur, D. A., Emanuel, T. W. J., and Phillips, S. I. 1999. Transfer of training effectiveness of a personal computer aviation training device. *International Journal of Aviation Psychology, 9*: 319–335.

Thompson, T. N., Bell, M. A., and Deaton, J. E. 2009. Justification for use of simulation. In D. Vincenzi, J. Wise, M. Mouloua and P. A. Hancock (Eds.). *Human Factors in Simulation and Training* (39–48). Boca Raton, FL: CRC Press.

Young, B., Sobhani, A., Lenné, M. G., and Sarvi, M. 2014. Simulation and safety: A review of the state of the art in road safety simulation models. *Accident Analysis & Prevention, 66*: 89–103.

Index

Page numbers followed by f and t indicate figures and tables, respectively.